16
Topics in Heterocyclic Chemistry

Series Editor: R. R. Gupta†

Topics in Heterocyclic Chemistry
Series Editor: R. R. Gupta†

Recently Published and Forthcoming Volumes

Aromaticity in Heterocyclic Compounds
Volume Editors: T. Krygowski, M. Cyrański
Volume 19, 2009

Heterocyclic Supramolecules I
Volume Editor: K. Matsumoto
Volume 17, 2008

Bioactive Heterocycles VII
Flavonoids and Anthocyanins in Plants, and Latest Bioactive Heterocycles II
Volume Editor: N. Motohashi
Volume 16, 2009

Bioactive Heterocycles VI
Flavonoids and Anthocyanins in Plants, and Latest Bioactive Heterocycles I
Volume Editor: N. Motohashi
Volume 15, 2008

Heterocyclic Polymethine Dyes
Synthesis, Properties and Applications
Volume Editor: L. Strekowski
Volume 14, 2008

Synthesis of Heterocycles via Cycloadditions II
Volume Editor: A. Hassner
Volume 13, 2008

Synthesis of Heterocycles via Cycloadditions I
Volume Editor: A. Hassner
Volume 12, 2008

Bioactive Heterocycles V
Volume Editor: M. T. H. Khan
Volume 11, 2007

Bioactive Heterocycles IV
Volume Editor: M. T. H. Khan
Volume 10, 2007

Bioactive Heterocycles III
Volume Editor: M. T. H. Khan
Volume 9, 2007

Bioactive Heterocycles II
Volume Editor: S. Eguchi
Volume 8, 2007

Heterocycles from Carbohydrate Precursors
Volume Editor: E. S. H. El Ashry
Volume 7, 2007

Bioactive Heterocycles I
Volume Editor: S. Eguchi
Volume 6, 2006

Marine Natural Products
Volume Editor: H. Kiyota
Volume 5, 2006

QSAR and Molecular Modeling Studies in Heterocyclic Drugs II
Volume Editor: S. P. Gupta
Volume 4, 2006

QSAR and Molecular Modeling Studies in Heterocyclic Drugs I
Volume Editor: S. P. Gupta
Volume 3, 2006

Heterocyclic Antitumor Antibiotics
Volume Editor: M. Lee
Volume 2, 2006

Microwave-Assisted Synthesis of Heterocycles
Volume Editors: E. Van der Eycken, C. O. Kappe
Volume 1, 2006

Bioactive Heterocycles VII

Flavonoids and Anthocyanins in Plants, and Latest Bioactive Heterocycles II

Volume Editor: Noboru Motohashi

With contributions by
J. J. Aaron · M. D. Gaye Seye · M. Ishihara · M. Kawase
N. Motohashi · H. Sakagami · Y. Shirataki · S. Trajkovska

Springer

The series *Topics in Heterocyclic Chemistry* presents critical reviews on "Heterocyclic Compounds" within topic-related volumes dealing with all aspects such as synthesis, reaction mechanisms, structure complexity, properties, reactivity, stability, fundamental and theoretical studies, biology, biomedical studies, pharmacological aspects, applications in material sciences, etc. Metabolism will be also included which will provide information useful in designing pharmacologically active agents. Pathways involving destruction of heterocyclic rings will also be dealt with so that synthesis of specifically functionalized non-heterocyclic molecules can be designed.
The overall scope is to cover topics dealing with most of the areas of current trends in heterocyclic chemistry which will suit to a larger heterocyclic community.
As a rule contributions are specially commissioned. The editors and publishers will, however, always be pleased to receive suggestions and supplementary information. Papers are accepted for *Topics in Heterocyclic Chemistry* in English.
In references *Topics in Heterocyclic Chemistry* is abbreviated *Top Heterocycl Chem* and is cited as a journal.

Springer WWW home page: springer.com
Visit the THC content at springerlink.com

ISSN 1861-9282 e-ISSN 1861-9290
ISBN 978-3-642-00335-6 e-ISBN 978-3-642-00336-3
DOI 10.1007/978-3-642-00336-3
Springer Dordrecht Heidelberg London New York

Library of Congress Control Number: PCN applied for

Cover design: WMXDesign GmbH, Heidelberg
Typesetting and Production: le-tex publishing services oHG, Leipzig

Printed on acid-free paper

Springer is part of Springer Science+Business Media (www.springer.com)

Topics in Heterocyclic Chemistry Also Available Electronically

For all customers who have a standing order to Topics in Heterocyclic Chemistry, we offer the electronic version via SpringerLink free of charge. Please contact your librarian who can receive a password or free access to the full articles by registering at:

springerlink.com

If you do not have a subscription, you can still view the tables of contents of the volumes and the abstract of each article by going to the SpringerLink Homepage, clicking on "Browse by Online Libraries", then "Chemical Sciences", and finally choose Topics in Heterocyclic Chemistry.

You will find information about the

- Editorial Board
- Aims and Scope
- Instructions for Authors
- Sample Contribution

at springer.com using the search function.

Color figures are published in full color within the electronic version on SpringerLink.

Preface

Fruits and vegetables contain abundant nutraceuticals or phytochemicals such as carbohydrates, proteins, lipids, vitamins, enzymes, dietary fibers, minerals, antioxidants, flavonoids, and anthocyanins and other diverse components including trace chemicals. Interestingly, the flavonoids and anthocyanins in phytochemicals could be significantly associated with reductions of risk of cardiovascular disease, macular degeneration, and improvements and preventions of many other diseases.

Actually, flavonoids and anthocyanins in plants could play a very important role in our daily health. Our ancestors used them as part of daily life for dyestuffs and folklore medicines. Evidentially, the benefits of colorful flavonoids and anthocyanins have been confirmed by the results of many researchers.

As part of the series *Topics in Heterocyclic chemistry*, this volume titled *Bioactive Heterocycles II* presents comprehensive and up-to-date reviews on selected topics concerning mainly the usefulness for human health of flavonoids and related compounds, *Sophora* flavonoids and their functions in *Sophora* species (*Leguminosae*), the relationship of biological activity with the diverse structures of heterocycles by quantum chemical calculation, advances in bioactive mesoionic heterocycles, and the spectroscopic properties and application of bioactive phenothiazines including also benzo[*a*]phenothiazines.

The first chapter, "Anthocyanins as Functional Food Colors" by Noboru Motohashi and Hiroshi Sakagami, presents the structures of flavonoids and anthocyanins and their functions, especially protective activity against age-related diseases such as hypertension, diabetes, heart defects and others. Anthocyanins have higher antioxidant activity against oxidative stress, and thus the human body might be protected from oxidative injury by anthocyanins. On the basis of these facts, this review presents the biosynthesis of plant flavonoids and their ability to scavenge oxidants, inhibit or activate enzymes, and the safety of proanthocyanidins and anthocyanidins present in daily food intake. This review might be beneficial for understanding the structures and functions of flavonoids and anthocyanins in relation to bioflavonoid nutraceuticals.

The second chapter, "Flavonoids in *Sophora* Species" by Yoshiaki Shirataki and Noboru Motohashi, presents the relationships of diverse *Sophora* flavonoids with their functions in *Sophora* species of *Leguminosae*, which abundantly exist in the natural kingdom. Many *Sophora* flavonoids with *Sophora* plant phenols exist in a remarkable position. Many flavonoids have been used since ancient times as important natural pigments or dyestuffs and as medicines. These flavonoids are useful for the health of daily human life because some colorful anthocyanins with their glucosides have effects related to color therapy and, nutraceutical and functional values for foodstuffs. This Chapter concentrates on flavonoids in *Sophora* plants, and the relationship of the chemical structures to the benefits of their nutraceuticals. For this purpose, soy-based infant formulas, osteoporosis, antitumor activity, antimicrobial activity, anti-HIV activity, radical generation and O_2^- scavenging activity, and enzyme inhibitory activity have been described. This review will be useful to researchers in the field of effectiveness of *Sophora* nutraceuticals in disease prevention and treatment.

The third chapter, "Quantitative Structure-Cytotoxicity Relationship of Bioactive Heterocycles by the Semi-empirical Molecular Orbital Method with the Concept of Absolute Hardness" by Mariko Ishihara, Hiroshi Sakagami, Masami Kawase, and Noboru Motohashi, presents the relationship between the cytotoxicity (defined as 50% cytotoxic concentration) of heterocycles such as phenoxazine, 5-trifluoromethyloxazoles, *O*-heterocycles such as 3-formylchromone and coumarins, and vitamin K_2 derivatives against some tumor cell lines and 15 chemical descriptors. The results suggest the importance of selecting the most appropriate descriptors for each cell type and compound. The review is of interest as it represents the relationship of the molecular structures with the cytotoxic activity of these heterocycles.

The fourth chapter, "The Chemistry of Bioactive Mesoionic Heterocycles" by Masami Kawase, Hiroshi Sakagami, and Noboru Motohashi, presents the latest developments in the biologically interesting mesoionic compounds, such as sydnones, sydnonimines, 1,2,3,4-oxatriazolium-5-aminates, and 1,3,4-thiadiazolium-2-aminides, mainly since 2000. Most mesoionics are generally unstable and are not stable enough to be assayed for their biological activities in vitro. However, the mesoionics are endowed with charged, yet net neutral electrical character and their chemical properties are valuable assets in their potential usefulness as medicinal agents. Several mesoionic compounds have shown a wide spectrum of biological activity. Especially, their properties as NO-releasing compounds have been extensively investigated and the biomedical mechanism of NO has been elucidated by several mesoionics utilized as a probe. This review could contribute to the development of novel drugs in medicinal chemistry.

The last chapter, "Bioactive Phenothiazines and Benzo[*a*]phenothiazines: Spectroscopic Studies, Biological and Biomedical Properties and Applications" by Jean-Jacques Aaron, Mame Diabou Gaye-Seye, Snezana Trajkovska, and Noboru Motohashi, presents recent advances in the spectroscopic, photophysical, photochemical, and analytical studies of bioactive phenothiazines and benzophenothiazines, including also their biological and biomedical properties, as well as recent results from the authors of this review Chapter. The electronic absorption and luminescence spectral properties, and the complexation and interactions of these compounds in organized media have been discussed. Moreover, various analytical and biomedical applications, relative to phenothiazines and benzophenothiazines have been described. Among the important biological and biomedical properties of these compounds, their neurological effects, their antibacterial, antifungal, antiviral, antiparasitic and antitumor activities, and their cytotoxicity have been particularly emphasized. This review especially shows the interest in studying the electronic absorption and luminescence spectroscopy of phenothiazines and benzophenothiazines, since some relationships between their spectral properties and their molecular structure as well as their biological activity have been observed.

It is hoped that this volume could lead to advances in the medicinal chemistry of flavonoids and anthocyanins in relation to human health and lead to higher interest among researchers in the field of phytochemical nutraceuticals, especially flavonoids and anthocyanins.

Tokyo, February 2009 Noboru Motohashi

Contents

Contents of Volume 15

Bioactive Heterocycles VI

Flavonoids and Anthocyanins in Plants, and Latest Bioactive Heterocycles I

Volume Editor: Motohashi, N.
ISBN: 978-3-540-79217-8

Top Heterocycl Chem (2009) 16: 1–40
DOI 10.1007/7081_2009_127

Published online: 5 March 2009

Anthocyanins as Functional Food Colors

Noboru Motohashi[1] (✉) · Hiroshi Sakagami[2]

[1]Meiji Pharmaceutical University, 2-522-1 Noshio, Kiyose-shi, 204-8588 Tokyo, Japan
noborumotohashi@jcom.home.ne.jp

[2]Division of Pharmacology, Department of Diagnostic and Therapeutic Sciences, Meikai University School of Dentistry, 1-1 Keyakidai, Sakado, 350-0283 Saitama, Japan

Abstract Anthocyanins, a proanthocyanidin-type of flavonoid, contain an abundance of functional phytochemicals and occur in fruits such as cranberry, blueberry, orange, apple and in vegetables such as tomato, sweet pepper, spinach, and radishes. Functional and essential diets have been ingested in daily life since the primitive era of history. When anthocyanins are coupled with some water-soluble sugar molecules, their color becomes red, yellow, violet, or blue. It is very intriguing that anthocyanins provide the colorful variety of pigments for pansies, petunias, plums, and other diverse flowers. Chlorophyll in various fruits and vegetables is the main green phyto-component, while anthocyanins are probably the most important visible plant pigments in the natural kingdom having specific colors. Anthocyanins have been clinically used in many folklore medicines worldwide. Anthocyanins could provide health benefits for age-related diseases as well as other diseases. Anthocyanins have higher antioxidant capacity against oxidative stress induced by excess reactive oxygen species (ROS) such as superoxide radicals, hydrogen peroxide, and thus the human body might be protected from oxidative injury by anthocyanins. On the basis of these facts, we review the synthesis of plant flavonoids and their ability to scavenge oxidants, inhibit or activate enzymes, and the safety of proanthocyanidins and anthocyanidins present in common foods.

Keywords Adsorption · Anthocyanins · Flavonoid Synthesis · Mutagenicity · Proanthocyanidins

Abbreviations

- Adriamycin (doxorubicin) (**76**)
- Afzelechin (**41**)
- Anthocyanins (**39**)
- Apigenin (flavone) (**67**)
- Ascorbic acid (organic acid) (**69**)
- Aurones (**28**)
- (+)-Catechin (**43**)
- Catechin-3-*O*-glucoside (anthocyanin) (**61**)
- Celebrex (celecoxib) (**75**)
- Chalcone (**57**)
- Cinnamic acid (**13**)
- Condensed tannins, proanthocyanidins (PA) ($n = 0, 1, 2, 3, -$) (**44**)
- Criodictyol (**30**)
- 4-Coumaroyl-CoA (**15**)
- Cyanidin (**51**)
- Cyanidin 3-pentose (**93**)
- Cyanidin 3-*O*-galactoside (**88**)
- Cyanidin 3-*O*-rutinoside (**89**)
- Cyanidin-3-*O*-glucoside (anthocyanin) (**60**)
- Cyanidin 3-*O*-xyloside (**95**)
- Daidzein (**21**)
- Delphinidin (**49**)
- Delphinidin 3-*O*-arabinoside (**91**)
- Dihydroflavonols (3-hydroxy-flavanones) (**32**)
- Dihydrokaempferol (3-hydroxy-flavanone) (**31**)
- Dihydromyricetin (**34**)
- Dihydroquercetin (**33**)

- Epiafzelechin (**53**)
- Epicatechin (**55**)
- (–)-Epicatechin (flavan-3-ol) (**62**)
- Epigallocatechin (**54**)
- (–)-Epigallocatechin (flavan-3-ol) (**63**)
- Etoposide (VP-16)(**8**)
- Flavan-3-ols (**40**)
- *epi*-Flavan-3-ols (**52**)
- Flavanone (**82**)
- Flavone (**77**)
- Flavones (**29**)
- Flavonoids (**58**)
- Flavonol glycosides (**36**)
- Flavonols (**35**)
- Gallocatechin (**42**)
- (+)-Gallocatechin (flavan-3-ol) (**64**)
- Genistein (**22**)
- 3-Hydroxy-anthocyanidins (**38**)
- 7β-Hydroxycholesterol (**5**)
- 25-Hydroxycholesterol (**6**)
- 4-Hydroxycinnamic acid (*p*-coumaric acid) (**14**)
- 2′-Hydroxyflavanone (**83**)
- 4′-Hydroxyflavanone (**84**)
- 6-Hydroxyflavanone (**85**)
- 7-Hydroxyflavanone (**86**)
- 3-Hydroxyflavone (flavonol) (**78**)
- 5-Hydroxyflavone (flavonol) (**79**)
- 7-Hydroxyflavone (flavonol) (**80**)
- 2′-Hydroxy-4′-methoxyisoflavones (**24**)
- 2′-Hydroxy-4′-methoxyisoflavonones (**25**)
- Ibuprofen (**72**)
- Isoflavones (**20**)
- Isoflavonoids (pterocarpans) (**26**)
- Isoorientin *C*-glycoside (flavone *C*-glycoside) (**66**)
- Kaempferol (**9**)
- Ketoconazole (**4**)
- Leucocyanidin (**48**)
- Leucodelphinidin (**46**)
- Leucopelargonidin (**47**)
- Liquiritigenin (7,4′-dihydroxyflavanone, a flavanone) (**19**)
- Luecoanthocyanidins (flavan-3,4-diols) (**37**)
- Luteolin (flavone) (**65**)
- Malonyl-CoA (**16**)
- Malvidin (**71**)
- Malvidin 3-*O*-glucoside (**90**)
- Medicarpin (**27**)
- 4′-Methoxyisoflavones (**23**)
- Morin (flavonol) (**81**)
- Myricetin (**11**)
- Naproxen (**73**)

- Naringenin (flavanone) (**1**)
- Naringin (flavanone *O*-glycoside) (**2**)
- Oxidized tannins (oxidized proanthocyanins, **45**)
- Pelargonidin (**50**)
- Pelargonidin 3-*O*-glucoside (**94**)
- Peonidin (**70**)
- Peonidin 3-*O*-glucoside (**92**)
- Peonidin 3-*O*-xyloside (**96**)
- Petunidin (**70′**)
- Phenylalanine (**12**)
- Resveratrol (stilbene) (**7**)
- Rutin (flavonol *O*-glycoside) (**3**)
- Quercetin (**10**)
- Quercetin-3-glucoside (flavonol *O*-glycoside) (**68**)
- Quercetin-3-*O*-rhamnoside
  (quercitrin, quercetin-3-L-rhamnoside, Flavonol *O*-glycoside) (**59**)
- Taxifolin (**87**)
- Tetrahydroxychalcone (**18**)
- Trihydroxychalcone (**17**)
- Triketo acid (**56**)
- Vioxx (refecoxib) (**74**)

1 Introduction

Anthocyanins are a proanthocyanidin-type of flavonoid distributed in various fruits. These anthocyanins are the most important visible plant pigments in the natural kingdom. Anthocyanins have been clinically used in many folklore medicines worldwide for the treatment of age-related diseases and other disease. This review presents the functionality of anthocyanins in relation to their chemistry, synthetic pathway, antioxidant activity, antitumor activity (including apoptosis-inducing activity), pharmacodynamics (absorption, metabolism, distribution, and excretion) and toxicity, and discusses their possible use as food and dietary supplements and usage in potential nutraceuticals.

2 Background

There are an abundance of colorful and multifunctional phytochemicals or nutraceuticals such as carotenoids, xanthophylls, flavonoids, and anthocyanins in nature. These colorful phytochemicals are present in our daily intake of fruits and vegetables, and sometimes from shrimp and sea fish. Additionally, the colorful phytochemicals are also distributed in processed foods such as juices, jams, teas, wines, infusions and decoctions, extracts, and other foods. The flavonoids belong to a group of polyphenolics such as gallic acid, chlorogenic acid, caffeic acid, betanin, ginkgetin, quercetin, and delphinidin. Generally, polyphenolics display various intriguing functionality originating from the phytochemicals (plant chemicals).

Phytochemicals with different colors such as anthocyanins and flavonoids could attract much attention for two main reasons: Firstly, these color-rich phytochemicals might generally be safer when compared to artificial colorants such as sudan red III, and other synthetic pigments [1]. Long-term intake of artificial food additives might be harmful to human health. Secondly, patents might not be necessary for the use of plant colorants.

Today, populations in many developed countries including the USA, Canada, France, UK, Japan, and others may think that three major nutritive elements—carbohydrates (sugar, glucide), proteins, and lipids—are sufficiently supplied in daily life. On the other hand, it has been also believed that many people need greater intake of phytochemicals such as vitamins, bio-minerals, and diverse colorful phytochemicals such as natural carotenoids and flavonoids, because age-related diseases such as hypertension, diabetes, heart failure, macular degeneration and cataracts, and cerebral infarction might be caused by deficiency of these colorful phytochemicals, especially plant-derived flavonoids. The flavonoids of minor phytochemicals are especially abundant in vegetables and fruits and might maintain human health.

Interestingly, an aglycon naringenin (**1**), a citrus flavonoid, could stimulate DNA repair against oxidative damage in LNCaP human prostate cancer cells (Fig. 1). It was suggested that the stimulation of the DNA base excision repair processes might prevent mutagenic changes in the prostate cancer cells [2, 3].

On the other hand, a neoangiogenesis of new blood vessels is necessary for the development and progression of the tumor. Many solid tumors might induce vascular proliferation by the production of angiogenic factors such as the vascular endothelial growth factor (VEGF). Glycosylated flavonoids such as naringin (**2**) and rutin (**3**) could work at the lowest concentration against MDA human breast cancer cells (Fig. 1). The inhibition of VEGF release by these flavonoids might be due in part to a novel mechanism for mammary cancer prevention. This means that anthocyanin flavanone glycosides such as naringin and rutin have been proven to be potent inhibitors of angiogenic peptide vascular endothelial growth factor, which is released from human tumor cells [3, 4].

Generally, it is known that there is a positive correlation between the risk of metabolic syndrome and fruit and vegetable intake, based on the serum concentration of inflammatory markers such as *C*-reactive protein (CRP). CRP is a globulin that forms a precipitate with the somatic *C*-polysaccharide of *Pneumococcus* in vitro and is the most predominant acute phase protein. Then, the appearance of CRP in the serum indicates whether the patients have infectious inflammation or not.

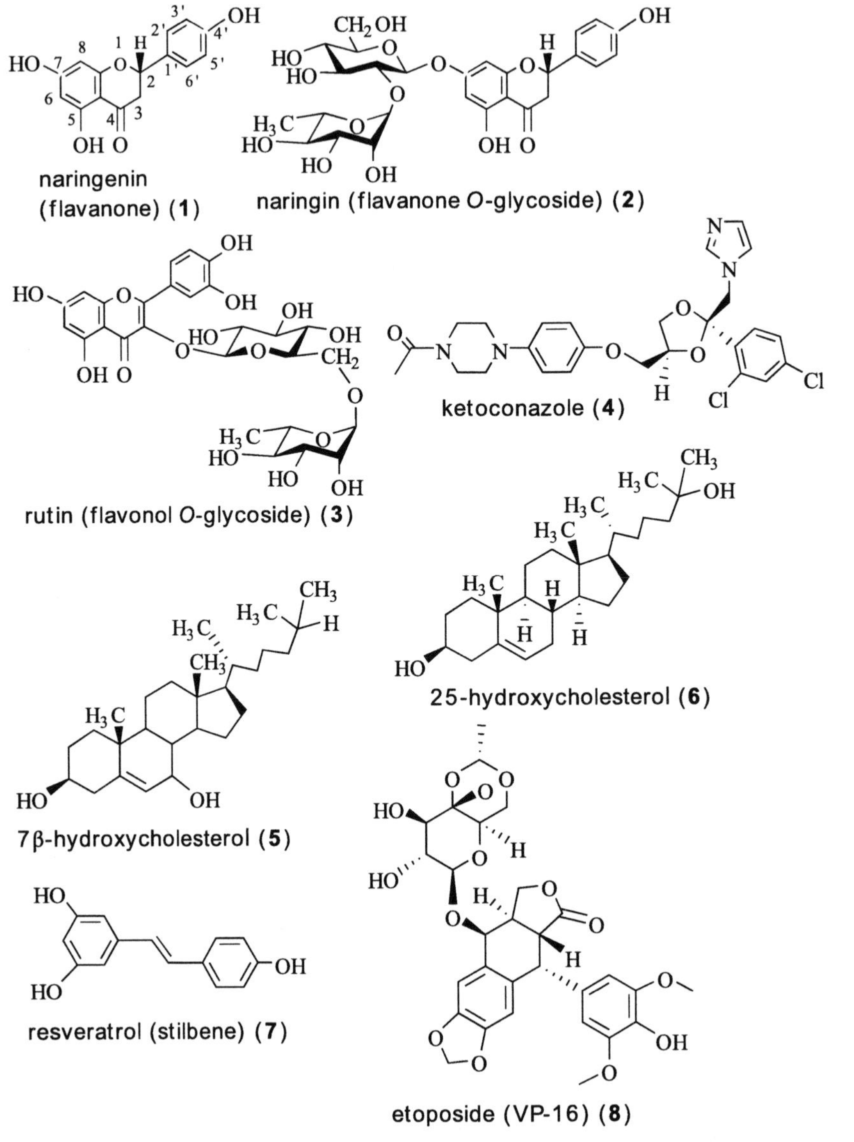

Fig. 1 Structures of naringenin (flavanone, **1**), naringin (flavanone *O*-glycoside, **2**) and rutin (flavonol *O*-glycoside, **3**), ketoconazole (**4**), 7β-hydroxycholesterol (**5**) and 25-hydroxycholesterol (**6**), resveratrol (**7**) and etoposide (**8**, VP-16)

The intake of fruit and vegetables was assessed with the use of a validated semiquantitative food-frequency questionnaire using 486 female teacher volunteers, aged between 40 and 60, in Tehran, Iran. The blood pressure of the volunteers was measured by standard methods. Metabolic syndrome was evaluated on the basis of the National Cholesterol Education Program Adult Treatment Panel III guidelines. Interestingly, daily fruit and vegetable intake by the two volunteer groups was 228 ± 79 and 186 ± 88 g/day, respectively. Both fruit and vegetable intake was inversely associated with plasma CRP concentrations. After correction for age, body mass index, and waist circumference, the mean plasma CRP concentrations across the five increasing quintile groups were determined: 1.94, 1.79, 1.65, 1.61, and 1.56 mg/L in the fruit intake group, and 2.03, 1.82, 1.58, 1.52, and 1.47 mg/L in the vegetable intake group. These inverse associations remained significant after additional correction for other potential confounding variables and dietary factors. After correction for potential confounders, persons with the highest quantity of fruit intake had a 34% lower chance of having metabolic syndrome (95% cardiac index (CI): 20%, 46%) and persons with the highest quantity of vegetable intake had a 30% lower chance (95% CI: 16%, 39%), when compared to persons with the lowest quantity. This result suggests that higher intake of fruit and vegetables might be associated with a lower risk of metabolic syndrome, possibly due to lower CRP concentrations. It might suggest that the daily intake of sufficient fruit and vegetables could prevent early possible cardiovascular disease [5].

Epidemiologically, it is known that antioxidative polyphenols in the human daily diet could protect us from age-related diseases such as cancer, diabetes and hypertension. The cytoprotective potential of the flavonoids such as rutin (**3**), kaempferol (**9**), quercetin (**10**) and myricetin (**11**) were examined against oxidative DNA damage in human colonocytes in vitro. Caco-2 cells, with specialized enterocyte/colonocyte cell functions, were used as an in vitro model for human colonocytes. Hydrogen peroxide (H_2O_2), one of the active oxygens, was used as an oxidant. The extent of DNA damage such as double-strand breakage of the DNA base pair and oxidization of purines and pyrimidines of DNA bases was determined by alkaline single-cell gel electrophoresis or comet assay. Both cell growth and viability were measured (Fig. 2).

Hydrogen peroxide dose-dependently increased the DNA strand breakage in human colonocytes, maybe via oxygen free radical generation. Two flavonoids of quercetin (**10**) and myricetin (**11**), protected Caco-2 cells from oxidative injury by active oxygen. Additionally, quercetin (**10**) decreased the hydrogen peroxide-mediated inhibition of growth. Two other flavonoids, rutin (**3**) and kaempferol (**9**), were ineffective. However, quercetin (**10**), which inhibits DNA double-strand breakage, could not suppress the oxidization of bases caused by hydrogen peroxide. Interestingly, ketoconazole (**4**) (Figs. 1 and 2), an antifungal agent, prevented quercetin (**10**) cytoprotection in Caco-2 cells, indicating that P450-mediated metabolism might alter the efficacy of the flavonoids against oxidative DNA damage. Quercetin (**10**), the most abundant flavonoid in our daily diet, may play an important role in the defense of human colonocytes against oxidative attack by active oxygens [6–8].

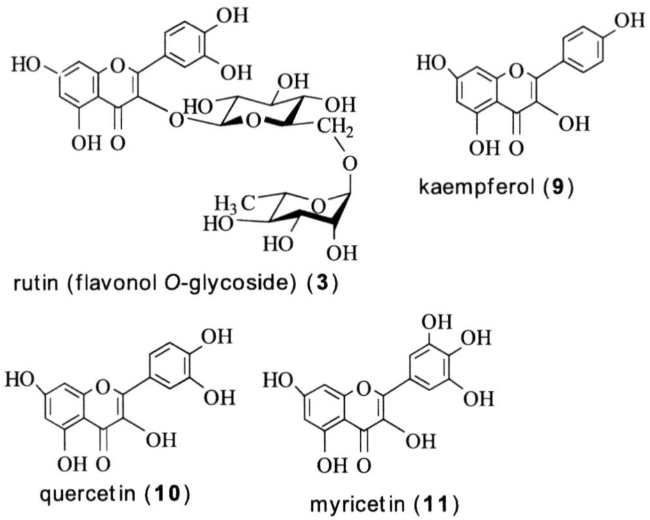

Fig. 2 Structures of four flavonols of rutin (flavonol *O*-glycoside) (**3**), kaempferol (**9**), quercetin (**10**) and myricetin (**11**)

Similarly, a high consumption of fruits and vegetables could contribute to a healthy life style because fruits and vegetables could decrease the incidence of cancers. The beneficial effects might be derived from many kinds of phytochemicals such as vitamins, dietary fibers and anthocyanins [9]. The modulatory effects of phytochemicals were examined for their cytotoxicity, genotoxicity and oxidative injury in cell systems. In this study, first, the potential beneficial effects of flavonoids are demonstrated. Phytochemicals such as flavonoids and β-carotenes might exhibit both antioxidant and free-radical scavenging activities [10, 11].

The protective effect of flavonoids was tested against H_2O_2-induced DNA damage in Caco-2 and HepG2 cells, with the comet assay. The two cells were first treated for 24 h with increasing concentrations of three flavonoids, rutin (**3**), quercetin (**10**), and myricetin (**11**) (Figs. 1 and 2), followed by exposure to H_2O_2 (50 μM) for 30 min. The exposure to H_2O_2 for 30 min at 37 °C induced significant DNA damage, but, pre-incubation with the flavonoids before H_2O_2 exposure significantly protected Caco-2 and HepG2 cells from H_2O_2-induced DNA damage. Second, the oxysterol-induced toxicity was examined by the cellular models. Oxysterols are generated during the cooking and processing of foods and might be also produced endogenously by the oxidation of membrane lipids. It is known that oxysterols might modulate their cytotoxicity by inducing apoptosis [12–14].

Two hydroxycholesterols, 7β-hydroxycholesterol (**5**) and 25-hydroxycholesterol (**6**) can commonly be found in daily foods (Fig. 1). The two hydroxycholesterols (**5**, **6**) were examined for their abilities to induce apoptosis in human monocytic leukemia cell line U937 and hepatoma cell line HepG2. The U937 and HepG2 cells were incubated for up to 48 h with 30 μM hydroxysterol. 7β-Hydroxycholesterol (**5**) induced apoptosis in U937 cells, as judged by the nonrandom DNA fragmentation, condensed and fragmented nuclei, and the generation of hypodiploid cells. However, contrastively, these two hydroxysterols (**5**, **6**) might possibly induce necrotic cell death by a different mechanism in the hepatoma cells [15] (Fig. 1).

The ability of these antioxidative phytochemicals in fruits, vegetables, plants, herbs and beverages to modulate chronic diseases was evaluated, based on the etiological point. The concentration of phytochemicals (flavonoids, stilbenes and plant extracts) that inhibits cell growth by 50% (IC_{50}) was determined by their antioxidant and genoprotective activities against U937 cells under oxidatively stressful conditions. U937 cells were treated with resveratrol (**7**) (Fig. 1), stilbene and citroflavan-3-ol, and four plant extracts, grapeseed polyphenols, olive leaf extract, bearberry, and *Echinacea purpurea*, and the viability was estimated by the fluorescein diacetate/ethidium bromide assay. U937 cells were pretreated with the test compounds at levels below the IC_{50}, and then exposed to oxidants such as 0.5 μM etoposide (VP-16) (**8**) (Fig. 1), a topoisomerase II inhibitor, 100 μM hydrogen peroxide (H_2O_2) or 400 μM *tert*-butylhydroperoxide (*t*BOOH). Their intracellular reduced glutathione levels were measured as an indicator of oxidative stress. The result showed that resveratrol (**7**) had the highest protective effect (IC_{50} = 13.7 μg/mL), and *Echinacea* had the lowest protective effect (IC_{50} = 9400 μg/mL). The etoposide (**8**)-induced oxidative stress was efficiently reduced by the olive leaf extract and bearberry. The grape seed polyphenols and bearberry were highly protective against H_2O_2- and *t*BOOH-induced DNA damage. These results suggest that non-nutrient dietary phytochemicals such as citroflavan-3-ol play a role as significant biofunctional phytochemicals and that the plant extracts such as bearberry, grapeseed polyphenols, and olive leaf extract protect the cells from the oxidative stress [16].

3
Flavonoid Synthesis in Plants

Flavonoids are abundantly distributed in the natural kingdom and have been known for a long time. Flavonoids might have evolved to protect plants from oxidative damage from sunlight (ultraviolet rays). When compared to alkaloids, flavonoids might not show specific toxicity like poisons. Humans eat daily colorful plants that contain various flavonoids and anthocyanins from birth. Each flavonoid has distinct properties, and these might help us to understand the relationships between their colorful structures and biological activity [17].

Studies have mainly concerned production and accumulation of diverse flavonoids in angiosperms [18–20]. At the same time, the major steps leading to various types of flavonoid subgroups by plant enzymes have also been examined in cell-free systems [21].

Plants have many enzymes involved in secondary metabolism, its precursor proteins and ultimate metabolites.

In addition, the initial source of variation in enzyme functions might depend on random mutations that occurred in their genes and the chromosomal rearrangements for enzymes which produce diverse flavonoid pigments as their final functional phytochemicals. The functional variations might occur in different parts of the plant, generating different evolutionary paths [22].

3.1 General Phenylpropanoid Pathway (12–16) to Stilbenes (7)

Phenylalanine ammonia-lyase (PAL) eliminates the amino group from phenylalanine (**12**) to produce cinnamic acid (**13**). Cinnamate-4-hydroxylase (C4H) hydroxidizes compound (**13**) to yield *p*-coumaric acid (**14**). 4-Coumaroyl:CoA-ligase (4CL) complex catalyzed the conversion of *p*-coumaric acid (**14**) and coenzyme A (CoA) to 4-coumaroyl-CoA (**15**) and 3 moles malonyl-CoA (**16**). Stilbene synthase (STS) converts these two compounds (**15**, **16**) into resveratrol of stilbene (**7**) (Fig. 3) [23, 24].

3.2 Isoflavonoid (26) Pathway via Isoflavones (20) from a General Phenylpropanoid Pathway (4-Coumaroyl-CoA (15) and Malonyl-CoA (16))

Chalcone synthase (CHS) and chalcone reductase (CHR) convert 4-coumaroyl-CoA (**15**) and 3 mol malonyl-CoA (**16**) to trihydroxychalcone (a chalcone) (**17**) via tetrahydroxychalcone (a chalcone) (**18**). Chalcone isomerase (CHI) converts trihydroxychalcone (a chalcone) (**17**) into liquiritigenin (7,4′-dihydroxyflavanone, a flavanone) (**19**). Isoflavone synthase (IFS) converted flavanone (**19**) to isoflavones (**20**) such as daidzein (**21**) and genistein (**22**). Isoflavone 2′-hydroxylase (I2′H) hydroxylated isoflavones (**20**) to 4′-methoxyisoflavones (**23**). Isoflavone 2′-hydroxylase (I2′H) hydroxylated 4′-methoxyisoflavones (**23**) into 2′-hydroxy-4′-methoxyisoflavones (**24**). Isoflavone reductase (IFR) reduced 2′-hydroxy-4′-methoxyisoflavones (**24**) to 2′-hydroxy-4′-methoxyisoflavonones (**25**). Finally, vestitone reductase (VR) and 4′-methoxyisoflavanol dehydrogenase (DMID) cyclized 2′-hydroxy-4′-methoxyisoflavonones (**25**) to form isoflavonoids (**26**) such as medicarpin (**27**) (Fig. 4) [23, 24].

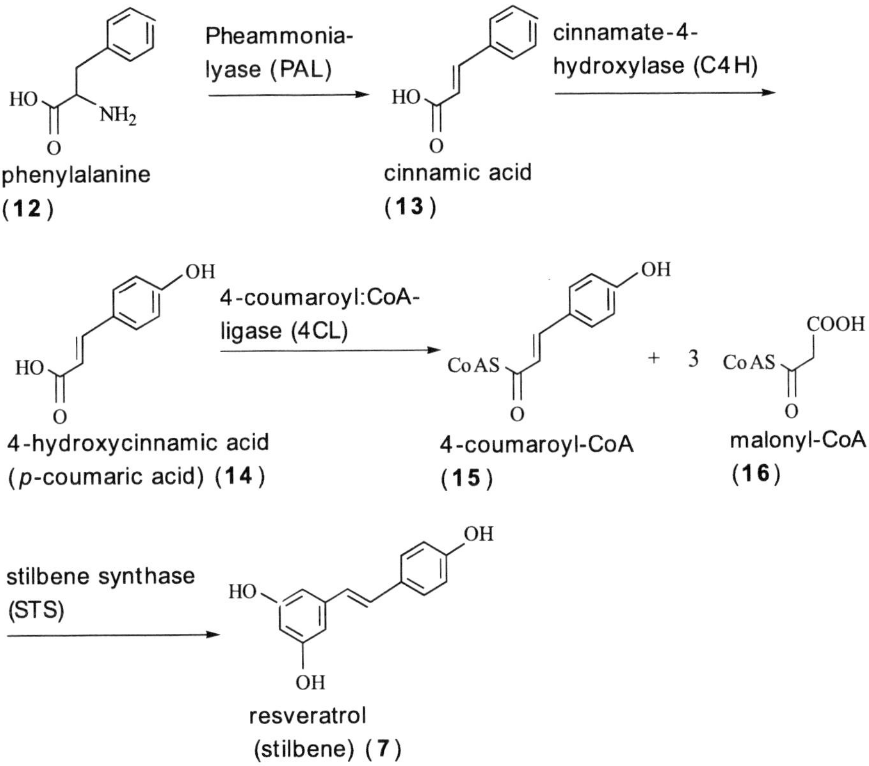

Fig. 3 Biosynthesis of stilbenes in plants

3.3 Aurones (28) from a General Phenylpropanoid Pathway (4-Coumaroyl-CoA (15) and Malonyl-CoA (16))

Aurones (**28**) were synthesized from 4-coumaroyl-CoA (**15**) and malonyl-CoA (**16**) via tetrahydroxychalcone (chalcone, **18**) and trihydroxychalcone (**17**)), using chalcone synthase (CHS) and chalcone reductase (CHR). Aureusidin synthase (AS) converted trihydroxychalcone (**17**) to aurones (**28**) (Fig. 5) [23, 24].

3.4 Flavones (29) from Tetrahydroxychalcone (18)

Tetrahydroxychalcone (**18**) was converted to naringenin (**1**) by chalcone isomerase (CHI), and then to flavones (**29**) by flavone synthase I (FSI) and flavone synthase II (FSII), respectively (Fig. 6).

Naringenin (**1**) was also converted to criodictyol (flavanone, **30**) by flavonoid 3′-hydroxylase (F3′H), and then to flavones (**29**) by FSI and FSII,

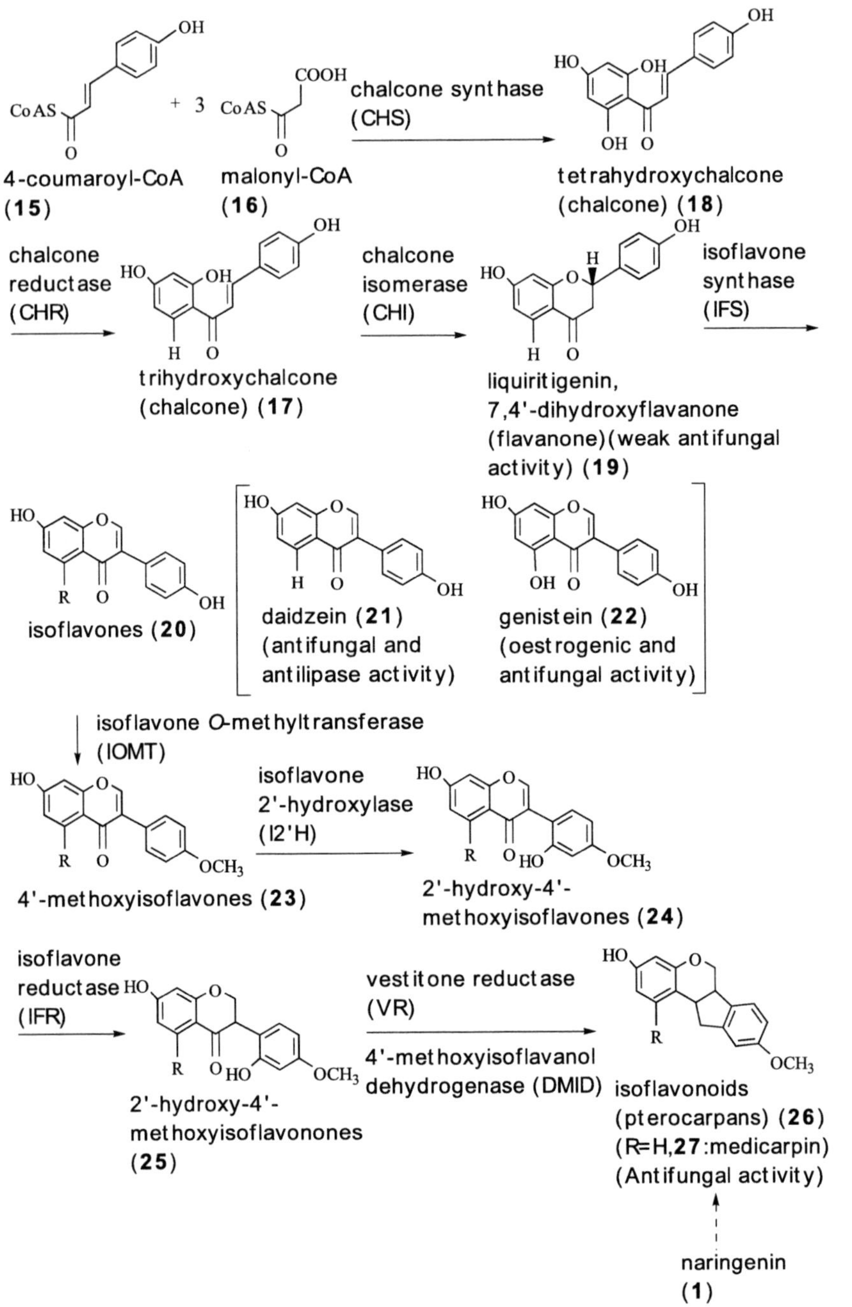

Fig. 4 Biosynthesis of isoflavonoids (pterocarpans) via isoflavones from phenylpropanoid pathway

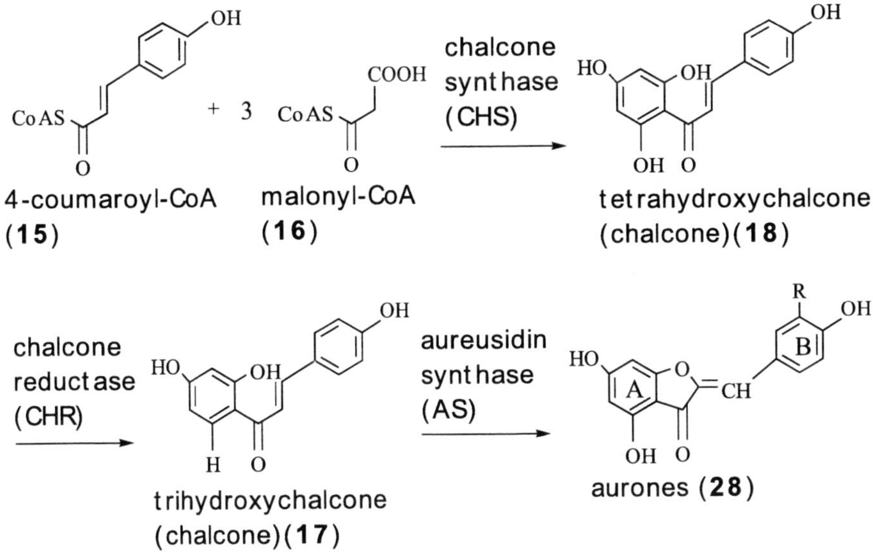

Fig. 5 Biosynthesis of aurones (28) from phenylpropanoid pathway

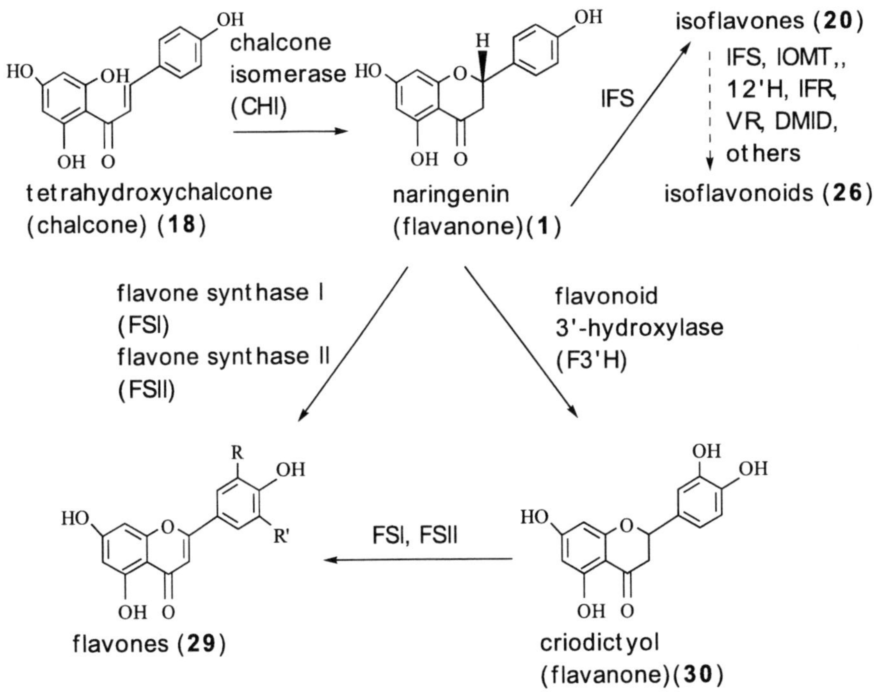

Fig. 6 Biosynthesis of flavones (29) from tetrahydroxychalcone (chalcone, 18)

respectively. On the other hand, naringenin (**1**) was converted to isoflavones (**20**) by isoflavone synthase (IFS) and then finally to isoflavonoids (**26**) by many enzymes (Fig. 6) [23].

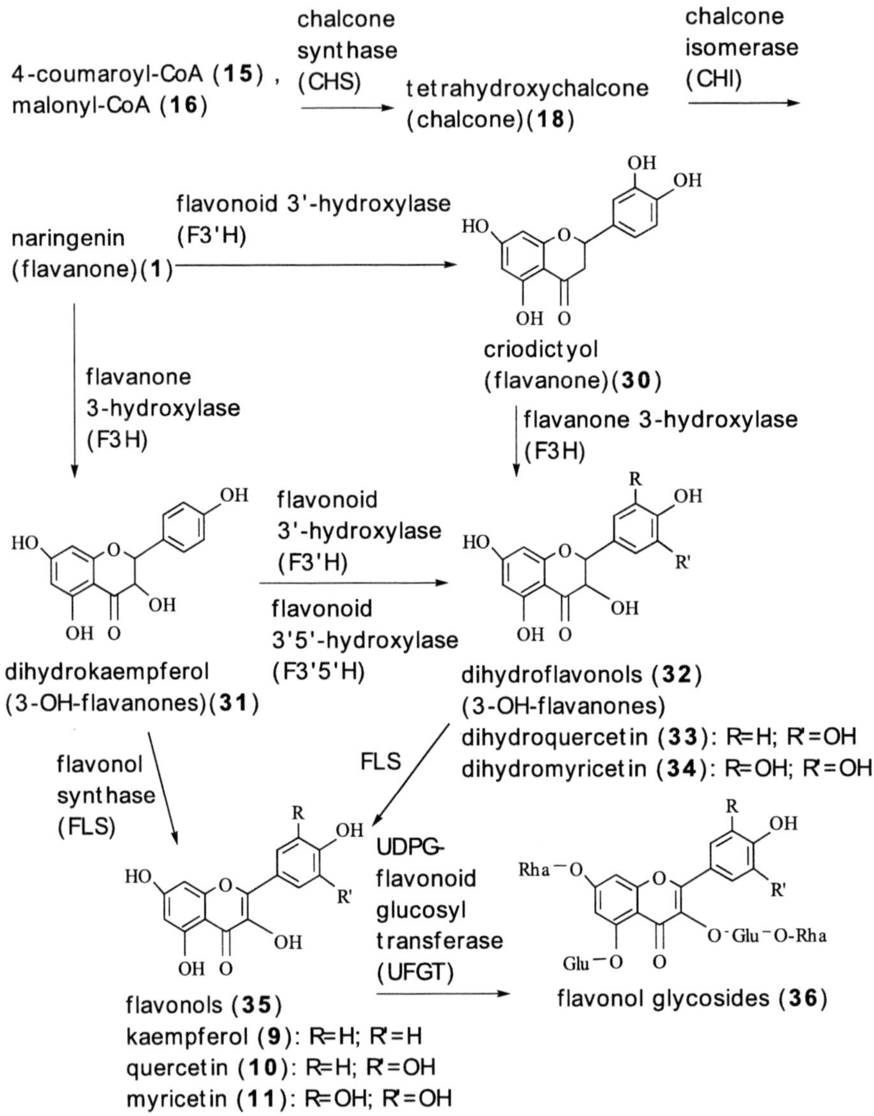

Fig. 7 Biosynthesis of flavonols (35) via naringenin (1) from tetrahydroxychalcone (chalcone, 18)

3.5 Biosynthesis of Flavonols (35) via Naringenin (1) from Tetrahydroxychalcone (Chalcone, 18)

Flavanone 3-hydroxylase (F3H) converted naringenin (**1**) to dihydrokaempferol (3-OH-flavanone) (**31**). Flavonoid 3′-hydroxylase (F3′H) and flavonoid 3′5′-hydroxylase (F3′5′H) converted dihydrokaempferol (**31**) to dihydroflavonols (3-OH-flavanones) (**32**) such as dihydroquercetin (**33**) and dihydromyricetin (**34**).

Flavonol synthase (FLS) converted dihydrokaempferol (**31**) to flavonols (**35**). Similarly, flavonol synthase (FLS) converted dihydroflavonols (3-OH-flavanones) (**32**) to flavonols (**35**) such as kaempferol (R = H; R′ = H) (**9**), quercetin (R = H; R′ = OH) (**10**) and myricetin (R = OH; R′ = OH) (**11**). Then, UDPG-flavonoid glucosyl transferase (UFGT) converted flavonols (**35**) to flavonol glycosides (**21**) (Fig. 7) [23, 24].

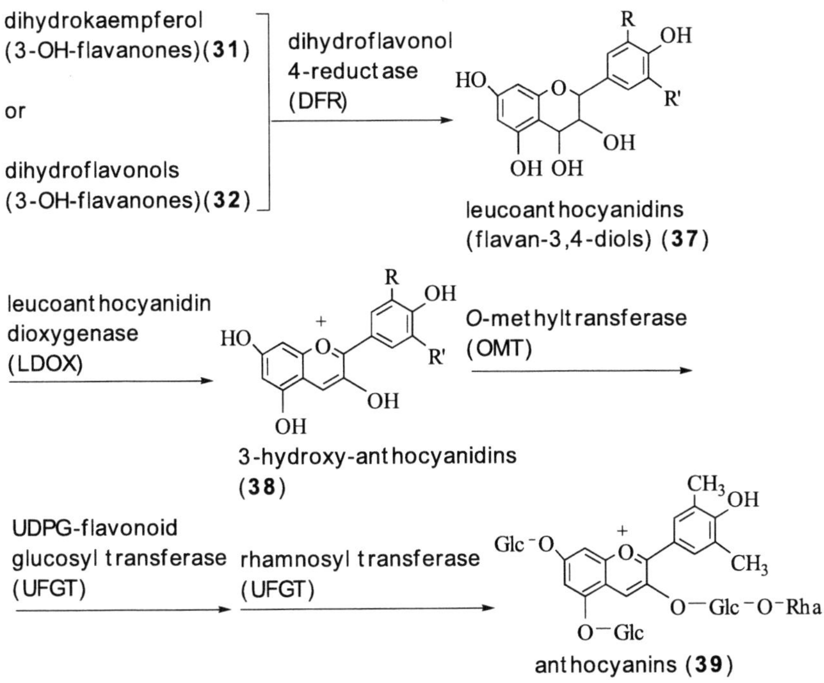

Fig. 8 Biosynthesis of anthocyanins (39) via leucoanthocyanidins (flavan-3,4-diols, 37) from dihydroflavonols (3-hydroxy-flavanones, 31, 32)

3.6
Biosynthesis of Anthocyanins (39) via Leucoanthocyanidins (Flavan-3,4-diols, 37) from Dihydroflavonols (3-OH-Flavanones, 31, 32)

Dihydroflavonol 4-reductase (DFR) converted dihydroflavonols (3-OH-flavanones, **32**) to leucoanthocyanidins (flavan-3,4-diols, **37**). The leucoanthocyanidins (flavan-3,4-diols, **37**) were converted by leucoanthocyanidin dioxygenase (LDOX) to 3-hydroxy-anthocyanidins (**38**). Finally, 3-hydroxy-anthocyanidins (**38**) were converted by three enzymes of *O*-methyltransferase (OMT), UDPG-flavonoid glucosyl transferase (UFGT) and rhamnosyl *trans*ferase (UFGT) to anthocyanins (**39**) (Fig. 8) [23, 24].

3.7
Biosynthesis of Flavan-3-ols (40) and Condensed Tannins (Proanthocyanidins, 44) from Leucoanthocyanidins (Flavan-3,4-diols, 37)

Leucoanthocyanidin reductase (LAR) converted leucoanthocyanidins (flavan-3,4-diols, **37**) to flavan-3-ols (**40**). The flavan-3-ols (**40**) were converted to condensed tannins (proanthocyanidins, PA, **44**) by condensing enzyme (CE). Following this condensed tannins (proanthocyanidins, PA, **44**) finally yielded their oxidized tannins (oxidized proanthocyanins, **45**) by proanthocyanidine oxidase (PRO) (Fig. 9) [23, 24].

3.8
Biosynthesis of 3-Hydroxy-anthocyanidins (38), *epi*-Flavan-3-ols (52) and Condensed Tannins (Proanthocyanidins, 44) from Leucoanthocyanidins (Flavan-3,4-diols, 37)

Leucoanthocyanidin dioxygenase (LDOX), an anthocyanidin synthase (ANS), converted luecoanthocyanidins (flavan-3,4-diols, **37**) to 3-hydroxy-anthocyanidins (**38**). The 3-hydroxy-anthocyanidins (**38**) were converted to *epi*-flavan-3-ols (**52**) by anthocyanidin reductase (ANR). Then, *epi*-flavan-3-ols (**52**) were converted to brownish condensed tannins (proanthocyanidins, PA, **44**) by condensing enzyme (CE). Similarly, condensed tannins (proanthocyanidins, PA, **44**) were converted to oxidized tannins (oxidized proanthocyanidins, **25**) by proanthocyanidine oxidase (PRO) (Fig. 10) [23, 24].

3.9
Biosynthesis of Flavonoids (58) from 4-Hydroxycinnamic Acid (14) and Acetic Acid

One mole 4-hydroxycinnamic acid (**14**) might react with 3 mol acetic acids to yield a precursor chalcone (**57**) and then flavonoids (**58**) (Fig. 11) [25, 26]. The proposed biosynthesis of flavonoids was confirmed in a *Antirrhinum majus*

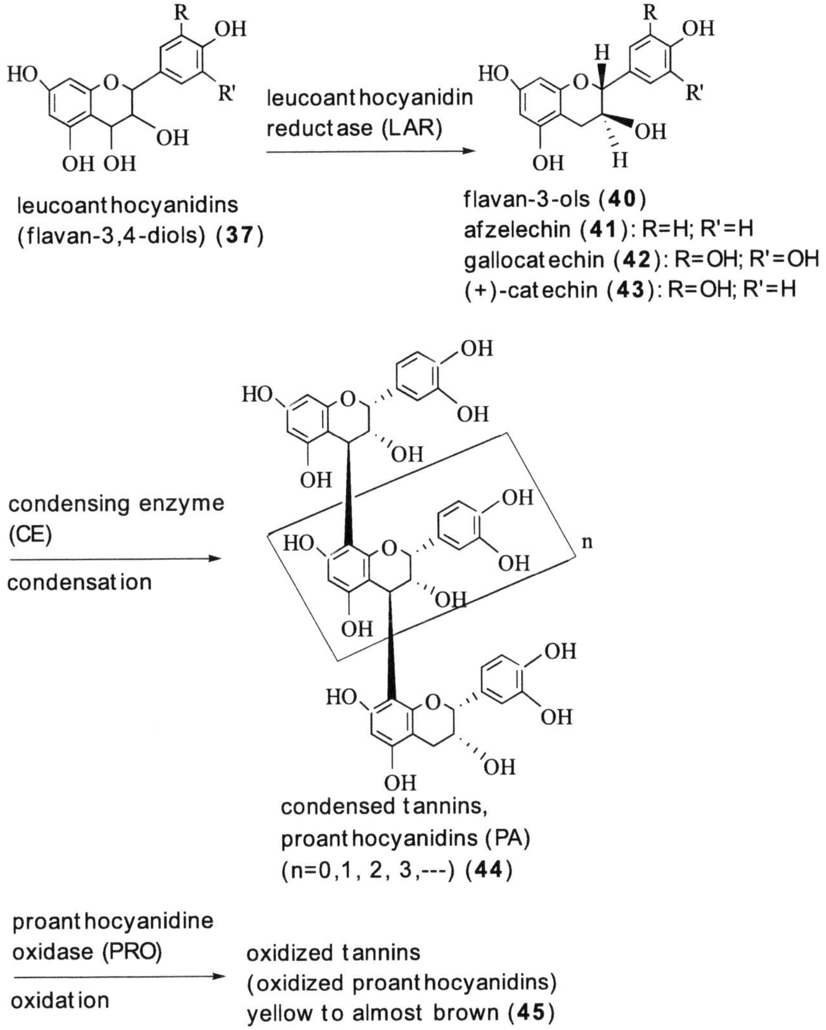

Fig. 9 Biosynthesis of flavan-3-ols (40), condensed tannins (proanthocyanidins, 44) and oxidized tannins (oxidized proanthocyanidines, 45) from leucoanthocyanidins (flavan-3,4-diols, 37)

mutant line that completely lacks these flavonoids, having only hydroxycinnamic acids [27].

The biosynthetic pathways of epicatechin (**55**), cyanidin (**51**) and proanthocyanidin (PA) (**44**) had the same intermediates to leucocyanidin (**48**). Next, leucoanthocyanidin reductase (LAR) converts the leucocyanidin (**48**) to epicathechin (**55**), whereas leucoanthocyanidin dioxygenase (LDOX) converted leucocyanidin (**48**) to cyanidin (**51**). Two cyanidin (**51**) and epicathechin (**55**) are a precursor of proanthocyanidins (PA) (**44**).

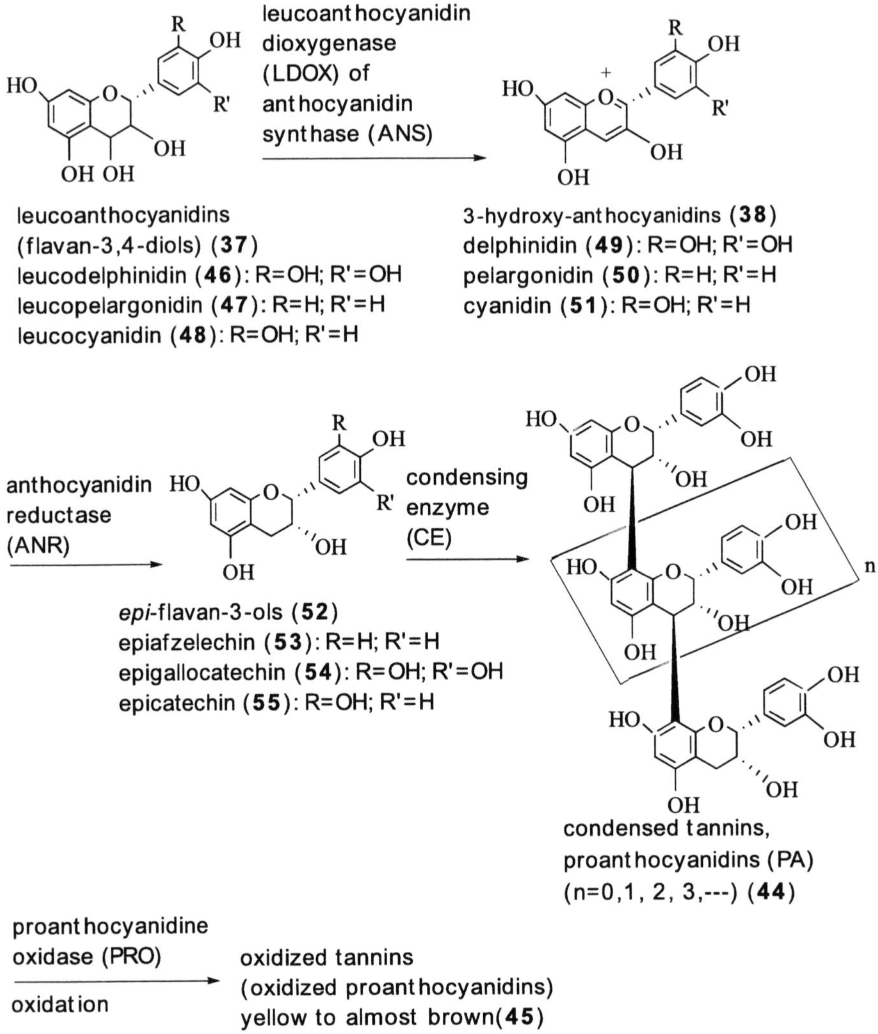

Fig. 10 Biosynthesis of 3-hydroxy-anthocyanidins (38), *epi*flavan-3-ols (52) and condensed tannins (proanthocyanidins, 44) from leucoanthocyanidins (flavan-3,4-diols, 37)

The *Arabidopsis* mutant tannin-deficient seed 4 (tds4-1) reduced the concentration and altered the accumulation pattern of proanthocyanidins (PA, **44**). By complementation of the tds4-1 mutation either with a cosmid encoding LDOX or a 35S, the LDOX construct was identified on the TDS gene. The independent *Arabidopsis* lines with a T-DNA insertion in the LDOX gene represented a similar phenotype, and a line was allelic to tds4-1. Anthocyanidin reductase encoded by the *BANYULS (BAN)* gene is the core enzyme in proanthocyanidin (PA, **44**) biosynthesis [28]. The seed pheno-

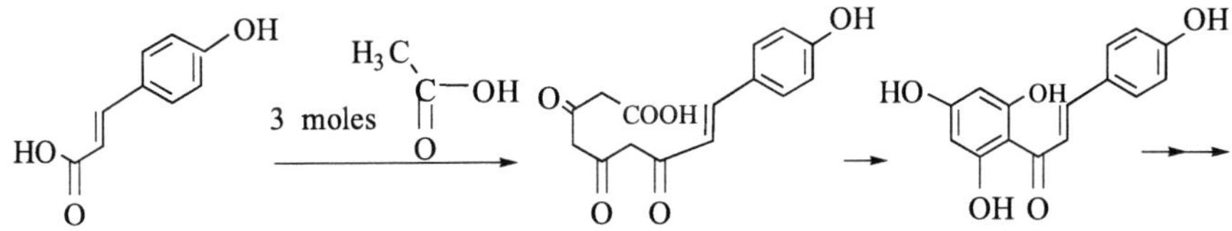

4-hydroxycinnamic acid (*p*-coumaric acid) (**14**)

triketoacid (**56**)

chalcone (**57**)

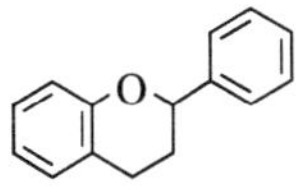

flavonoids (**58**)

Fig. 11 Biosynthesis of flavonoids (58) from 4-hydroxycinnamic acid (14) and acetic acid

type of ban tds4 double mutants showed that LDOX precedes the *BANYULS (BAN)* gene in the proanthocyanidin (PA, **44**) pathway by the identification of the *BANYULS (BAN)* gene as an anthocyanidin reductase. Double-mutant analysis was also used to order the other TDS genes. Analysis of the proanthocyanidin (PA, **44**) intermediates in tds4-1 revealed the accumulation of three dimethylaminocinnamaldehyde (DMACA) reacting compounds in extracts from developing seeds. *Arabidopsis*, unlike many other plants, exclusively use the epicatechin pathway for the synthesis of proanthocyanidins (PA, **44**), and not the catechin pathway. Transmission electron microscopy (TEM) showed that when seeds of tds4 were stained with DMACA, proanthocyanidin (PA, **44**) intermediates were accumulated in the cytoplasm of endothelial cells. The fluorescent marker dyes were used to show that the tds4 endothelial cells had multiple small vacuoles, instead of a large central vacuole observed in the wild types (WT). These results show that in addition to its established role in the formation of anthocyanin, LDOX is also part of the proanthocyanidin (PA, **44**) biosynthesis pathway (Fig. 12) [28, 29].

Through the phenotypic characterization of the *Arabidopsis thaliana* transparent *testa12 (tt12)* mutant encoding a membrane protein of the multidrug and toxic efflux transporter family, *tt12* is suggested to be involved in the vacuolar accumulation of proanthocyanidin precursors in the seed. The metabolite analysis of *tt12* seeds revealed the absence of flavan-3-ols and proanthocyanidins, and the reduction of quercetin-3-*O*-rhamnoside (**59**), the major flavonol. The *tt12* promoter is active in cells synthesizing proanthocyanidins. By using translational fusions between *tt12* and green fluorescent protein, the transporter was found to be localized in the tonoplast. Yeast vesicles expressing *tt12* could transport the anthocyanin cyanidin-3-*O*-glucoside (**60**) in the presence of MgATP, but not the aglycones, cyanidin and epicatechin. The inhibitor studies demonstrate that *tt12* could act in vitro as a cyanidin-3-*O*-glucoside (**60**)/H^+-antiporter. The *tt12* does not transport glycosylated flavonols and procyanidin dimers, and a direct transport ac-

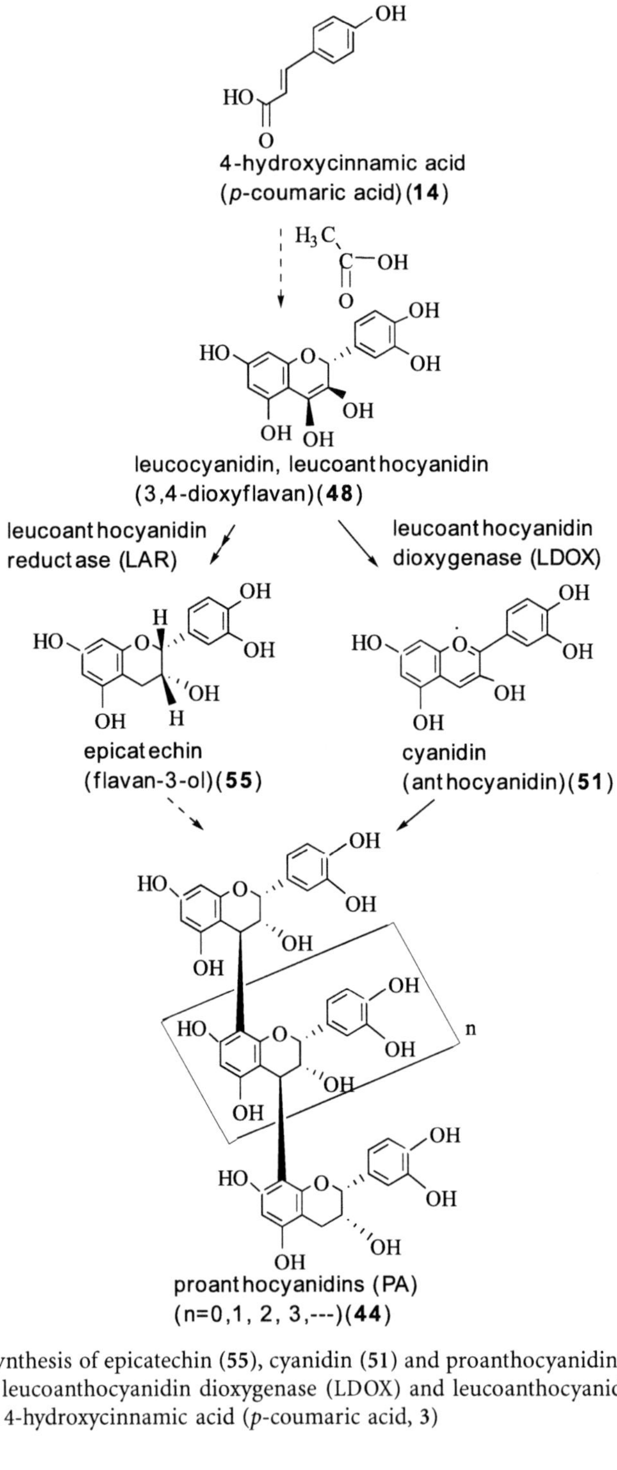

Fig. 12 Biosynthesis of epicatechin (55), cyanidin (51) and proanthocyanidins (44) by two enzymes of leucoanthocyanidin dioxygenase (LDOX) and leucoanthocyanidin reductase (LAR) from 4-hydroxycinnamic acid (*p*-coumaric acid, 3)

tivity for catechin-3-*O*-glucoside (**61**), a glucosylated flavan-3-ol, was not detectable. However, catechin-3-*O*-glucoside (**61**) inhibited *tt12*-mediated transport of cyanidin-3-*O*-glucoside (**60**) in a dose-dependent manner, while flavan-3-ol aglycones and glycosylated flavonols had no effect on anthocyanin transport. It is proposed that *tt12* could transport the glycosylated flavan-3-ols in vivo. *Mutant banyuls (BAN)* seeds could accumulate the anthocyanins instead of proanthocyanidins, yet the *ban tt12* double mutant showed the reduced anthocyanin accumulation. These results suggest that *tt12* might be involved in anthocyanin transport in vitro (Fig. 13) [30].

Camellia sinensis L. such as green tea leaves contains abundantly functional catechins such as (+)-catechin (**43**), (–)-epicatechin (**62**), (–)-epigallocatechin (**63**) and (+)-gallocatechin (**64**). Interestingly, these flavan-3-ols might play a crucial role in the defense of the human body against various bacteria (Fig. 14).

The anthocyanidin reductase enzyme recently described in *Arabidopsis* and *Medicago* was shown to be present in tea with very high activity and could produce epicatechin as well as epigallocatechin from the respective anthocyanidins, thus explaining very high contents of flavan-3-ols. Especially, two enzymes, dihydroflavonol 4-reductase and leucoanthocyanidin 4-reductase selectively catalyze key steps in their biosynthesis of catechins (**43**, **62**) and gallocatechins (**63**, **64**), respectively (Fig. 14) [31].

Condensed tannins (CTs) are flavonoid oligomers, many of which have beneficial effects on animal and human health. (–)-Epicatechin (**62**) of the

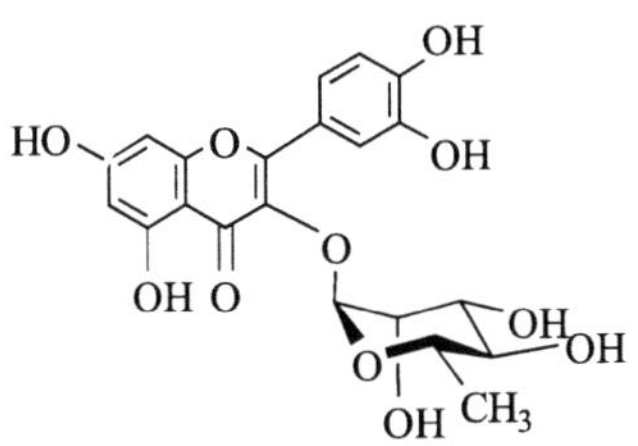

quercetin-3-*O*-rhamnoside
(Flavonol *O*-glycoside) (**59**)

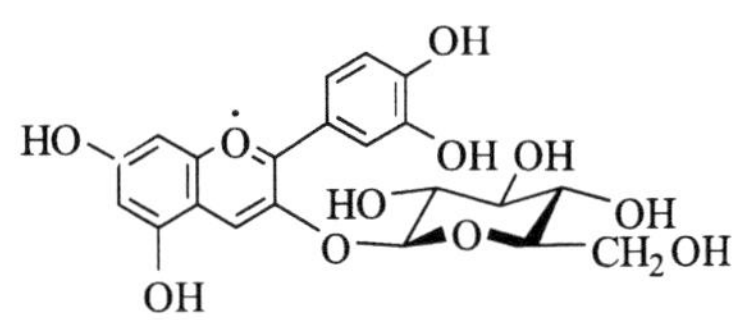

cyanidin-3-*O*-glucoside
(anthocyanin) (**60**)

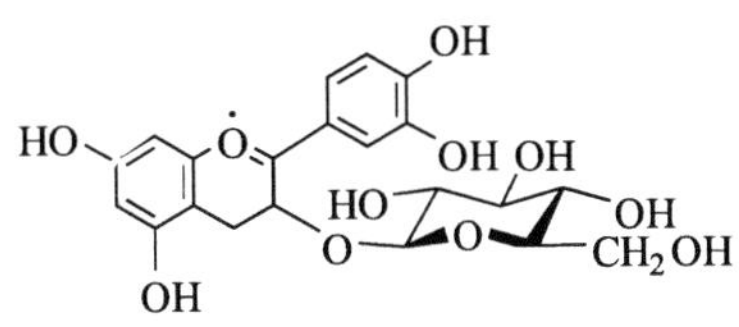

catechin-3-*O*-glucoside
(anthocyanin) (**61**)

Fig. 13 Structures of quercetin-3-*O*-rhamnoside (**59**), cyanidin-3-*O*-glucoside (**60**) and catechin-3-*O*-glucoside (anthocyanin) (**61**)

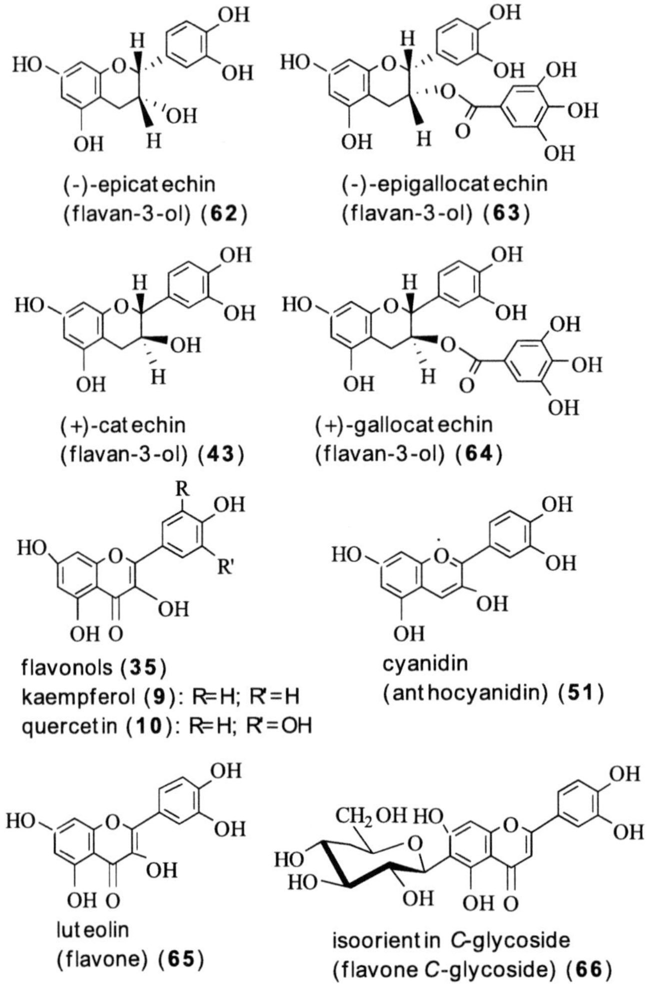

Fig. 14 (–)-Epicatechin (flavan-3-ol, **62**), (–)-epigallocatechin (flavan-3-ol, **63**), (+)-catechin (flavan-3-ol, **43**) and (+)-gallocatechin (flavan-3-ol, **64**) in leaves of tea (*Camellia sinensis* L.), and **analogs** of luteolin (flavone, **65**) and isoorientin *C*-glycoside (flavone *C*-glycoside, **66**) in angiosperms

flavan-3-ol is a component of diverse condensed tannins and gives particular flavor and astringency to teas and wines.

Then, *BANYULS (BAN)* genes from *Arabidopsis thaliana* and *Medicago truncatula* encode anthocyanidin reductase, which could convert anthocyanidins to the corresponding 2,3-*cis*-flavan-3-ols (**43, 62, 63, 64**). Interestingly, the ectopic expression of *BAN* in tobacco flower petals and *Arabidopsis* leaves induced the accumulation of condensed tannins without production of 2,3-*cis*-flavan-3-ols (**43, 62, 63, 64**) (Fig. 14) [32].

4
Oxidant Scavenging Capacity

Considering the importance and putative modes of action of specific flavonoids as bioactive components of the diet in in vivo and in vitro models, the consumed fruit and vegetables of mixed varieties available in the UK were analyzed for the composition of major phenolic components. Total phenolic content and the vitamin C levels were also determined. The antioxidant capacities of aqueous and methanolic extracts were comparatively examined using the TEAC (trolox equivalent antioxidant capacity), FRAP (ferric reducing ability of plasma), and ORAC (oxygen radical absorbance capacity) assays. These assays estimate polyphenols, simple phenols, and ascorbate components, respectively. The antioxidant activity was calculated using 100 g (raw weight) of each uncooked material. Among many fruits and vegetables, strawberry, raspberry, and red plum were the richest in anthocyanins (which had the highest antioxidant activities), followed by orange and grapefruit (rich in flavanones) and onion, leek, spinach, and green cabbage (rich in flavonols). On the other hand, apple, tomato, pear, and peach rich in hydroxycinnamates such as 4-hydroxycinnamic acid (*p*-coumaric acid) (**14**) showed consistently lower antioxidant activities. The TEAC, FRAP, and ORAC values of the individual extracts were similar to each other in good relationship with the total phenolic and vitamin C contents. The antioxidant activity measured by TEAC on a 100 g raw uncooked portion was in the following order: strawberry $\gg$ raspberry = red plum $\gg$ red cabbage $\ggg$ grapefruit = orange > spinach > broccoli > green grape $\approx$ / = onion > green cabbage > pea > apple > cauliflower > tomato $\approx$ / = peach = leek > banana $\approx$ / = lettuce. This means that the antioxidative activity of fruits and vegetables might mainly be total activity of polyphenols, ascorbic acid, β-carotene, and other components in the carotenoids (Fig. 3) [33].

Fourteen daily common European fruit and vegetable juices such as apple, beetroot, blueberry, carrot, elderberry, lemon, lingonberry, multivitamin, orange, pink grapefruit, sauerkraut, and tomato juices including sour cherry nectar were investigated for their total oxidant scavenging capacity (TOSC) against three reactive oxygen species (ROS)—peroxyl radical, hydroxyl radical, and peroxynitrite. These three ROS showed both physiological and technological relevance. An apparent concentration-dependent antioxidant capacity could not be observed due to the complicated assay systems [34].

The antioxidative role of anthocyanins and flavones in providing stable blue flower colors in the angiosperms has been studied. The leaf anthocyanins and flavones showed a protective effect against UV-B radiation. These protective UV-B flavonoids in angiosperms were mainly kaempferol (**9**), quercetin (**10**), cyanidin (**51**), luteolin (**65**), and isoorientin *C*-glycoside (flavone *C*-glycoside, **66**) (Fig. 14) [35].

The antioxidant potencies of several common dietary flavonoids were examined when compared with vitamin C as a positive control. Antioxidant effects of flavonoids and vitamin C on the oxygen radical-generated DNA damage in human lymphocytes were examined by the comet assay with single-cell gel electrophoresis. Pretreatment with all flavonoids and vitamin C dose-dependently inhibited the oxidative DNA damage. At 279 μmol/L, luteolin (**65**) showed the highest antioxidant potency (**65**, 9% of damage caused by hydrogen peroxide), followed by myricetin (**11**, 10%), quercetin (**10**, 22%), kaempferol (**9**, 32%), quercitrin (**59**, quercetin-3-L-rhamnoside, 45%), apigenin (**67**, 59%), quercetin-3-glucoside (**68**, 62%), ascorbic acid (control, **69**, 78%), and rutin (quercetin-3-β-D-rutinoside, **3**, 82%). This indicated that kaempferol (**9**), quercetin (**10**), myricetin (**11**), and uteolin (**32**) except rutin (**3**), apigenin (**67**), and quercetin-3-glucoside (**68**) had higher antioxidant activity when compared to ascorbic acid (**69**).

The extent of DNA damage was co-related with ED_{50} (concentration to produce 50% protection) values. The protective effect of quercetin (**10**) and ascorbic acid (**69**) at a concentration of 23.2 μmol/L was found to be additive (quercetin (**10**) alone: 71% of maximal DNA damage; ascorbic acid (**69**): 83%; in combination: 62%). Therefore, free flavonoids are more protective than the conjugated flavonoids, i.e. quercetin (**10**) vs. its conjugate quercetin-3-glucoside (**68**) ($P < 0.001$). In addition, the antioxidant activity of free flavonoids might be also related to both the number and position of hydroxyl groups in the flavonoid structures (Figs. 14 and 15) [36].

Furthermore, preventing initial DNA damage may be protective against cardiovascular disease, certain forms of cancer, and aging. The protective effects of quercetin (**10**) against oxidative DNA damage and the formation of bulky DNA adducts were studied in human lymphocytes both in vitro and ex vivo. First, human lymphocytes were pre-incubated with various concentrations of quercetin (**10**), followed by incubation with hydrogen peroxide. Their protection against oxidative DNA damage was measured by the single-cell gel electrophoresis (comet) assay. Second, the quercetin (**10**)-treated human lymphocytes were treated with carcinogenic benzo[*a*]pyrene, and a benzo[*a*]pyrene-DNA adduct formation was measured by ^{32}P-postlabeling. Third, in a pilot study, lymphocytes from female volunteers who consumed a quercetin (**10**)-rich blueberry/apple juice mixture for four weeks were treated ex vivo with an effective dose of hydrogen peroxide and benzo[*a*]pyrene, respectively, at three different time points, i.e. before ($t = 0$ weeks), during ($t = 2$ weeks), and after ($t = 4$ weeks) the intervention. Quercetin (**10**) dose-dependently protected the cells from oxidative DNA damage ($p < 0.01$) and the formation of benzo[*a*]pyrene-DNA adducts ($p < 0.05$). Administration of juice in vivo for 4 weeks significantly increased the total antioxidant capacity of plasma, and increased the plasma quercetin (**10**) concentration from 5.0 to 10.6 nM ($p = 0.03$). After the intervention for 4 weeks, the level of the oxidative damage upon ex vivo exposure to hydrogen

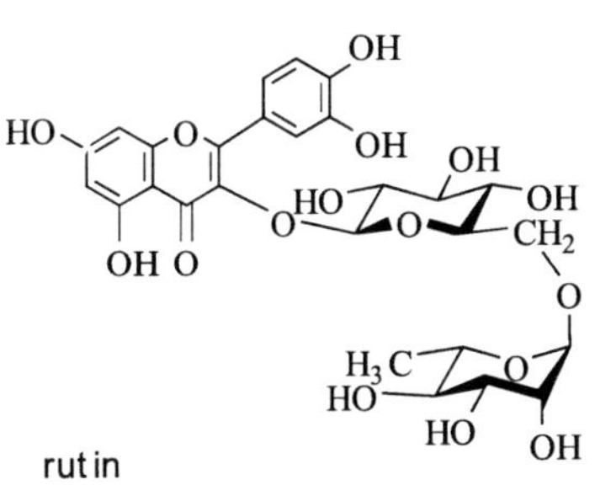

rutin
(flavonol *O*-glycoside) (**3**)

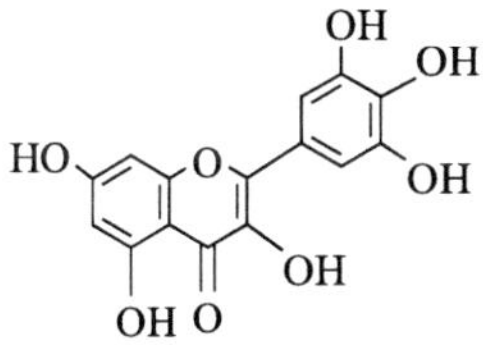

myricetin
(flavonol) (**11**)

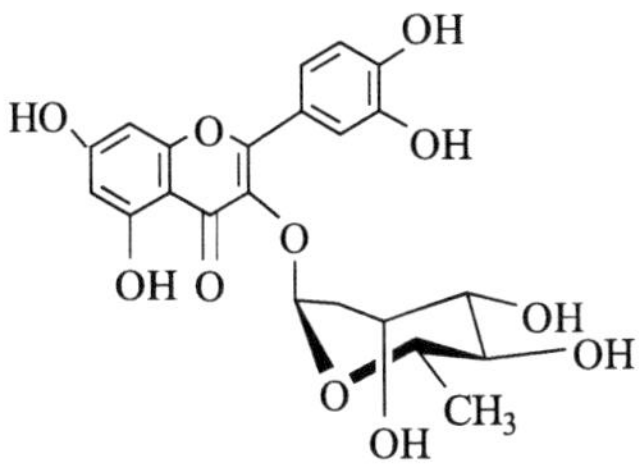

quercitrin (quercetin-3-L-rhamnoside)
(flavonol *O*-glycoside) (**59**)

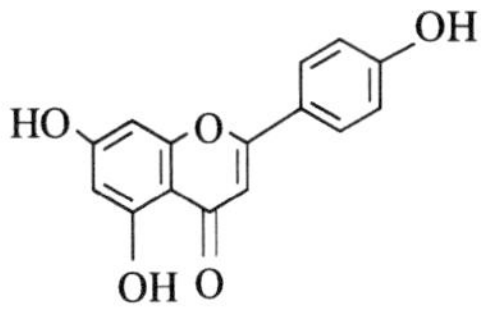

apigenin
(flavone) (**67**)

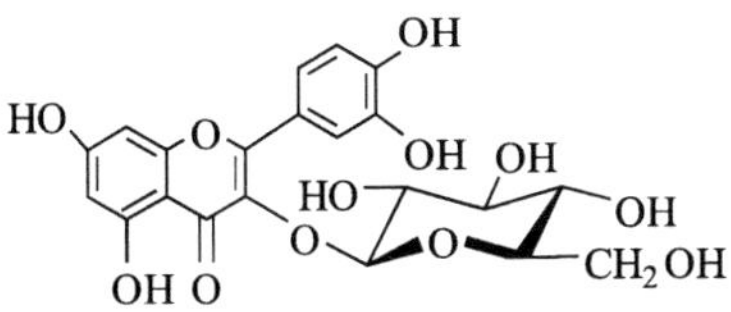

quercetin-3-glucoside
(flavonol *O*-glycoside) (**68**)

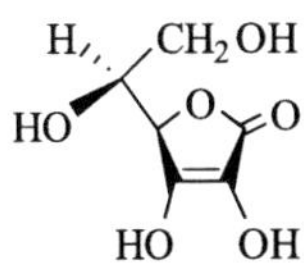

ascorbic acid
(organic acid) (**69**)

Fig. 15 Some dietary antioxidative flavonoids that induce DNA damage (control: ascorbic acid (**69**))

peroxide declined by 41%, and the benzo[*a*]pyrene-DNA adduct level induced ex vivo also diminished by 11%. In vitro and ex vivo tests suggest that quercetin (**10**) could protect human lymphocytes from chemically induced DNA damage [37].

5 Inhibition or Activation of Enzyme

From recent widespread consumption of diets rich in anthocyanin and catechin, the evaluation of their in vitro inhibitory effects on two cyclooxygenase

(COX) enzymes and on the proliferation of human cancer cell lines were examined.

Five anthocyanidins, cyanidin (**51**), delphinidin (**49**), pelargonidin (**50**), peonidin (**70**), and malvidin (**71**) at 40 μM concentration were tested for their ability to inhibit COX-1 and COX-2 activity. Eleven epimer catechins, (+)-catechin (**43**), (–)-catechin, (±)-catechin, (+)-epicatechin, (–)-epicatechin (**62**), (–)-epigallocatechin, (–)-gallocatechin, (–)-epicatechin gallate, (–)-catechin gallate, (–)-epigallocatechin gallate, and (–)-gallocatechin gallate were tested at 80 μM. Among these flavonoids, six galloyl derivatives of the catechins, cyanidin (**51**) and malvidin (**71**) showed the highest COX inhibitory activities when compared to the anti-inflammatory drugs ibuprofen (**72**, 10 μM), naproxen (**73**, 10 μM), vioxx (rofecoxib, **74**, 1.67 ppm), and celebrex (celecoxib, **75**, 1.67 ppm) (Fig. 16).

Antitumor activity against four human cancer cell lines, MCF-7 (breast), SF-268 (central nervous system, CNS), HCT-116 (colon), and NCI-H460 (lung) was investigated at 6.25–100 μM, in comparison with a popular anticancer drug, adriamycin (doxorubicin) (**76**) (6.25 μM). Five anthocyanidins, cyanidin (**51**), delphinidin (**49**), pelargonidin (**50**), peonidin (**70**), malvidin (**71**) and five catechins, (+)-catechin (**43**), (–)-catechin, (±)-catechin, (+)-epicatechin, and (–)-epicatechin (**62**) were inactive at 100 μM. Three galloyl-catechins, (–)-gallocatechin, (–)-epigallocatechin gallate, and (–)-gallocatechin gallate inhibited the growth of MCF-7 cells by 95, 100, and 97%, respectively, at 50 μM. Two galloyl-catechins, (–)-gallocatechin and (–)-gallocatechin gallate, were the most effective against HCT-116 cells, inhibiting their growth by 85% and 93%, respectively, and against NCI-H460 cells, 87% and 67%, respectively, at 50 μM. The CNS cells were the most sensitive among the test cell lines, and total growth inhibition was obtained with two catechins of (–)-gallocatechin and (–)-gallocatechin gallate at 100 μM. Overall, only six galloyl derivatives of six catechins, (–)-epigallocatechin, (–)-gallocatechin, (–)-epicatechin gallate, (–)-catechin gallate, (–)-epigallocatechin gallate, and (–)-gallocatechin gallate inhibited the proliferation of these cancer cell lines (Fig. 16) [38].

Seven structurally related flavones, kaempferol (**9**), quercetin (**10**), flavone (**77**), 3-hydroxyflavone (3-OH flavone, **78**), 5-hydroxyflavone (5-OH flavone, **79**), 7-hydroxyflavone (7-OH flavone, **80**), and morin (**81**) were investigated for their antitumor activity against three human malignant colorectal carcinoma cells, HT29, COLO205, and COLO320-HSR. Among seven flavones (flavonols), flavone (**77**) showed the highest cytotoxic activity against these three cell types, as judged by MTT assay. The cytotoxicity induced by flavone (**77**) was mediated by induction of apoptosis characterized by the appearance of the DNA ladders, apoptotic bodies, and hypodiploid cells.

The activation of caspase-3 assessed by the cleavage of poly ADP-ribose polymerase (PARP), a caspase-3 substrate, was confirmed in flavone-treated cells. Ac-DEVD-FMK (an inhibitory peptide for caspase-3), but not Ac-YVAD-

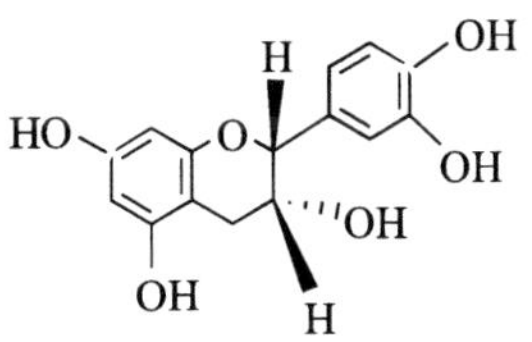

(-)-epicatechin (**62**)

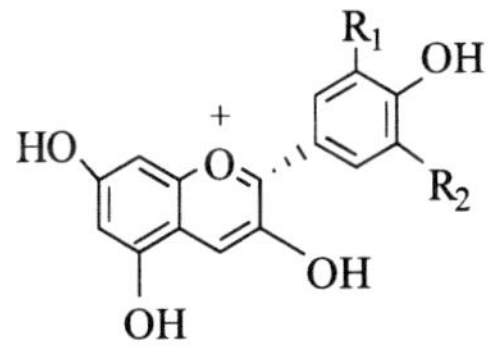

anthocyanidins
cyanidin (**51**): R_1=OH; R_2=H
peonidin (**70**): R_1=OCH_3; R_2=H
delphinidin (**49**): R_1=OH; R_2=OH
petunidin (**70'**): R_1=OCH_3; R_2=OH
malvidin (**71**): R_1=OCH_3; R_2=OCH_3

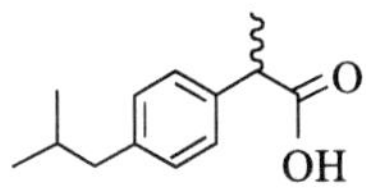

ibuprofen (**72**) (anti-inflammatory drug)

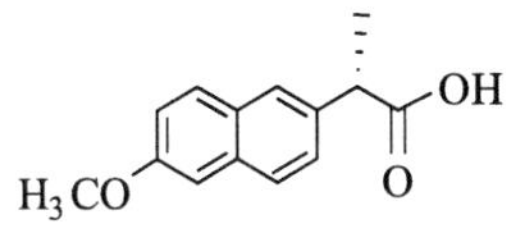

naproxen (**73**) (anti-inflammatory drug)

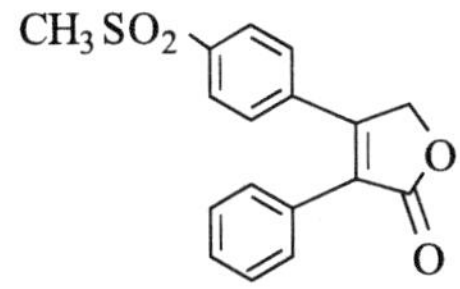

vioxx (refecoxib) (**74**) (anti-inflammatory drug)

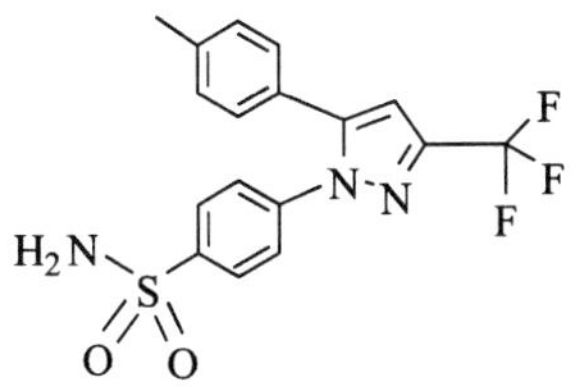

celebrex (celecoxib) (**75**) (anti-inflammatory drug)

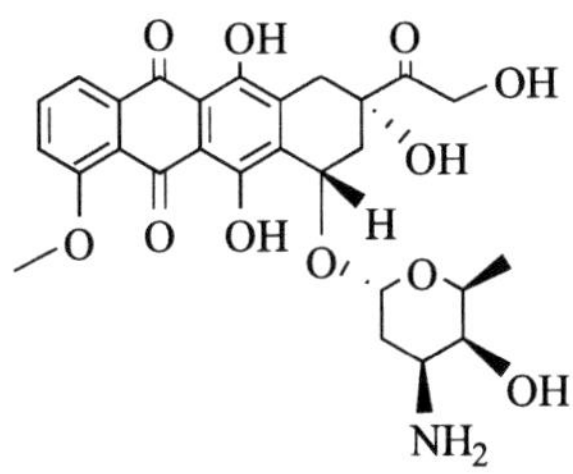

adriamycin (doxorubicin) (**76**) (anthracycline antibiotic)

Fig. 16 Antioxidative anthocyanidins, catechins, and anti-inflammatory and an antitumor agents

FMK (inhibitor for caspase-1), decreased the cytotoxic effect of flavones against COLO205 and HT29 cells. The elevation of p21, but not p53, could be observed in flavone-treated cells. Intracellular peroxide level (detected by DCHF-DA assay) was elevated in flavone-treated cells. Antioxidants such as tiron, catalase, SOD, PDTC, but not DPI, suppressed the flavone-induced cytotoxic effect. The in vivo antitumor test showed that the flavone inhibited their tumorigenesis by eliciting the apoptotic cells and an increase in p21 (not p53) in s.c. injection of COLO205 cells in nude mice. The p21 protein expression was induced in the tumor tissues derived from the flavone-treated group. Then, the flavones induced the apoptosis in the primary colon carcinoma cells

COLO205-X with appearance of DNA ladders, caspase-3 activation, and an increase in p21 (not p53) protein. The results indicate that these flavones might effectively induce apoptosis in colorectal carcinoma cells in vitro and in vivo (Fig. 17) [39].

Similarly, eight flavanones, naringenin (**1**), flavanone (**82**), 2′-hydroxyflavanone (2′-OH flavanone, **83**), 4′-hydroxyflavanone (4′-OH flavanone, **84**), 6-hydroxyflavanone (6-OH flavanone, **85**), 7-hydroxyflavanone (7-OH flavanone, **86**), and taxifolin (**87**) were investigated for their antitumor activity against three colorectal carcinoma cells—HT29, COLO205, and COLO320HSR—with the MTT assay. 2′-Hydroxyflavanone (**83**) flavanone showed highly potent cytotoxicity against these three cells, inducing DNA ladders, apoptotic bodies, and hypodiploid cells. The induction of caspase-3 protein processing and enzyme activity associated with the cleavage of PARP was identified in 2′-hydroxyflavanone (**83**)-treated cells. A peptidyl inhibitor (Ac-DEVD-FMK) of caspase-3 reduced the cytotoxicity of 2′-hydroxyflavanone (**83**) in COLO205 and HT-29 cells. The elevation of p21 (but not p53) and a decrease in Mcl-1 protein were found in 2′-hydroxyflavanone (**83**)-treated COLO205 and HT-29 cells. The elevation of intracellular ROS was detected in 2′-hydroxyflavanone (**83**)-treated cells by the 2′,7′-dichlorodihydrofluorescein diacetate (DCHF-DA) assay. ROS scavengers such as 4,5-dihydro-1,3-benzene disulfonic acid (tiron), catalase, superoxide dismutase (SOD), and pyrrolidine dithiocarbamate (PDTC) apparently suppressed 2′-hydroxyflavanone (**83**)-induced cytotoxicity. 2′-Hydroxyflavanone (**83**) significantly inhibited the growth of the tumor cells implanted in nude mice by s.c. injection of COCO205 cells. The HE staining method with immunohistochemical investigation revealed the increase of p21 (but not

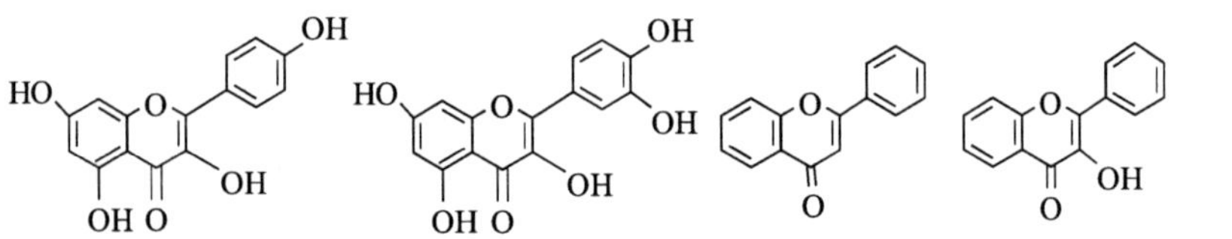

kaempferol (flavonol) (**9**)   quercetin (flavonol) (**10**)   flavone (flavone) (**77**)   3-hydroxyflavone (flavonol) (**78**)

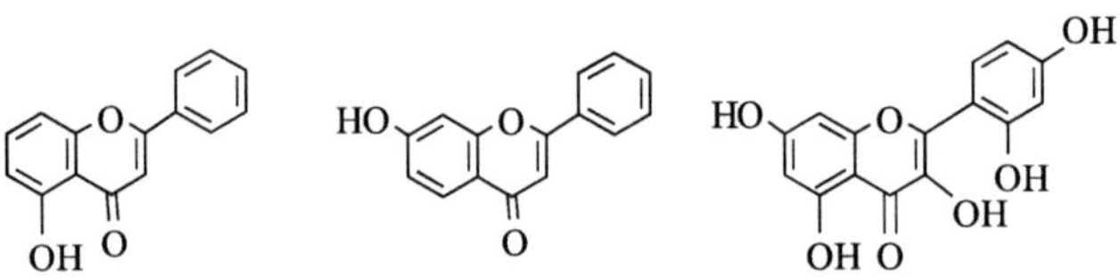

5-hydroxyflavone (flavonol) (**79**)   7-hydroxyflavone (flavonol) (**80**)   morin (flavonol) (**81**)

Fig. 17 Inhibition of tumor growth via apoptosis in vitro and in vivo by flavones (flavonols)

p53) protein in the tumor tissues by 2′-hydroxyflavanone (**83**) treatment. The primary tumor cells (COLO205-X) derived from COLO205 were confirmed. 2′-Hydroxyflavanone (**83**) induced apoptosis in COLO205-X cells, as judged from the appearance of DNA ladders, caspase-3 protein processing, PARP protein cleavage, and increasing p21 protein. These results revealed in vitro, ex vivo, and in vivo antitumor activities of 2′-hydroxyflavanone (**83**) via apoptosis induction. Another seven flavanones also induced apoptosis (Fig. 18) [40].

The development of androgen independence and resistance to apoptosis in prostate cancer might be often correlated with high levels of serum tumor necrosis factor (TNF)-α in these patients. The loss of sensitivity to TNF-α-induced apoptosis in androgen-insensitive prostate carcinoma cells might be due in part to the constitutive activation of Rel/nuclear factor (NF)-κB transcription factors that regulate several cell survival and antiapoptotic genes. The growth inhibitory and apoptotic activities of apigenin (**67** , a flavone) (Fig. 15) against a variety of human prostate carcinoma cells was studied by NF-κB expression in androgen-insensitive human prostate carcinoma cells exhibiting high constitutive levels of NF-κB. Then, the effect of apigenin (**67**) on androgen-insensitive human prostate carcinoma PC-3 cells was estimated by the NF-κB activation, using elec-

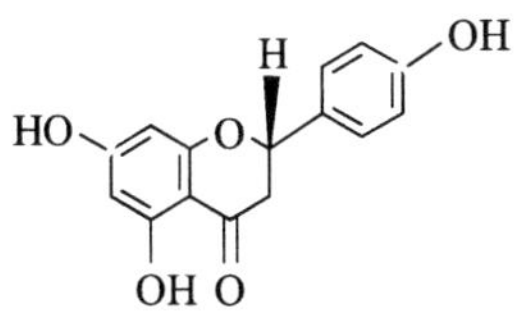

naringenin (flavanone) (**1**)

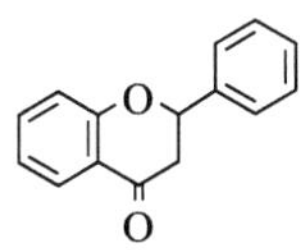

flavanone (**82**)

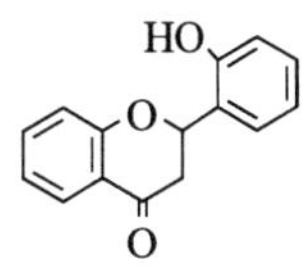

2′-hydroxyflavanone (**83**)

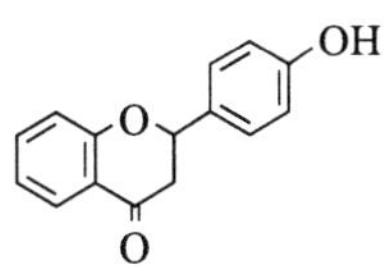

4′-hydroxyflavanone (**84**)

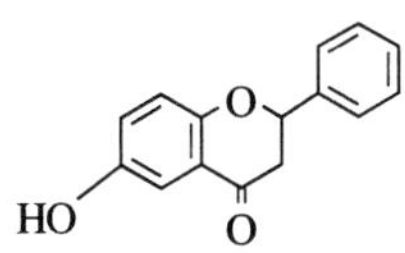

6-hydroxyflavanone (**85**)

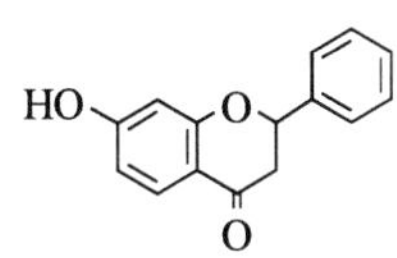

7-hydroxyflavanone (**86**)

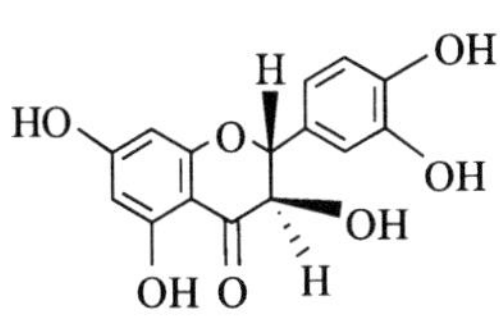

taxifolin (**87**)

Fig. 18 Some flavanones that inhibit the tumor growth

trophoretic mobility shift assay (EMSA) and reporter gene assay. The expression of NF-κB subunits p65 and p50, IκBα, p-IκBα, in-beads kinase assay and NF-κB-regulated genes were determined by Western blot analysis. Apoptosis was determined by fluorescence-activated cell-sorting analysis after annexin V/propidium iodide staining. Treatment of cells with 10–40 μM apigenin (**67**) inhibited the DNA binding and reduced the nuclear levels of the p65 and p50 subunits of NF-κB. Apigenin (**67**) inhibited the IκBα degradation and IκBα phosphorylation and significantly decreased IKKα kinase activity. Apigenin (**67**) also inhibited the TNF-α-induced activation of NF-κB via the IκBα pathway, thereby sensitizing the cells to the TNF-α-induced apoptosis. The inhibition of NF-κB activation might correlate with the decreased expression of NF-κB-dependent reporter gene and a suppressed expression of NF-κB-regulated genes including specifically Bcl2, cyclin D1, COX-2, matrix metalloproteinase 9, nitric oxide synthase-2 (NOS-2), and vascular endothelial growth factor (VEGF). These results suggest that the inhibition of NF-κB by apigenin (**67**) might suppress prostate cancer by transcriptional repression of the NF-κB-responsive genes as well as selective sensitization of the prostate carcinoma cells to the TNF-α-induced apoptosis (Fig. 15) [41].

6 Adsorption of Flavonoids

6.1 Rat Test Examples

Anthocyanin glycosides with different colors are the daily dietary plant pigments; they might prevent many diseases such as hypertension, diabetes, cardiac infarction, and eye insufficiency. The preventive and therapeutic properties of anthocyanins and flavonoids highly depend on absorption and metabolism in the living body after intake.

Generally, these anthocyanins are quickly absorbed as the glycoside forms in bodies of animals and humans [42–47].

Interestingly, administrated intact anthocyanins (glycosides) could be found in plasma in rats or humans only a few minutes after their oral administration without any structural changes [42, 44, 45].

6.1.1 Absorption of Anthocyanin Glycosides into the Stomach

It could be suggested that such rapid detections in animals or humans might come from their absorption through the gastric wall. Gastric absorptions of blackberry anthocyanins (14 and 750 μmol/L), bilberry anthocyanins

(88 μmol/L), and purified anthocyanins (14 μmol/L) as a control sample were compared 30 min after administration. Approximately 25% (average of 23–26.2%) of anthocyanin glycosides, such as cyanidin 3-*O*-glucoside (**60**) and malvidin 3-*O*-glucoside (**90**), and a galactoside of cyanidin 3-*O*-galactoside (**88**), were directly absorbed from the stomach. The absorption of cyanidin 3-*O*-rutinoside (**89**) was relatively lower, whereas that of bilberry anthocyanins was relatively higher (32%) (Fig. 19).

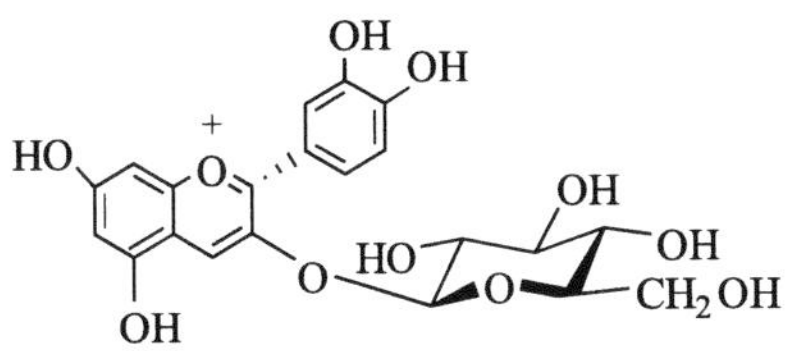

cyanidin 3-*O*-glucoside (**60**)

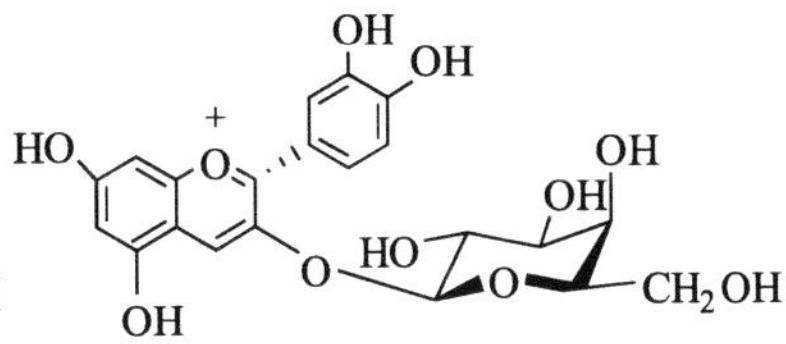

cyanidin 3-*O*-galactoside (**88**)

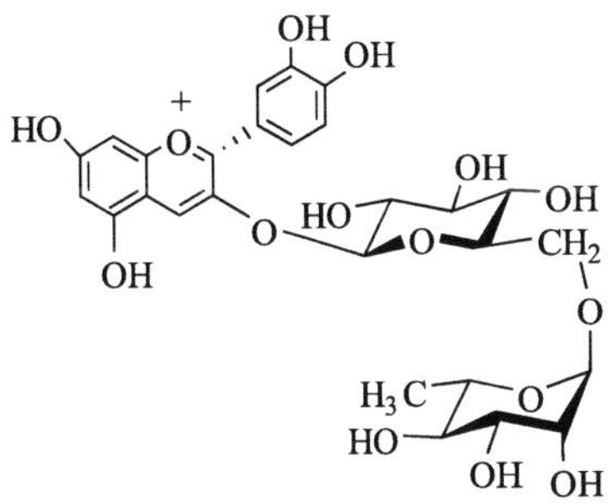

cyanidin 3-*O*-rutinoside (**89**)

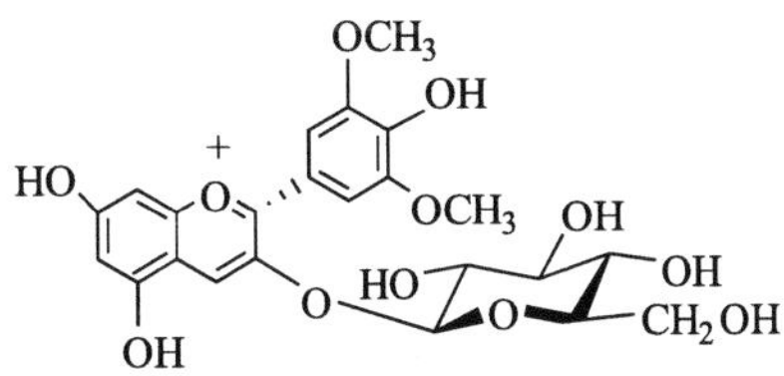

malvidin 3-*O*-glucoside (**90**)

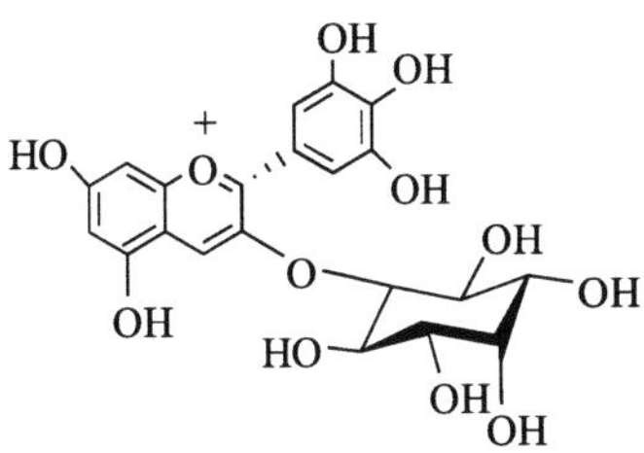

delphinidin 3-*O*-arabinoside (**91**)

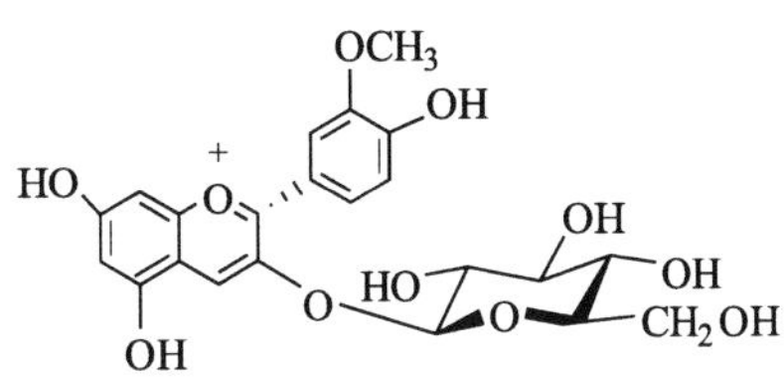

peonidin 3-*O*-glucoside (**92**)

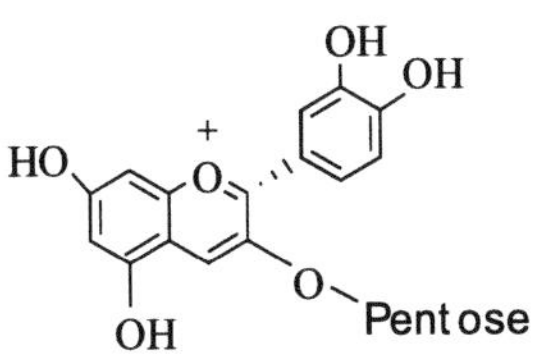

cyanidin 3-pentose (**93**)

Fig. 19 Structures of main anthocyanin glycosides (**60, 88–93**) in blackberry and bilberry extracts

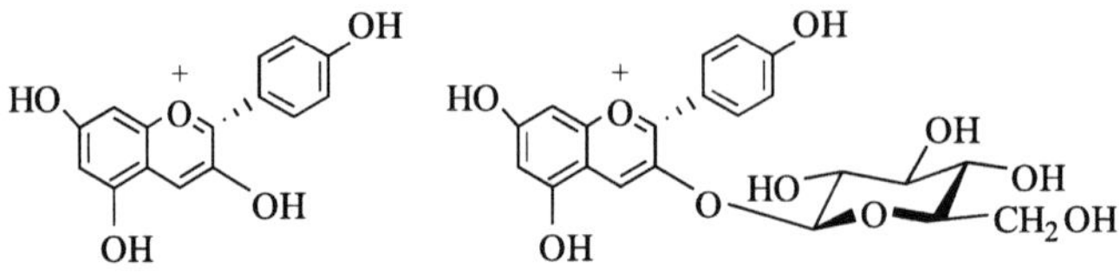

pelargonidin (**50**) pelargonidin 3-*O*-glucoside (**94**)

Fig. 20 Pelargonidin (**50**) and pelargonidin 3-*O*-glucoside (**94**) in strawberry

Surprisingly, the absorption varied widely from 19 to 37%, depending on the individual anthocyanin structure in the bilberry extract. Delphinidin 3-*O*-arabinoside (**91**) showed the highest absorption rate (37%).

Interestingly, the absorption rate of cyanidin 3-*O*-galactoside (**88**) (24%) after direct adminstration of a high concentration (750 μmol/L) of blackberry anthocyanins into the gastric lumen was lower than that after administration at lower concentration (14 μmol/L) (26%). After administration of a high concentration of blackberry anthocyanins, the blackberry anthocyanins were detected in the plasma of both the gastric vein and aorta; however, the aglycones and metabolites were undetectable in both. The cyanidin 3-*O*-glucoside (**60**) was detected in the bile approximately 20 min later. Peonidin 3-*O*-glucoside (**92**), a methylated form of cyanidin 3-*O*-glucoside (**60**), and unknown anthocyanin metabolites were also detected in the bile. These results suggest that anthocyanin glycosides are quickly and efficiently absorbed through the stomach, and rapidly excreted into bile as both intact and metabolized forms (Fig. 19) [48].

The gastric and intestinal absorption of pelargonidin 3-*O*-glucoside (**94**) was examined by using in situ rat. A high proportion of pelargonidin 3-*O*-glucoside (**94**) was rapidly absorbed through both the stomach (23%) and small intestine (24%). Then, the metabolism of pelargonidin 3-*O*-glucoside (**94**) was examined by feeding the rats for 8 days with a diet enriched in freeze-dried strawberries. Low amounts of anthocyanins (only 0.013–0.163% of ingested anthocyanins) were recovered from the urine after 24 h ($n = 8$). When strawberry anthocyanins were analyzed, comparable amounts of intact glycosides (around 53%) and glucuronidated metabolites (about 47%) were found in the urine. Pelargonidin 3-*O*-glucoside (**94**) was thus glucuronized to a larger extent than cyanidin 3-*O*-glucoside (**60**). The results suggest that aglycone structure might strongly affect anthocyanin metabolism (Fig. 20) [49].

6.1.2
Absorption of Anthocyanin Glycosides into the Small Intestine

The intestine is also a typical and important site for anthocyanin absorption. Anthocyanin absorption through the intestine was examined. The jejunum and ileum of male Wistar rats were perfused for 45 min with a physiological buffer supplemented with various anthocyanins (glycosides). The purified anthocyanin glycosides [cyanidin 3-*O*-glucoside (**60**), cyanidin 3-*O*-galactoside (**88**), cyanidin 3-*O*-rutinoside (**89**), malvidin 3-*O*-glucoside (**90**)] (9.2 nmol/min), blackberry (9.0 nmol/min) or bilberry (45.2 nmol/min) anthocyanins were perfused. A high rate of absorption of anthocyanin glycosides through the small intestine after perfusion was observed. The absorption rate basically depends on chemical structure of the anthocyanins, ranging from 10.7% for malvidin 3-*O*-glucoside (**90**) to 22.4% for cyanidin 3-*O*-glucoside (**60**). Surprisingly, though anthocyanins were even perfused, only anthocyanin glycosides were recovered in the intestinal lumen. After the perfusion of a large amount of blackberry anthocyanins (600 nmol/min), native cyanidin 3-*O*-glucoside (**60**) was recovered from the urine and plasma of the aorta and mesenteric vein. Methylated and glucuronized derivatives of their metabolites were also identified. Cyanidin 3-*O*-glucoside (**60**) and its methylated derivatives such as peonidin 3-*O*-glucoside (**92**) and peonidin glucuronide appeared rapidly in the bile. These results suggested that anthocyanin glycosides might be rapidly and efficiently absorbed through the small intestine. Following the rapid metabolism, anthocyanins were excreted into both bile and urine as the intact glycosides such as methylated and glucuronized forms (Fig. 19) [50].

6.1.3 Absorption of Anthocyanin Glycosides into Organs such as Stomach, Jejunum, Liver, Kidney, Brain, and Other Organs

Anthocyanins are metabolized and distributed into digestive organs such as stomach, jejunum, liver, kidney, and the brain. To investigate anthocyanin metabolism, rats were fed with a blackberry (*Rubus fruticosus* L.), anthocyanin-enriched diet, for 15 days. During the identification and quantification of blackberry anthocyanin metabolites, only intact blackberry anthocyanins with cyanidin 3-*O*-glucoside (**60**) and cyanidin 3-pentose (**93**) as blackberry metabolites were detected in the stomach. However, in other organs such as jejunum, liver, and kidney, all of the intact, methylated and conjugated blackberry anthocyanidins (such as cyanidin monoglucuronides and peonidin monoglucuronides) were found. The proportions of anthocyanin derivatives differed according to the organ considered, with the liver presenting the highest proportion of methylated forms. Jejunum and plasma also contained aglycone forms. In the brain, total anthocyanin content measured as blackberry anthocyanins and peonidin 3-*O*-glucoside (**92**) was 0.25 ± 0.05 nmol/g tissue ($n = 6$). The urinary excretion of total anthocyanins was as low as $0.19 \pm 0.02\%$ of ingested amounts. These results suggest that the digestive organs including jejunum might play a role in the metabolism of anthocyanins such as methylation and/or glucurono-conjugation. Additionally, through consumption in an anthocyanin-rich diet, anthocyanins could enter even the brain (Fig. 19) [51].

6.2 Human Test Examples

First, blackberry contains anthocyanins such as cyanidin (**51**), peonidin (**70**), cyanidin 3-*O*-glucoside (**60**) of major pigments and also cyanidin 3-*O*-xyloside (**95**), peonidin 3-*O*-glucoside (**92**), peonidin 3-*O*-xyloside (**96**), pelargonidin 3-*O*-glucoside (**94**), and pelargonidin (**50**) (Fig. 21) [52].

Second, to evaluate anthocyanin urinary excretion after intake of a meal with blackberry, a total of five healthy volunteers were orally administered 200 g blackberries that contain 960 μmol anthocyanins. The urine samples were analyzed by solid-phase extraction for anthocyanin metabolites. Among a total of 12 anthocyanin metabolites (1537 nmol/24 h), peonidin glucuronide (glucuronide of **92**) and cyanidin 3-*O*-xyloside (**95**) were the major metabolites (529 nmol), followed by cyanidin glucuronide (a glucuronide of **60**, 355 nmol), cyanidin 3-*O*-glucoside (**60**, 193 nmol), cyanidin (**51**, 112 nmol), peonidin 3-*O*-glucoside (**92**, 94 nmol), unknown anthocyanin metabolites (86 nmol), peonidin (**70**, 43 nmol), peonidin glucuronide (glucuronide of **92**, 40 nmol), cyanidin glucuronide (glucuronide of **60**, 32 nmol), peonidin glucuronide (a glucuronide of **92**, 30 nmol), and cyanidin diglucuronide (a glucuronide of **60**, 19 nmol) (Fig. 21) [52].

Minor anthocyanin metabolites such as acylated derivatives of cyanidin 3-*O*-glucoside (**60**), methylated glycosides, glucuronides of anthocyanidins and anthocyanins, a sulfoconjugate of cyanidin (**51**), and anthocyanidins were identified in the urine. The total urinary excretion of blackberry anthocyanin metabolites was $0.160 \pm 0.020\%$ of anthocyanins ingested in the five volunteers. Monoglucuronides of anthocyanidin metabolites represented over 60% of the excreted metabolites. The urinary excretion of anthocyanins was maximal between 2 and 4 h after the administration. Interestingly, the urinary excretion of anthocyanins continued over a 24 h test period. The results suggest that anthocyanins are not only methylated, glucuronized, but also sulfonated in humans, and that the main metabolites of the blackberry anthocyanins excreted in human urine are monoglucuronides [52].

Similarly, the bioavailability of anthocyanins in humans who had intaken strawberries was examined, and possible urinary metabolites were identified. Six healthy volunteers comprising of three women and three men were orally administered a meal with 200 g strawberries containing 179 μmol pelargonidin-3-*O*-glucoside (**94**) as a daily diet. The anthocyanins in the urine were rapidly determined before and after the meal. Then, in addition to pelargonidin-3-*O*-glucoside (**94**), five anthocyanin metabolites including three monoglucuronides of pelargonidin (**50**), one sulfoconjugate of pelargonidin, and pelargonidin (**50**), a nonmetabolite, were identified in

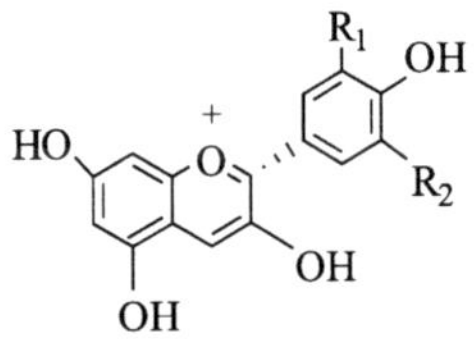

anthocyanidins in blackberry
cyanidin (**51**): R_1=OH; R_2=H
peonidin (**70**): R_1=OCH_3; R_2=H

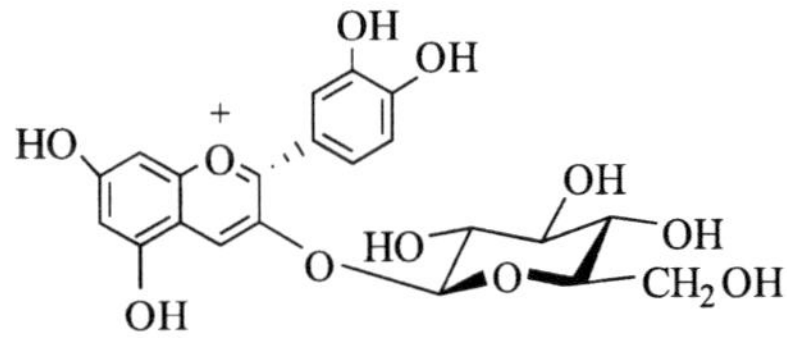

cyanidin 3-*O*-glucoside (**60**)

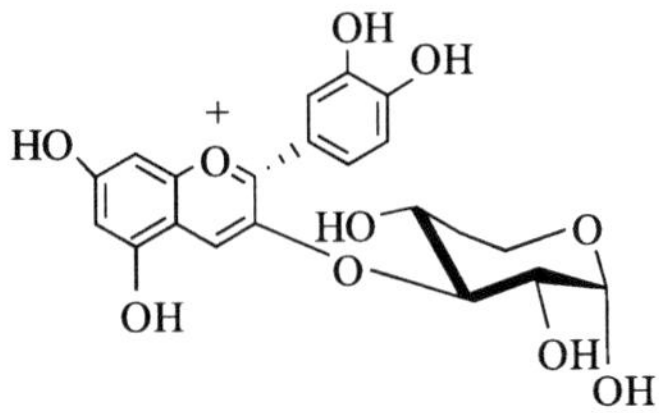

cyanidin 3-*O*-xyloside (**95**)

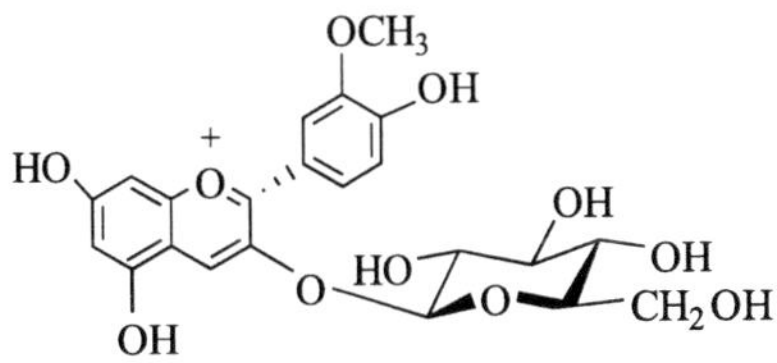

peonidin 3-*O*-glucoside (**92**)

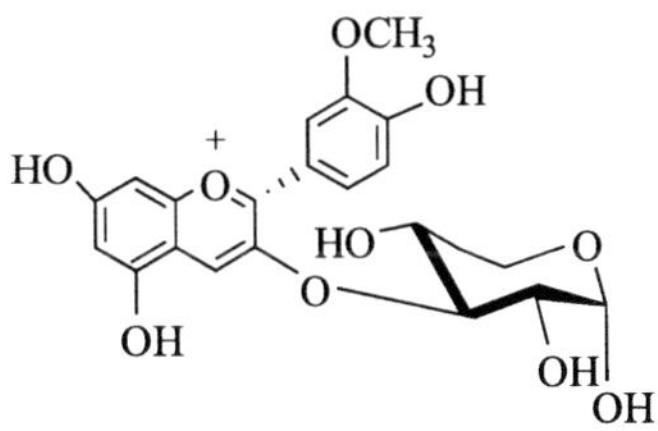

peonidin 3-*O*-xyloside (**96**)

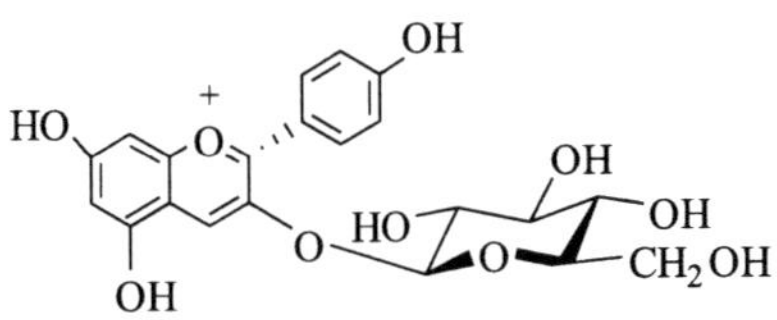

pelargonidin 3-*O*-glucoside (**94**)

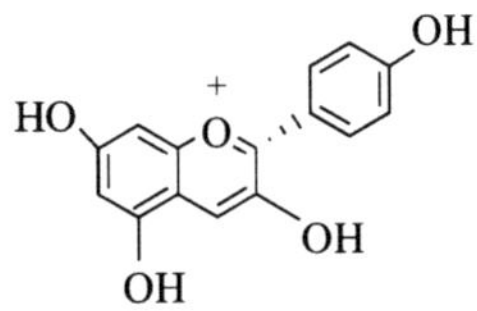

pelargonidin (**50**)

Fig. 21 Structures of cyanidin (51), cyanidin 3-*O*-glucoside (60),peonidin 3-*O*-glucoside (92), peonidin 3-*O*-xyloside (96), pelargonidin 3-*O*-glucoside (94) and pelargonidin (50) found mainly in metabolites of blackberry and strawberry

the urine. The total strawberry anthocyanin metabolites including monoglucuronides in the urinary excretion corresponded to $1.80 \pm 0.29\%$ ($n = 6$) of pelargonidin-3-*O*-glucoside (**94**) that had been administered to the six volunteers. Interestingly, more than 80% of the excreted metabolites corresponded to pelargonidin monoglucuronide. More than two-thirds of the pelargonidin-3-*O*-glucoside (**94**) metabolites were excreted within 4 h of the meal. However, the urinary excretion of the metabolites continued over the 24 h experiment. It was shown that the orally administered pelargonidin-3-*O*-glucoside (**94**) was metabolized to glucuronized anthocyanins and sulfonated anthocyanins in humans, and that the main metabolite of strawberry anthocyanins in human urine was a mono-glucuronized pelargonidin (**50**) (Fig. 21) [53].

The dose response and metabolism of strawberry anthocyanins were examined with 12 healthy adults who had consumed each of three different doses of strawberry anthocyanin. The four main anthocyanins in pureed strawberries were pelargonidin 3-*O*-glucoside (**94**, 87.8%), cyanidin 3-*O*-glucoside (**60**, 7.7%), pelargonidin-3-rutinoside (4.5%), and pelargonidin-3-(6″-acetoyl)glucoside (trace) (Fig. 21).

Oral administration of 100, 200, or 400 g pureed strawberries corresponded with 15, 30, or 60 μmol anthocyanin, respectively. The urine metabolites of strawberry anthocyanins were collected 24 h after each anthocyanin administration and the urine samples were analyzed.

The major compounds collected were pelargonidin 3-*O*-glucoside (**94**, 3.6–7.3%), a nonmetabolite, and pelargonidin glucuronide (Pg-Monoglucuronide B, 77.9–82.3%), pelargonidin glucuronide (Pg-Monoglucuronide A, 7.6–8.3%), pelargonidin glucuronide C (Pg-Monoglucuronide C, 6.4–6.9%) in the urine after the ingestion. Interestingly, pelargonidin glucuronide B was detected in the urine at 10-times higher concentrations than pelargonidin glucuronide A and pelargonidin glucuronide C. When volunteers were administrated three different doses (100, 200, or 400 g pureed strawberries), the amount of pelargonidin monoglucuronide excreted in the urine was shown to increase with increasing dose. These results suggest that the absorption and metabolism of pelargonidin 3-*O*-glucoside (**94**) increase with increasing dose (Fig. 21) [54].

7 Safety of Proanthocyanidins

Procyanidins with different degrees of polymerization such as dimers, trimers, and polymers from extracts of different natural sources were found to be nonmutagenic in the *Salmonella* mutagenesis assay. Moreover, the mutagenic impurity in procyanidin B-4 was isolated by reversed-phase high-performance liquid chromatography (HPLC). Procyanidin B-4 was identified as rutin (flavonol *O*-glycoside, **3**) (Fig. 22) [55].

Edible berry extracts rich in anthocyanins possess a broad spectrum of therapeutic, pharmacologic, and anticarcinogenic properties. Six kinds of berry extract such as wild blueberry, bilberry, cranberry, elderberry, raspberry, and strawberry were examined for their antioxidant efficacy, cytotoxicity, cellular uptake, and anti-angiogenicity. Combinations of six edible berry extracts were synergic. OptiBerry with many berry extracts exhibited a higher oxygen radical absorbance capacity (ORAC) value, low cytotoxicity, and superior anti-angiogenic properties when compared to other combination administrations. OptiBerry was safe and antioxidative against broad

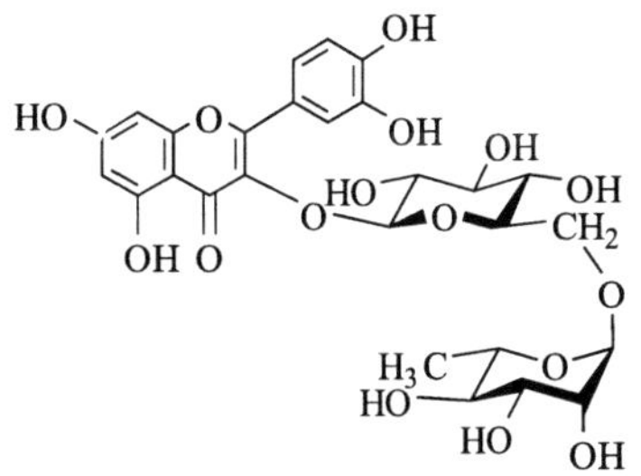

Fig. 22 Rutin (flavonol *O*-glycoside) (**3**)

spectra in vivo. The acute oral 50% lethal dose (LD_{50}) of OptiBerry was greater than 5 g/kg in rats. The acute dermal LD_{50} of OptiBerry was greater than 2 g/kg. No changes in the body weight or adverse effects were observed following the necropsy.

OptiBerry was found to be slightly irritating to the skin and minimally irritating to the eye of New Zealand albino rabbits. The antioxidative ability of OptiBerry was examined. OptiBerry feeding for 8 weeks prevented apparently hyperbaric oxygen (HBO)-induced GSH oxidation in the lung and liver of vitamin E-deficient Sprague Dawley rats. Furthermore, OptiBerry feeding protected the mice from whole body HBO-induced oxidation. Daily foods together with OptiBerry might be reasonably safe with antioxidant properties [56].

Both grape seed extract (GSE) and grape skin extract (GSKE) that contain proanthocyanidins were examined for their safety. GSE and GSKE were added into daily foods and fed to rats for at least 3 months, according to the GLP-compliant subchronic toxicity test. On the basis of the histologic observations on all tissues from control and high-dose GSE and GSKE groups, any toxicity could not be found, indicating that a dietary concentration of 2.5% GSE or 2.5% GSKE is nontoxic. The administration dose of GSE or GSKE was equivalent to a time-weighted average dose over the course of the test: approximately 1.78 g/kg body weight/day GSE or GSKE in male rats and 2.15 g/kg body weight/day in female rats, respectively [57].

Similarly, proanthocyanidin extract of grape seeds was examined systematically for toxicity after oral administration. The grape seed extract was examined for acute and subchronic toxicity. For the reverse mutation test, Fischer 344 rats and the test of *Salmonella typhimurium* were used; also were used the chromosomal aberration test for CHL cells, and the micronucleus test for ddY mice. No evidence of acute oral toxicity at dosages of 2 and 4 g GSE/kg, or of mutagenicity could be found. Combination administration of a diet with 0.02, 0.2, or 2% (w/w) GSE to the rats for 90 days produced no toxicity. This result suggests that the intakes of GSE proanthocyanidin-rich extract for human daily food is safe enough for the effective administration [58].

The oligonol containing catechin-type monomers and lower oligomers of proanthocyanidin could artificially convert the polyphenol polymers into oligomers. Using 30 healthy volunteers, the toxicity of a single-dose administration was examined by the administration of oligonol (2 g/kg body weight) via a gavage administration for 4 weeks. The oligonol was found to be safe without any side effects such as abnormal behavior and alopecia. Both body weight gain and food consumption showed the normal levels. Oligonol showed no toxicity at the dose with 1/25 of LD_{50} for 6 months. Therefore, oligonol might be safe even with repeated intake of 200 mg oligonol/day. The highest permissible dose is 12 g oligonol daily for an adult man with 60 kg body weight.

The 50% lethal dose (LD_{50}) was 5.0 g/kg body weight (95% confidence limit: 3.5–6.4 g/kg). Following administration of oligonol at doses of 100 mg/day and 200 mg/day for 92 days, all 30 healthy volunteers showed good bioavailability, maintaining normal liver and kidney functions, and hematological parameters. The mutagenicity of oligonol to induce gene mutation was examined with six bacteria of *Salmonella typhimurium* TA98, TA100, TA104, TA1535, TA153, and *Escherichia coli* WP2uvrA. Oligonol was found to be nonmutagenic to these strains. The results indicate that oligonol has no toxicity, and therefore might be effective as both food and dietary supplements, and also for usage in nutraceuticals [59].

References

1. Francis FJ (1989) Crit Rev Food Sci Nutr 28:273
2. Gao K, Henning SM, Niu Y, Youssefian AA, Seeram NP, Xu A, Heber D (2006) J Nutr Biochem 17:89
3. Mata-Bilbao L, Andrés-Lacueva C, Roura E, Jáuregui O, Escribano E, Torre C, Lamuela-Raventós RM (2007) Br J Nutr 98:86
4. Schindler R, Mentlein R (2006) J Nutr 136:1477
5. Esmaillzadeh A, Kimiagar M, Mehrabi Y, Azadbakht L, Hu FB, Willett WC (2006) Am J Clin Nutr 84:1489
6. Duthie SJ, Johnson W, Dobson VL (1997) Mutat Res 390:141
7. Duthie SJ, Dobson VL (1999) Eur J Nutr 38:28
8. Collins AR, Duthie SJ, Fillion L, Gedik CM, Vaughan N, Wood SG (1997) Biochem Soc Trans 25:326
9. Nichenametla SN, Taruscio TG, Barney DL, Exon JH (2006) Crit Rev Food Sci Nutr 46:161
10. Glei M, Matuschek M, Steiner C, Böhm V, Persin C, Pool-Zobel BL (2003) Toxicol In Vitro 17:723
11. Loft S, Poulsen HE (1996) J Mol Med 74:297. Review and Erratum in: (1997) J Mol Med 75:67
12. Lordan S, Mackrill JJ, O'Brien NM (2008) Int J Toxicol 27:279
13. Patel RP, Diczfalusy U, Dzeletovic S, Wilson MT, Darley-Usmar VM (1996) J Lipid Res 37:2361

14. Gesquière L, Loreau N, Blache D (1997) Free Radic Biol Med 23:541
15. O'Brien NM, Woods JA, Aherne SA, O'Callaghan YC (2000) Biochem Soc Trans 28:22
16. O'Brien NM, Carpenter R, O'Callaghan YC, O'Grady MN, Kerry JP (2006) J Med Food 9:187
17. Harborne JB (1986) Prog Clin Biol Res 213:15
18. Harborne JB (ed) (1988) The Flavonoids, Advances in Research Since 1980. Chapman & Hall, UK
19. Markham KR (1988) In: Harborne JB (ed) The Flavonoids, Advances in Research Since 1980. Chapman & Hall, UK, p 427
20. Markham KR (1990) Proc Phytochem Soc 29:143
21. Swan T (1986) The evolution of flavonoids. In: Cody V, Middleton E, Harborne JB (eds) Plant Flavonoids in Biology and Medicine, I: Biochemical, Pharmacological and Structure-Activity Relationships. Alan R Liss, New York, p 1
22. Stafford HA (1991) Plant Physiol 96:680
23. Winkel-Sfirley B (2001) Plant Physiol 126:485
24. Lepiniec L, Debeaujon I, Routaboul J-M, Baudry A, Pourcel L, Nesi N, Caboche M (2006) Ann Rev Plant Biol 57:405
25. Robinson R (1936) Nature 137:172
26. Heller W (1986) Prog Clin Biol Res 213:25
27. Geissmann TA, Harborne JB (1955) Arch Biochem Biophys 55:447
28. Debeaujon I, Nesi N, Perez P, Devic M, Grandjean O, Caboche M, Lepiniec L (2003) Plant Cell 15:2514
29. Abrahams S, Lee E, Walker AR, Tanner GJ, Larkin PJ, Ashton AR (2003) Plant J 35:624
30. Marinova K, Pourcel L, Weder B, Schwarz M, Barron D, Routaboul JM, Debeaujon I, Klein M (2007) Plant Cell 19:2023
31. Punyasiri PA, Abeysinghe IS, Kumar V, Treutter D, Duy D, Gosch C, Martens S, Forkmann G, Fischer TC (2004) Arch Biochem Biophys 431:22
32. Xie DY, Sharma SB, Paiva NL, Ferreira D, Dixon RA (2003) Science 299:396 (and Comment in: (2003) Science 299:352)
33. Proteggente AR, Pannala AS, Paganga G, Van Buren L, Wagner E, Wiseman S, Van De Put F, Dacombe C, Rice-Evans CA (2002) Free Radic Res 36:217
34. Lichtenthäler R, Marx F (2005) J Agric Food Chem 53:103
35. Harborne JB, Williams CA (2000) Phytochemistry 55:481
36. Noroozi M, Angerson WJ, Lean ME (1998) Am J Clin Nutr 67:1210
37. Wilms LC, Hollman PC, Boots AW, Kleinjans JC (2005) Mutat Res 582:155
38. Seeram NP, Zhang Y, Nair MG (2003) Nutr Cancer 46:101
39. Chen YC, Shen SC, Chow JM, Ko CH, Tseng SW (2004) Int J Oncol 25:661
40. Shen SC, Ko CH, Tseng SW, Tsai SH, Chen YC (2004) Toxicol Appl Pharmacol 197:84
41. Shukla S, Gupta S (2004) Clin Cancer Res 10:3169
42. Tsuda T, Horio F, Osawa T (1999) FEBS Lett 449:179
43. Miyazawa T, Nakagawa K, Kudo M, Muraishi K, Someya K (1999) J Agric Food Chem 47:1083
44. Matsumoto H, Inaba H, Kishi M, Tominaga S, Hirayama M, Tsuda T (2001) J Agric Food Chem 49:1546
45. Cao G, Muccitelli HU, Sánchez-Moreno C, Prior RL (2001) Am J Clin Nutr 73:920
46. Felgines C, Texier O, Besson C, Fraisse D, Lamaison JL, Rémésy C (2002) J Nutr 132:1249
47. Galvano F, La Fauci L, Vitaglione P, Fogliano V, Vanella L, Felgines C (2007) Ann Ist Super Sanita 43:382

48. Talavéra S, Felgines C, Texier O, Besson C, Lamaison JL, Rémésy C (2003) J Nutr 133:4178
49. Felgines C, Texier O, Besson C, Lyan B, Lamaison JL, Scalbert A (2007) Br J Nutr 98:1126
50. Talavéra S, Felgines C, Texier O, Besson C, Manach C, Lamaison JL, Rémésy C (2004) J Nutr 134:2275
51. Talavéra S, Felgines C, Texier O, Besson C, Gil-Izquierdo A, Lamaison JL, Rémésy C (2005) J Agric Food Chem 53:3902
52. Felgines C, Talavera S, Texier O, Gil-Izquierdo A, Lamaison JL, Remesy C (2005) J Agric Food Chem 53:7721
53. Felgines C, Talavéra S, Gonthier MP, Texier O, Scalbert A, Lamaison JL, Rémésy C (2003) J Nutr 133:1296
54. Carkeet C, Clevidence BA, Novotny JA (2008) J Nutr 38:897
55. Yu CL, Swaminathan B (1987) Food Chem Toxicol 25:135
56. Bagchi D, Roy S, Patel V, He G, Khanna S, Ojha N, Phillips C, Ghosh S, Bagchi M, Sen CK (2006) Mol Cell Biochem 281:197
57. Bentivegna SS, Whitney KM (2002) Food Chem Toxicol 40:1731
58. Yamakoshi J, Saito M, Kataoka S, Kikuchi M (2002) Food Chem Toxicol 40:599
59. Fujii H, Sun B, Nishioka H, Hirose A, Aruoma OI (2007) Food Chem Toxicol 45:378

Top Heterocycl Chem (2009) 16: 41–91
DOI 10.1007/7081_2009_128

Published online: 5 March 2009

Flavonoids in *Sophora* Species

Yoshiaki Shirataki[1] (✉) · Noboru Motohashi[2]

[1]Faculty of Pharmaceutical Sciences, Josai University, 1-1 Keyakidai, Sakado-shi, 350-0295 Saitama, Japan
shiratak@josai.ac.jp

[2]Meiji Pharmaceutical University, 2-522-1 Noshio, Kiyose-shi, 204-8588 Tokyo, Japan

Abstract *Sophora* species of *Leguminosae* are abundantly present in the natural kingdom. Today, among *Sophora* plants, the flavonoids of the plant phenols occupy a remarkable position. For a very long time flavonoids have been used as natural pigments and dyes. Some of the colorful anthocyanins of the glucosides are used for color and flavor in foodstuffs. Therefore, these flavonoids are beneficial to daily human life. Herein we concentrate on flavonoids in *Sophora* plants, and the relationship between their chemical structures and nutraceutical effect. For this purpose, soy-based infant formulas, osteoporosis, antitumor activity, antimicrobial activity, anti-HIV activity, radical generation and O_2^- scavenging activity, and enzyme inhibitory activity have been described.

Keywords Enzyme inhibitory activity · Flavonoids · Radical scavengers · *Sophora* species

Abbreviations

- Acetyldaidzin (**33**)
- Acetylgenistin (**34**)
- Acetylglycitin (**35**)
- Aminoguanidine (**131**)
- Ampicillin (**113**)
- 1-(6-Benzyloxy-2-hydroxyphenyl)-3-(4-hydroxyphenyl)propenone (**132**)
- Biochanin A (**78**)
- Butein (**109**)
- Budesonide (**142**)
- Calycosin (**80**)
- *cis*-Octadecyl ferulic acid (**116**)
- Clarithromycin (**85**)

- Cyclokuraridin (**97**)
- Daidzein (**27**)
- Daidzin (**30**)
- Desmethylanhydroicaritin (**119**)
- 7,4′-Dihydroxyflavone (**61**)
- 7,4′-Dihydroxy-5-methoxy-8-(γ,γ-dimethylallyl)-flavanone (**5**)
- (2*S*)-7,4′-Dihydroxy-5-methoxy-8-(γ,γ-dimethylallyl)-flavanone (**121**)
- 7,4′-Dihydroxy-3′-methoxyisoflavone (**81**)
- 3′,7-Dihydroxy-4′-methoxyisoflavone (**1**)
- 2-(2,4-Dihydroxy-5-prenylphenyl)-5,6-methylenedioxybenzofuran (**98**)
- 6,8-Diprenylgenistein (**84**)
- Epalrestat (**130**)
- Estradiol (**40**)
- Estrogen (**Hor1**, oestrone) (**41**)
- Euchrestaflavanone A (**75**)
- Fisetin (**107**)
- Formononetin (equol) (**4**)
- Genistein (**28**)
- Genistin (**31**)
- Gentamicin (**114**)
- 7-*O*-β-D-Glucopyranoside-4′-*O*-β-D-glucopyranoside (**52**)
- 7-*O*-β-D-Glucopyranoside-4′-*O*-[(β-D-glucopyranosyl)-(1 $\rightarrow$ 2)-β-D-glucopyranoside] (**New2**) (**49**)
- 7-*O*-β-D-Glucopyranoside-4′-*O*-[(α-L-rhamnopyranosyl)-(1 $\rightarrow$ 2)-β-D-glucopyranoside] (**New1**) (**48**)
- Glycitein(**29**)
- Glycitin (**32**)
- Hesperetin (**70**)
- (–)-4-Hydroxy-3-methoxy-(6a*R*,11a*R*)-8,9-methylenedioxypterocarpan (**118**)
- Irisolidone (**79**)
- Isokurarinone (**12**)
- Isoliquiritigenin (**108**)
- Isosophoranone (**87**)
- Isoxanthohumol (**7**)
- Kaempferol (**106**)
- Kaempferol 3-*O*-β-D-glucopyranoside (**57**)
- Kaempferol 3-*O*-β-rutinoside (**55**)
- Kojic acid (**126**)
- Kosamol Q (**19**)
- Kosamol R (**24**)
- Kuraridin (**8**)
- Kuraridinol (**9**)
- Kurarinol (**3**)
- Kurarinone (**13**)
- (–)-Kurarinone (**68**)
- Kushecarpin A (**25**)
- Kushenol A (**26**)
- Kushenol C (**22**)
- Kushenol D (**14**)
- Kushenol E (**124**)

- Kushenol F (**127**)
- Kushenol H (**112**)
- Kushenol I (**20**)
- Kushenol K (**111**)
- Kushenol N (**21**)
- Kushenol O (**15**)
- Kushenol X (**64**)
- 8-Lavandulylkaempferol = 8-*C*-Lavandulylkaempferol (**96**)
- Leachianone A (**66**)
- Leachianone G (**125**)
- Licoisoflavone A (**82**)
- Licoisoflavone B (**84**)
- Lupeol (**59**)
- Maackiain = (–)-maackiain = *l*-maackiain (**6**)
- (–)-Maackiain 3-sulfate (**58**)
- Malonyldaidzin (**36**)
- Malonylgenistin (**37**)
- Malonylglycitin (**38**)
- 2′-Methoxykurarinone (**23**)
- (2*S*)-2′-Methoxykurarinone (**67**)
- 5-Methylsophoraflavanone B (**144**)
- Naringenin (**69**)
- Norkurarinol (**18**)
- Norkurarinone (**65**)
- Ononin (**60**)
- Orobol (**46**)
- Oxymatrine (ammothaminine) (**128**)
- Oxyphenylbutazone (**47**)
- Penicillamine (**123**)
- Piceatannol (**104**)
- 6-Prenylnaringenin (**71**)
- 8-Prenylkaempferol (**63**)
- Prunetin 4′-*O*-β-D-glucopyranoside (**53**)
- Quercetin (**43**)
- Quercetin 3-*O*-β-D-glucopyranoside (**56**)
- Resveratrol (**103**)
- 7-*O*-α-D-Rhamnopyranoside-4′-*O*-[(β-D-glucopyranosyl)-(1→2)-β-D-glucopyranoside] (**51**)
- 7-*O*-α-L-Rhamnopyranoside-4′-*O*-[(α-L-rhamnopyranosyl)-(1→2)-β-D-glucopyranoside](**New3**) (**50**)
- Rhaponticin (**105**)
- Rutin (**42**)
- Secundifloran (**88**)
- Secundiflorol A (**89**)
- Secundiflorol D (**90**)
- Secundiflorol E (**91**)
- Secundiflorol F (**92**)
- β-Sitosterol (**129**)
- Sophoflavescenol (**110**)
- Sophorabioside (**45**)

- Sophoradione (**99**)
- Sophoraflavanone A (**73**)
- Sophoraflavanone B (**72**)
- Sophoraflavanone D (**74**)
- Sophoraflavanone G (**10**)
- Sophoraflavanone H (**76**)
- Sophoraflavanone I (**77**)
- Sophoraflavanone K (**94**)
- Sophoraflavanone L (**95**)
- Sophoraisoflavanone A (**17**)
- Sophoraisoflavone A (**83**)
- Sophorastilbene A (**100**)
- Sophoricoside (**44**)
- Sophororicoside (**54**)
- Sophoronol (**86**)
- (+)-Syringaresinol (**62**)
- Tetrapterol G (**93**)
- *trans*-Hexadecyl ferulic acid (**115**)
- *trans*-Hexadecyl sinapic acid (**117**)
- Trifolirhizin (pterocarpan) (**2**)
- Trifolirhizin-6″-malonate (**16**)
- 2,4,4′-Trihydroxydeoxybenzoin (**39**)
- (2*S*)-3*β*,7,4′-trihydroxy-5-methoxy-8-(*γ*,*γ*-dimethylallyl)-flavanone (**120**)
- 3,7,4′-Trihydroxy-5-methoxy-8-prenylflavanone (**143**)
- Umbelliferone (**122**)
- (+)-*α*-Viniferin (**101**)
- (−)-*ε*-Viniferin (**102**)
- Xanthohumol (**11**)

1 Introduction

Some flavonoids from *Sophora* plants have been used as natural materials in tanning in the leather industry, fermentation of various herb teas, and industrial cocoa processing for their flavor qualities. Very recently, many studies have been carried out to determine the functionality of flavonoids. Many flavonoids from the *Sophora* plants have thus been discovered and their functionalities described. For instance, 24 flavonoids were identified from *Sophora flavescens* Ait (Figs. 1–4). Among the 24 flavonoids, 14 were completely identified including 3′,7-dihydroxy-4′-methoxy-isoflavone (**1**), trifolirhizin (**2**), kurarinol (**3**), formononetin (**4**), 7,4′-dihydroxy-5-methoxy-8-(*γ*,*γ*-dimethylallyl)-flavanone (**5**), maackiain (**6**), isoxanthohumol (**7**), kuraridin (**8**), kuraridinol (**9**), sophoraflavanone G (**10**), xanthohumol (**11**), isokurarinone (**12**), kurarinone (**13**), and kushenol D (**14**). Additionally, ten of the flavonoids were tentatively identified as kushenol O (**15**), trifolirhizin-6″-malonate (**16**), sophoraisoflavanone A (**17**), norkurarinol/kosamol Q [norku-

Fig. 1 Flowers of *Sophora flavescens*, a typical *Sophora* species in Asia. Photographed by Y. Shirataki

Fig. 2 Leaves of *Sophora flavescens*, a typical *Sophora* species in Asia. Photographed by N. Motohashi

Fig. 3 Young fruits of *Sophora flavescens*, a typical *Sophora* species in Asia. Photographed by N. Motohashi

Fig. 4 Fruits of *Sophora flavescens*, a typical *Sophora* species in Asia. Photographed by Y. Shirataki

rarinol (**18**)/kosamol Q (**19**)], kushenol I/N [kushenol I (**20**)/kushenol N (**21**)], kushenol C (**22**), 2′-methoxykurarinone (**23**), kosamol R (**24**), kushecarpin A (**25**), and kushenol A (**26**) (Fig. 5) [1].

Isoflavones of phytoestrogens such as daidzein (**27**), genistein (**28**), and glycitein (**29**) are widely distributed in the natural kingdom. The glucoside

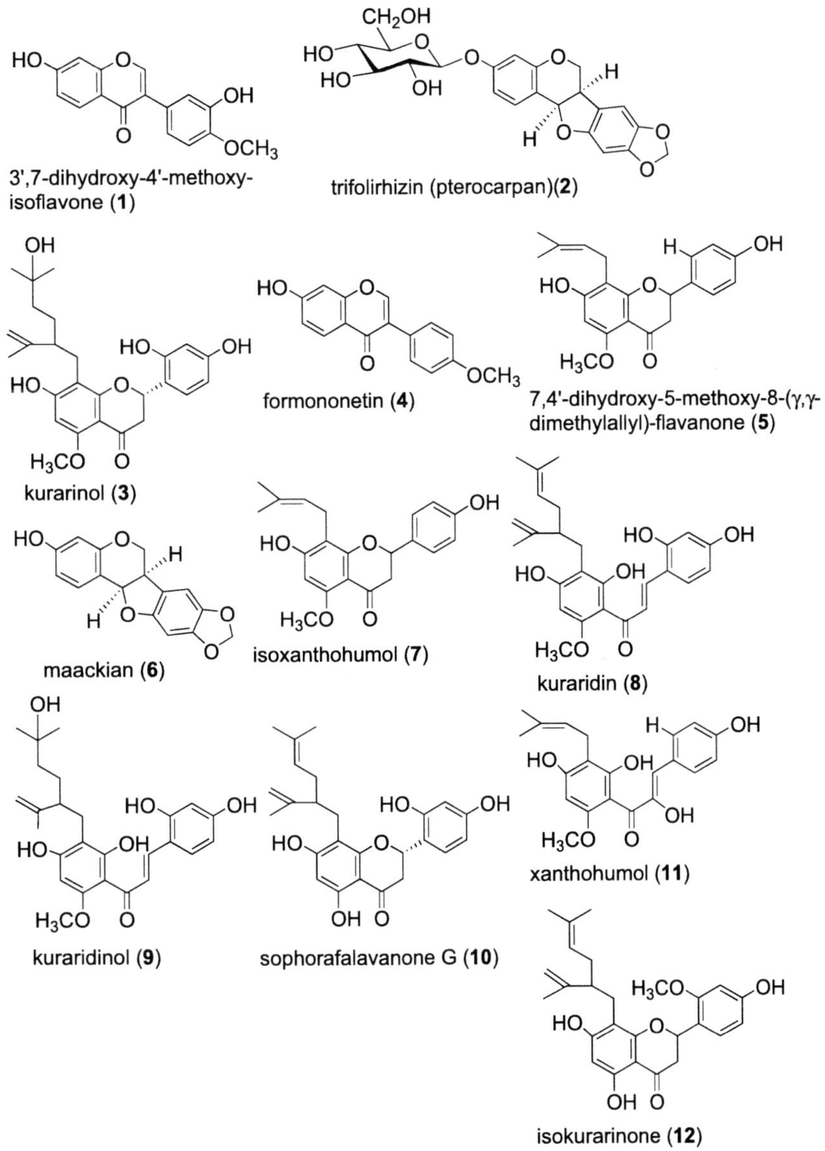

Fig. 5 Flavonoids from *Sophora flavescens* Ait

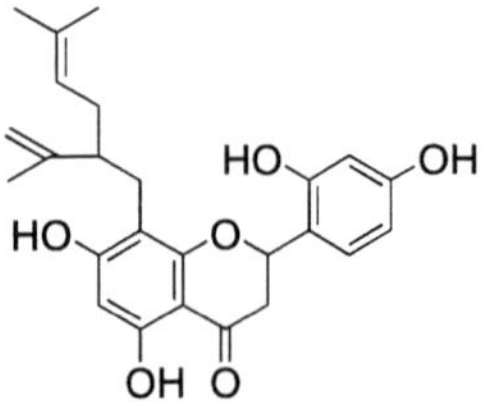

kurarinone (**13**)

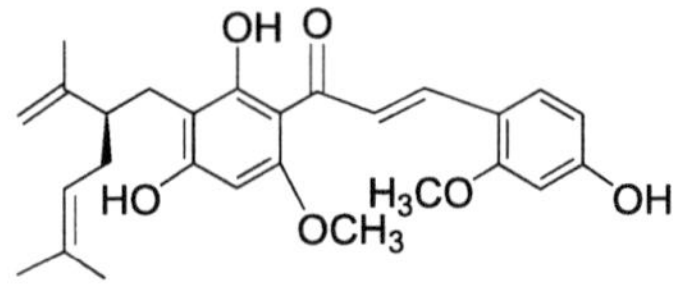

kushenol D (**14**)

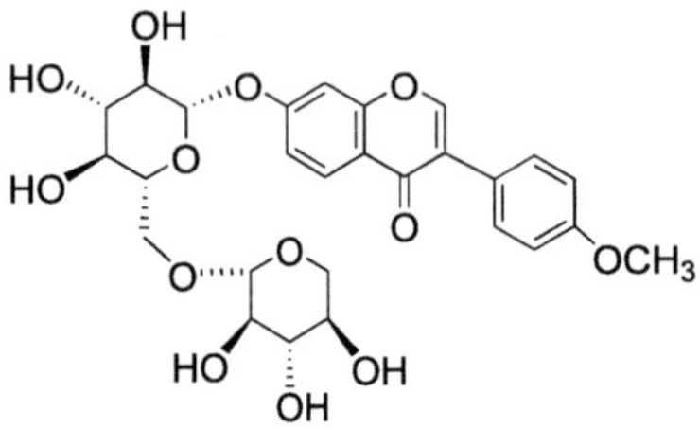

kushenol O (**15**)

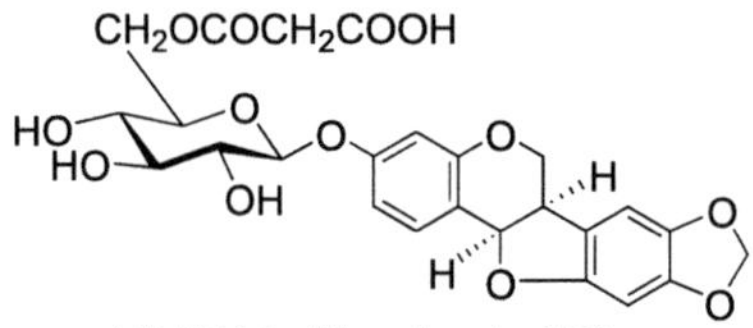

trifolirhizin-6"-malonate (**16**)

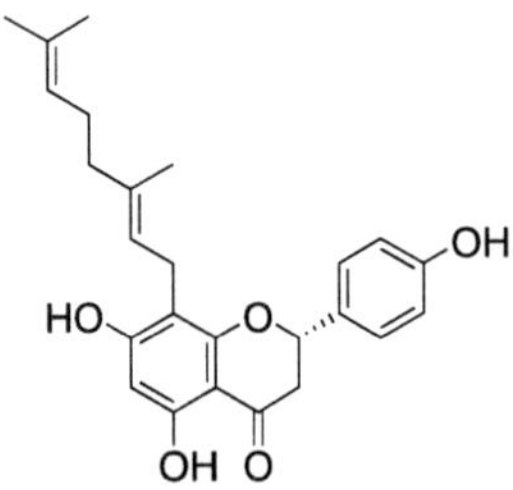

sophoraflavanone A (**17**)

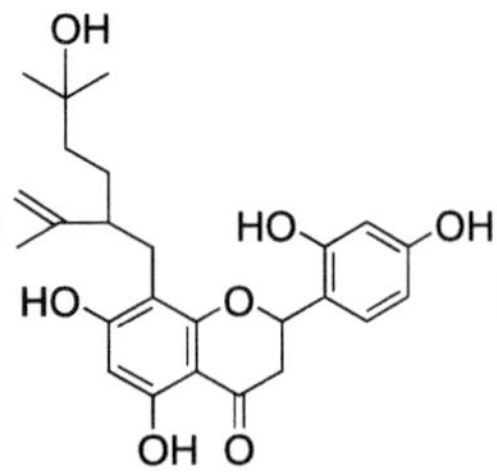

norkurarinol (**18**)

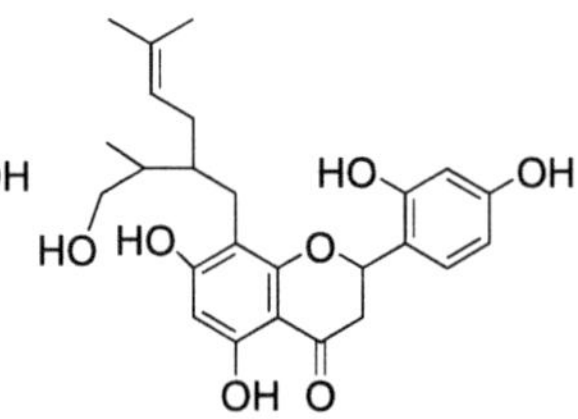

kosamol Q (**19**)

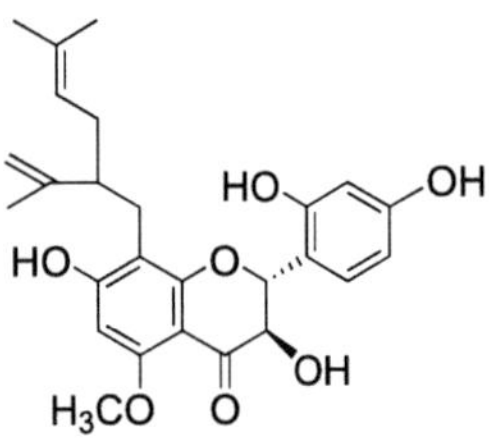

kushenol I (**20**)

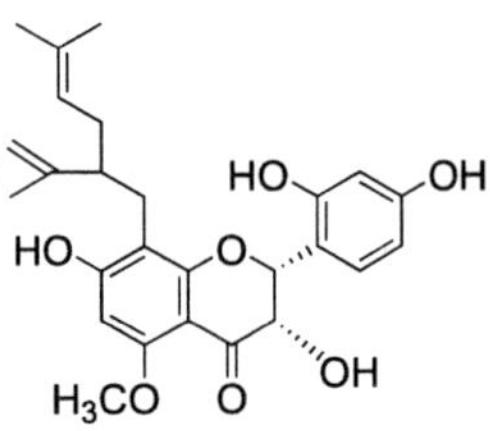

kushenol N (**21**)

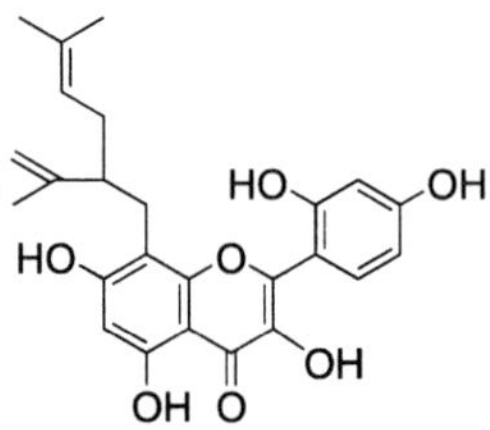

kushenol C (**22**)

Fig. 5 (continued)

forms of isoflavones could be converted to aglycones of free forms by enzymes in the gut and then provide protection and prevention for age-related diseases such as hypertension, diabetes, and heart disease.

To determine the soy isoflavone content of foods, the concentrations of 12 isoflavone isomers, three aglycones including daidzein (**27**), genistein (**28**), and glycitein (**29**), and nine glucosides including daidzin (**30**), genistin (**31**), glycitin (**32**), acetyldaidzin (**33**), acetylgenistin (**34**), acetylglycitin (**35**), mal-

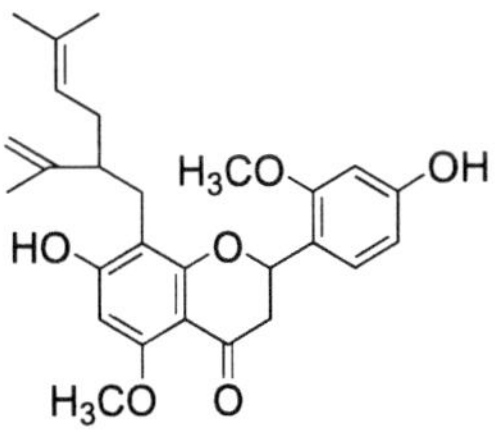

2'-methoxykurarinone (**23**)

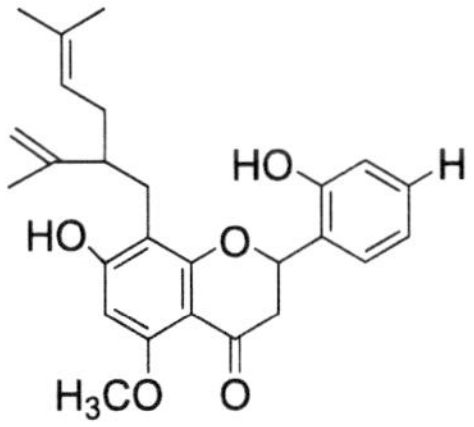

kosamol R (**24**)

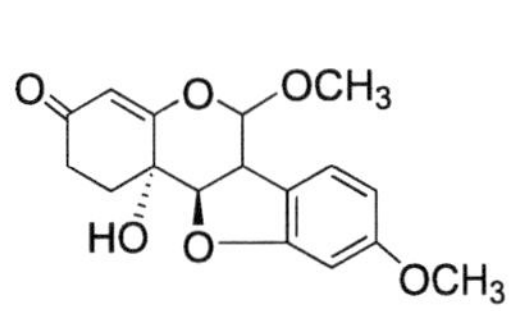

kushecarpin A (**25**)

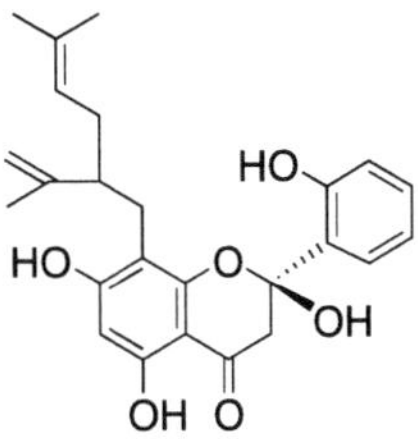

kushenol A (**26**)

Fig. 5 (continued)

onyldaidzin (**36**), malonylgenistin (**37**), and malonylglycitin (**38**) were measured in a variety of soybeans and soyfoods (Fig. 6) [2].

2
Functions of *Sophora* Flavonoids

2.1
Soy-based Formulas

Soy-based infant formulas have been used for more than 30 years. These formulas are manufactured from soy protein isolates, which contain abundant phytoestrogen isoflavones. In terms of isoflavone composition soy protein contains genistein (**28**) which is found predominantly in the form of glycosidic conjugates, and accounts for more than 65% of the isoflavones in soy-based formulas. Total isoflavone concentrations of soy-based formulas prepared for infant feeding range from 32 to 47 mg/l. However, isoflavone concentrations in human breast milk are only 5.6 ± 4.4 μg/l ($n = 9$). Therefore, for infants fed soy-based formulas it might be necessary to restrict intake to 22–45 mg isoflavones/day (6–11 mg × kg body weight/day). Moreover, the intake of phytoestrogens from human milk might be negligible (< 0.01 mg/day).

Regarding the metabolic components of isoflavones from soy-based infant formula, it is thought that the plasma isoflavone concentrations for 4-month-

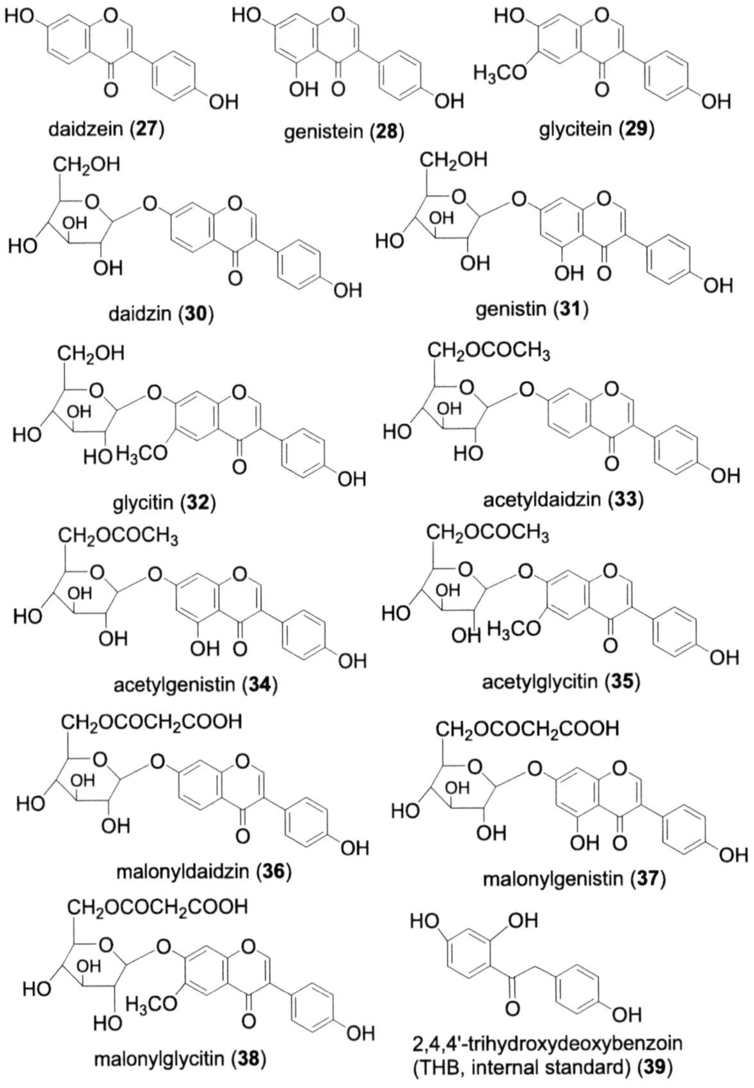

Fig. 6 Isoflavones and related compounds from *Soy* species

old infants fed the soy-based formula were 654–1775 μg/l (mean: 979.7 μg/l) higher when compared to the plasma concentrations of the infants fed either cow-milk formula (9.4 ± 1.2 μg/l) or human breast milk (4.7 ± 1.3 μg/l) [3].

The high plasma concentration of isoflavones in the infants fed the soy-based formula might be due to reduced intestinal biotransformation, as evidenced by low or undetectable concentrations of formononetin (equol) (**4**) of isoflavone and other metabolites, and could be maintained by the constant daily exposure due to frequent feeding. Moreover, isoflavones circulate at con-

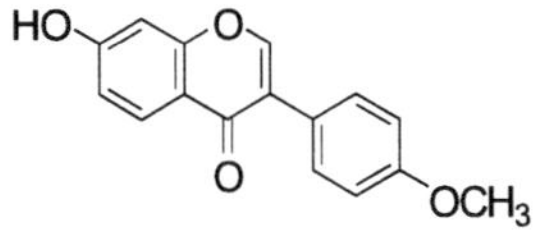

formononetin (equol) (**4**)

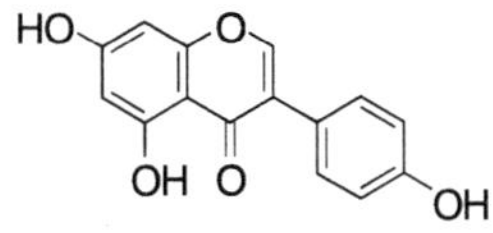

genistein (**28**)

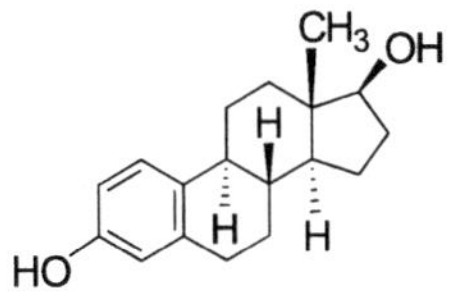

estradiol
(17β-estradiol) (**40**)

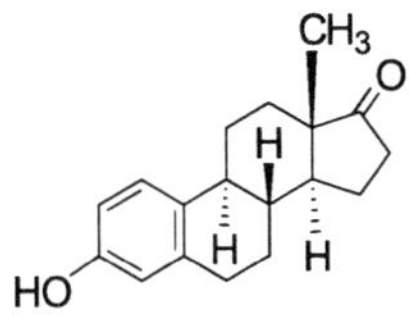

estrogen
(oestrone) (**41**)

Fig. 7 Isoflavones with activity in soy-based infant formulas

centrations that are 13 000–22 000-times higher when compared to plasma estradiol (**40**) concentrations in early life. The phytoestrogen isoflavones from early life might maintain long-term health benefits for humans against hormone-dependent diseases (Fig. 7) [4].

Moreover, safe and easy use by women of isoflavones has been increasing worldwide as an alternative medicine to estrogen hormone therapy against an increased risk of breast cancer, stroke, and heart attacks in response to estrogen and progesterone intervention. Generally, isoflavones might also be selective estrogen (oestrone) (**41**) receptor modulators like azaarene-phenols with a structural similarity to estradiol (17β-estradiol, estradiol-17β, oestradiol) (**40**). Actions at the cellular level of isoflavones might depend on the target tissue, receptor status of the tissue, and the level of the endogenous estrogen. Isoflavone hormone therapy might be beneficial as an alternative to estrogen (**41**) for hormone replacement in postmenopausal women (Fig. 7) [5].

Sophora japonica L. also contains diverse phytochemicals such as genistein (**28**), rutin (**42**), quercetin (**43**), sophoricoside (**44**), and sophorabioside (**45**). Interestingly, *Sophora* flower buds contain a great deal of rutin (**42**), while the immature fruits contain sophoricoside (**44**) and sophorabioside (**45**) (Fig. 8) [6].

Interleukin-5 (IL-5) is mainly produced from type 2 T helper (Th2) cells activated by allergen. From the fresh fruits of *Sophora japonica* L., isoflavonoids such as sophoricoside (**44**), genistein (**28**), orobol (**46**), and genistin (**31**) were found as the IL-5 inhibitors. These isoflavonoids exhibited their inhibitory effect in the IL-5 bioassay of mIL-5-dependent Y16 proliferation. Among the IL-5 inhibitors, sophoricoside (**44**) was the most effective with 89% inhibition at 12.5 μM; the 50% inhibitory concentration (IC_{50}) of sophoricoside (**44**) was 1.5 μM when compared to 31.7 μM of oxyphenylbutazone (oxypb) (**47**)—a nonsteroidal anti-inflammatory agent as a positive

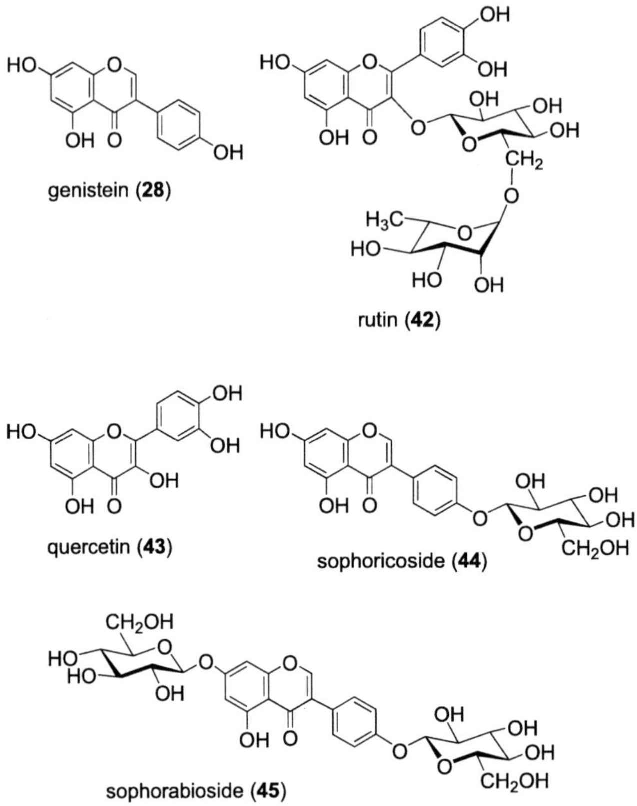

Fig. 8 Flavonoids in *Sophora japonica* L.

control. The order of IC_{50} values in terms of inhibitory potency in the IL-5 bioassay was as follows: sophoricoside (**44**) > orobol (**46**) (9.8 μM) approximately equal to genistin (10.6 μM) (**31**) > genistein (51.9 μM) (**28**). The position of *O*-glycosylation and number of hydroxy groups in the isoflavonoids seems to play an important role in the inhibitory effect in the IL-5 bioassay (Fig. 9) [7].

From the pericarps of *Sophora japonica*, four new isoflavone triglycosides, genistein 7-*O*-*β*-D-glucopyranoside-4′-*O*-[(*α*-L-rhamnopyranosyl)-(1→2)-*β*-D-glucopyranoside] (**New1**) (**48**), genistein 7-*O*-*β*-D-glucopyranoside-4′-*O*-[(*β*-D-glucopyranosyl)-(1→2)-*β*-D-glucopyranoside] (**New2**) (**49**), genistein 7-*O*-*α*-L-rhamnopyranoside-4′-*O*-[(*α*-L-rhamnopyranosyl)-(1→2)-*β*-D-glucopyranoside](**New3**) (**50**), and genistein 7-*O*-*α*-L-rhamnopyranoside-4′-*O*-[(*β*-D-glucopyranosyl)-(1→2)-*β*-D-glucopyranoside] (**New4**) (**51**), together with nine known isoflavonoids including genistein 7-*O*-*β*-D-glucopyranoside-4′-*O*-*β*-D-glucopyranoside (**52**), sophorabioside (**45**), prunetin 4′-*O*-*β*-D-glucopyranoside (**53**), sophororicoside(**54**), genistin (**31**), rutin (**42**), kaemp-

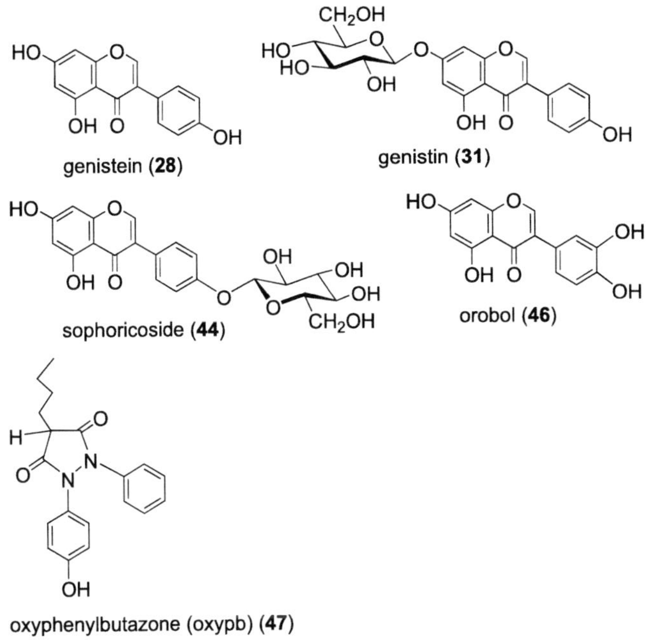

Fig. 9 Isoflavones from *Sophora japonica* L. and a control

ferol 3-*O*-β-rutinoside (**55**), quercetin 3-*O*-β-D-glucopyranoside (**56**), and kaempferol 3-*O*-β-D-glucopyranoside (**57**) were isolated (Fig. 10) [8].

A new pterocarpan, (–)-maackiain 3-sulfate (**58**) was isolated from the methanol extract of roots of *Sophora subprostarata* together with (–)-maackiain (**6**), trifolirhizin (**2**), lupeol (**59**), ononin (**60**), 7,4′-dihydroxyflavone (**61**), and (+)-syringaresinol (**62**). Three flavonoids (**60–62**) were isolated from this plant for the first time (Fig. 11) [9].

For prenylflavonoids and lavandulylflavonoids from the roots of *Sophora flavescens* Aiton (*Leguminosae* = *Fabaceae*), the ability of 8-prenylkaempferol (**63**), kushenol X (**64**), norkurarinone (**65**), leachianone A (**66**), kushenol C (**22**), maackiain (**6**) and a root-extract of *Sophora flavescens* to displace 17β-estradiol (**40**) from the rat uterine estrogen (**41**) receptor (ER) was determined. The relative binding affinities (RBA) of prenylated flavonoids were weak with RBA values between 0.004 and 0.072. However, a lavandulyl or prenyl group at position 8 on the flavonoids could enhance binding to the rat uterine ER (Fig. 12) [10].

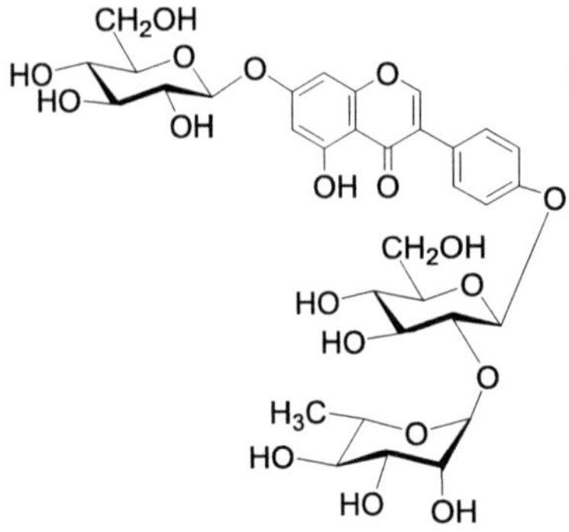

New1 (48)

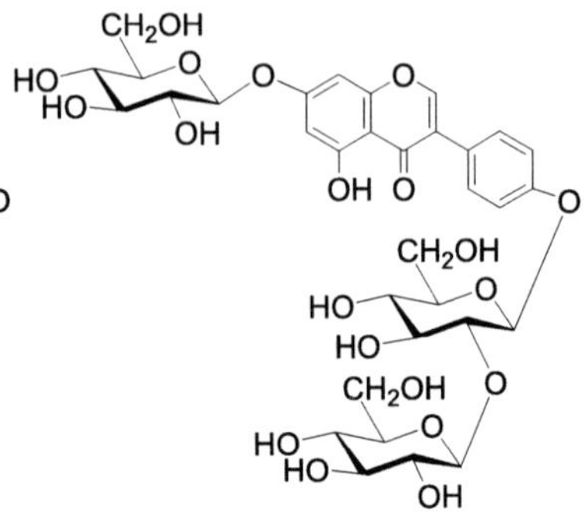

New2 (49)

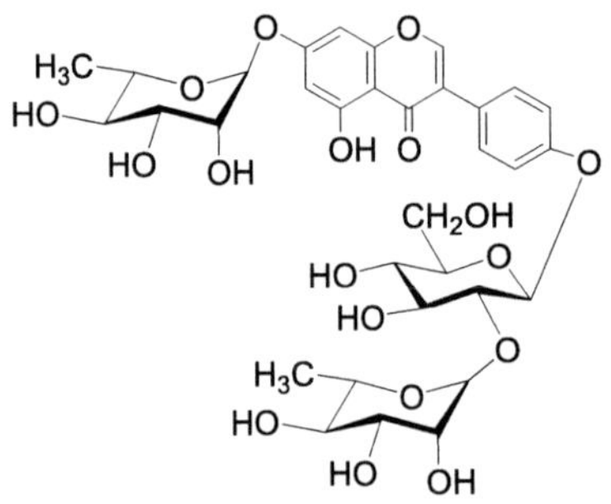

New3 (50)

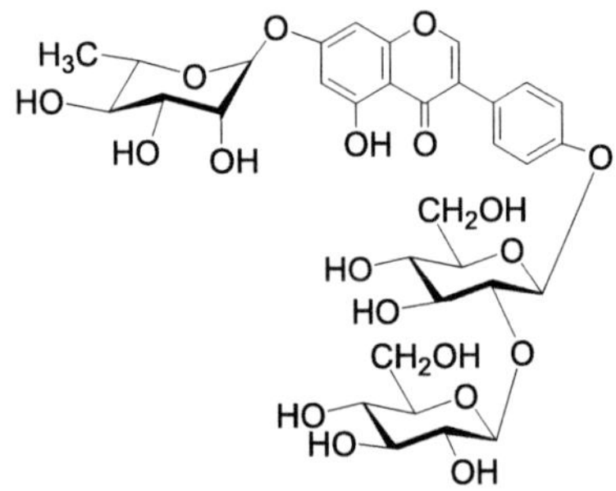

New4 (51)

Fig. 10 Isoflavone triglycosides from *Sophora japonica*

2.2 Osteoporosis

Generally, osteoporosis may be caused by hormone deficiency in post-menopausal women and as an age-related disease in the elderly. When estrogen (**41**) is reduced in the body, local factors such as interleukin-1β (IL-1β) and interleukin-6 (IL-6) which might be related to bone resorption may be increased and promote osteoclastogenesis, which is responsible for bone resorption (Fig. 13). One study looked at whether the glucosidic isoflavones of Isocal, PIII extracted from *Sophorae fructus* could affect the proliferation of osteoblasts and prevent osteoclastogenesis in vitro by attenuating the upstream cytokines such as IL-1β and IL-6 in a human osteoblastic cell line (MG-63) and in a primary osteoblastic culture from Sprague Dawley rat femurs. Interestingly, interleukin-1β (IL-1β) and interleukin-6 (IL-6) mRNA were significantly suppressed in osteoblast-like cells treated with 17β-estradiol (**40**) and PIII (glucosidic isoflavones extracted from *Sophorae fructus*) when compared to a positive control of two soybean isoflavones [genistin (**31**), daidzin (**30**)], and this suppression was more effective at 10^{-8}%

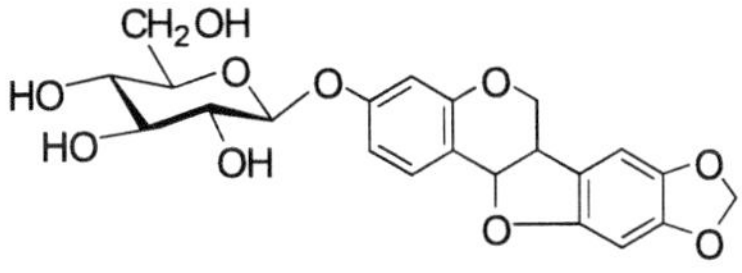

trifolirhizin (**2**)

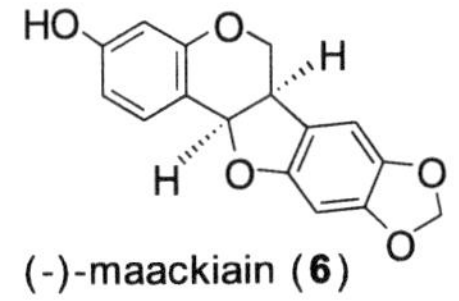

(-)-maackiain (**6**)

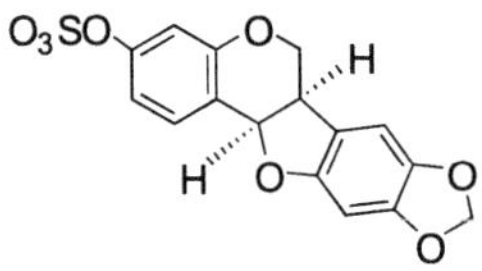

(-)-maackiain 3-sulfate (**58**)

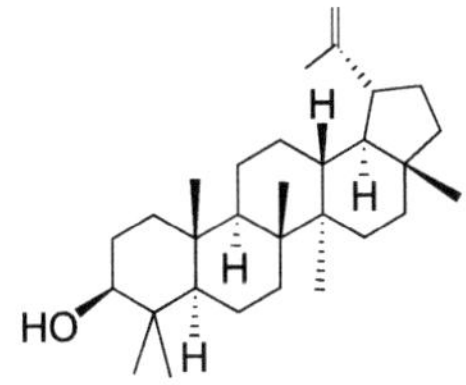

lupeol (**59**)

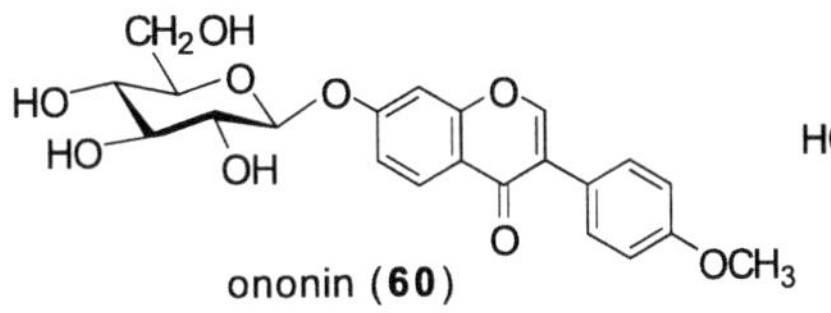

ononin (**60**)

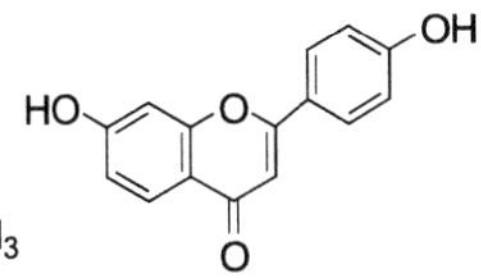

7,4'-dihydroxyflavone (**61**)

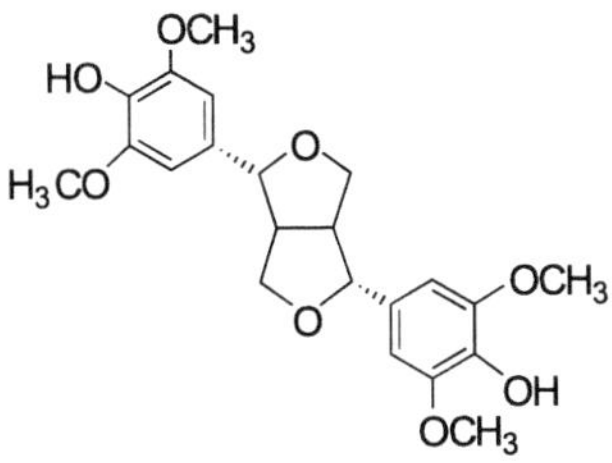

(+)-syringaresinol (**62**)

Fig. 11 Flavonoids in *Sophora subprostrata*

than at the highest concentration of 10^{-4}% (Fig. 13). In the cell line, the cells showed that 17β-estradiol (**40**) was the most effective in osteoblastic proliferation over the whole range of concentration (10^{-4}–10^{-12}%), even though PIII also showed the second greatest effectiveness at 10^{-8}%. The nitric oxide (NO) was significantly upregulated in PIII and 17β-estradiol (**40**) over the concentration range 10^{-6}%–10^{-8}% when compared to two soybean isoflavones [genistin (**31**), daidzin (**30**)], without showing any dose dependency. The bone marrow primary culture showed that PIII effectively suppressed osteoclastogenesis next to 17β-estradiol (**40**) when compared to two soybean isoflavones [genistin (**31**), daidzin (**30**)]. These results might suggest that local bone-resorbing cytokines could be regulated by PIII at lower concentrations. Therefore, PIII also might preferentially induce an anti-osteoporosis response by attenuating osteoclastic differentiation and by upregulating the NO (Fig. 13) [11].

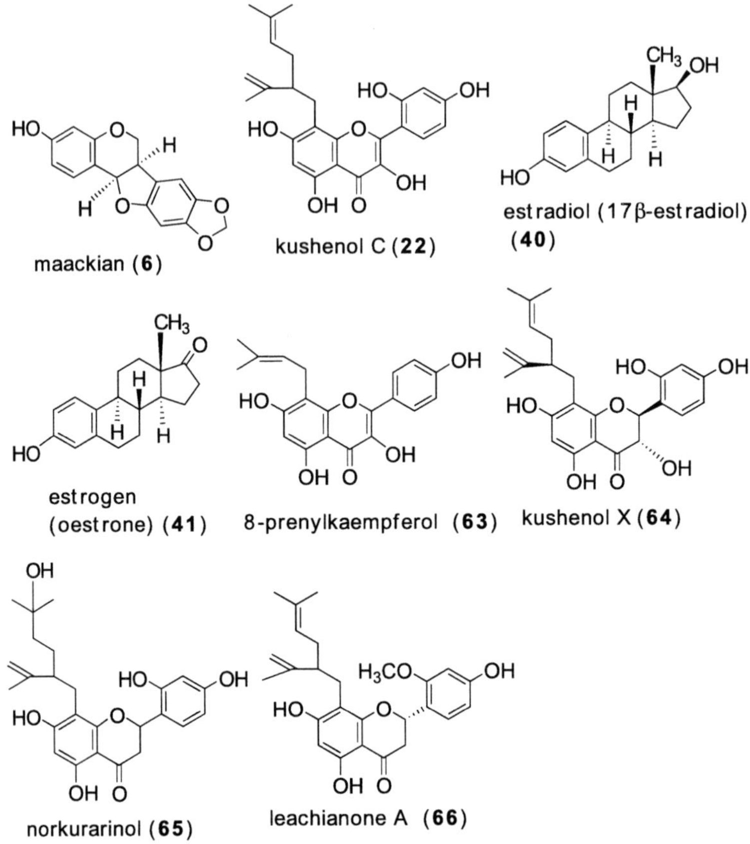

Fig. 12 Flavonoids for test of estrogen receptor binding from *Sophora flavescens*

2.3 Antitumor Activity

Two new lavandulylated flavanones (2*S*)-2′-methoxykurarinone (**67**) and (–)-kurarinone (**68**)—found in the root of *Sophora flavescens*—were determined together with two lavandulyl flavanones sophoraflavanone G (**10**) and leachianone A (**66**), and two isoflavonoids formononetin (**4**) and l-maackiain (**6**). Four flavanones (**3**, **66**, **67**, and **68**) exhibited significant cytotoxic activity against human myeloid leukemia HL-60 cells (Fig. 14) [12, 13].

Ten prenylflavanones (**10**, **69**–**77**) including sophoraflavanone G (**10**), naringenin (**69**), hesperetin (**70**), 6-prenylbaringenin (**71**), sophoraflavanone B (**72**), sophoraflavanone A (**73**), sophoraflavanone D (**74**), euchrestaflavanone A (**75**), sophoraflavanone H (**76**), sophoraflavanone I (**77**) were extracted from *Sophora tomentosa* L., and *Sophora moorcroftiana* Benth. ex Baker (*Leguminosae*) (Fig. 15).

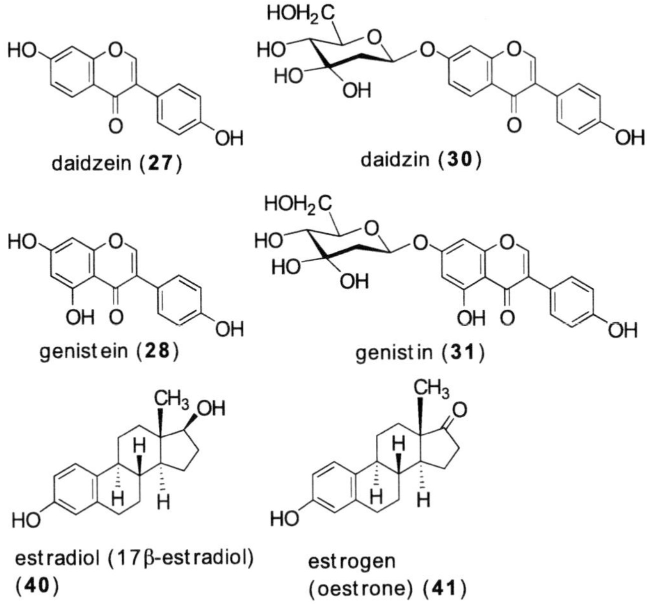

Fig. 13 Two glucosidic isoflavones of daidzin (**30**) and genistin (**31**) from *Sophorae fructus* and related compounds

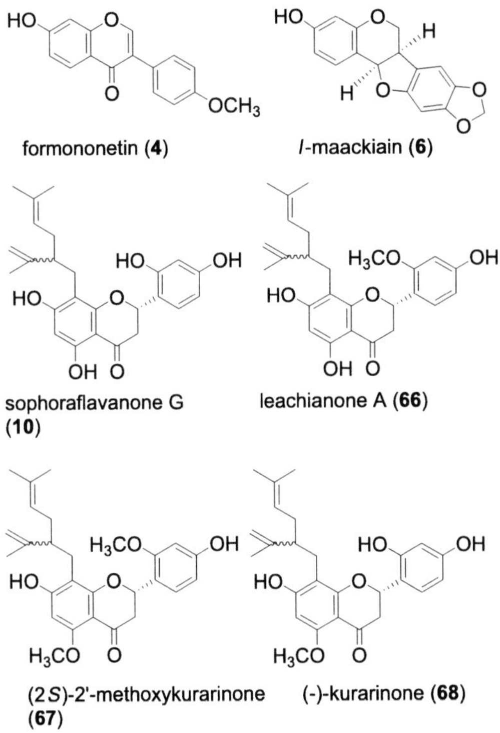

Fig. 14 Isoflavonoids from the root of *Sophora flavescensnes*

Looking at two oral tumor cells (HSC-2, HSG) and normal gingival fibroblast (HGF) in this study, sophoraflavanone G (**10**), sophoraflavanone B (**72**), sophoraflavanone A (**73**), and euchrestaflavanone A (**75**) were cytotoxic against the two tumor cells. HSG cells as a control were more resistant when compared to HSC-2 cells. Interestingly, sophoraflavanone H (**76**) and

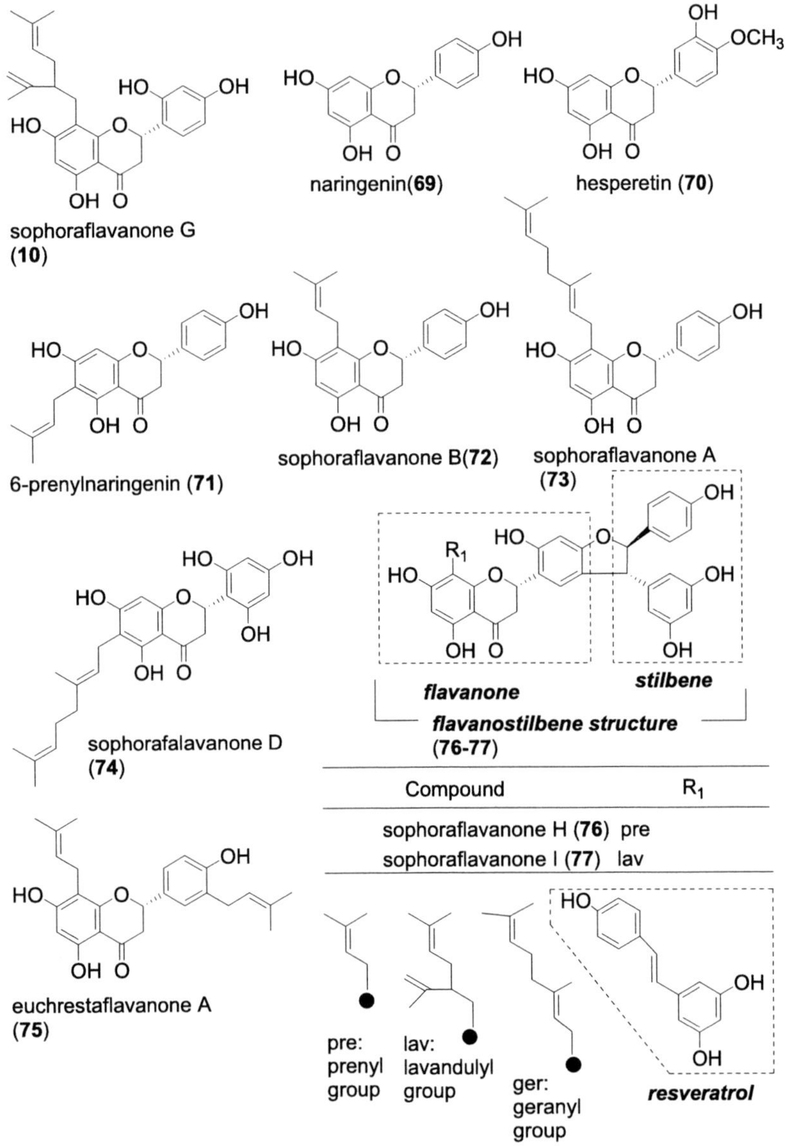

Fig. 15 Prenylflavanones (**10, 69–77**) from *Sophora* plants and its related compounds

Table 1 Antitumor activity, radical generation, and O_2^- scavenging activity of prenylflavanones (**10**, **69–77**)

Compd	Cytotoxic activity[1] (CC_{50}: μg/ml) (mM) Human oral tumor cell line HSC-2		HSG		Human gingival fibroblast (HGF)		Radical generation (at pH 10.5)	O_2^- scavenging activity at pH 7.8 SOD (U/1.5 mg)
69	425	(1.560)	> 500	(> 1.840)	> 500	(> 1.840)	< 0.03	1.5
70	370	(1.230)	480	(1.590)	189	(0.630)	< 0.03	2.1
71	22	(0.065)	32	(0.094)	35	(0.103)	0.25	17.1
72	19	(0.056)	20	(0.059)	23	(0.068)	< 0.03	3.9
73	< 16	(< 0.039)	16	(0.039)	13	(0.032)	< 0.03	4.8
10	< 8	(< 0.019)	8	(0.019)	8	(0.019)	< 0.03	5.6
74	24	(0.055)	58	(0.132)	45	(0.102)	< 0.03	48.0
75	17	(0.042)	31	(0.076)	27	(0.066)	0.04	23.5
76	38	(0.065)	91	(0.156)	100	(0.172)	1.14	160.1
77	15	(0.023)	50	(0.077)	45	(0.069)	0.85	72.6
Dextran sulfate (μg/ml)								
AZT (μM)								
ddC (μM)								

[1] Near confluent HSC-2, HSG, and HGF cells were incubated for 24 h with various concentrations of **10**, **69–77**, and the relative viable cell number (A_{540}) was determined by MTT method. Each value represents mean from duplicate determinations. Control A_{540} values of HSC-2, HSG, and HGF cells were 0.97, 0.54, and 0.31, respectively.

sophoraflavanone I (**77**) of flavanostilbene structure were more selective for the two human oral tumor cell lines HSC-2 and HSG when compared to normal human gingival fibroblast (HGF) (Table 1).

Prenylflavanones having either prenyl-, lavandulyl-, or geranyl groups on the A ring, two bioactive flavonostilbenes on the B ring and stilbene (resveratrol) were studied for tumor-specific cytotoxic activity, radical generation, and O_2^- scavenging activity. A good relationship of tumor-specific cytotoxic activity with radical generation and O_2^- scavenging activity in these prenylflavanones was observed [14] (Table 1).

Eleven isoflavones were also examined for their cytotoxicity (Fig. 16).

Among these 11 isoflavones (**4**, **27**, **28**, **78–84**, **84′**), genistein (**28**) produced the higher cytotoxic activity against two human oral tumor cell lines (HSC-2, HSG) when compared to normal human gingival fibroblast cells (HGF), suggesting its tumor-specific action. Interestingly, by electron-spin resonance (ESR) spectroscopy, genistein (**28**) was shown not to produce radicals, nor scavenge the O_2^- radicals generated by the hypoxanthine–xanthine oxidase

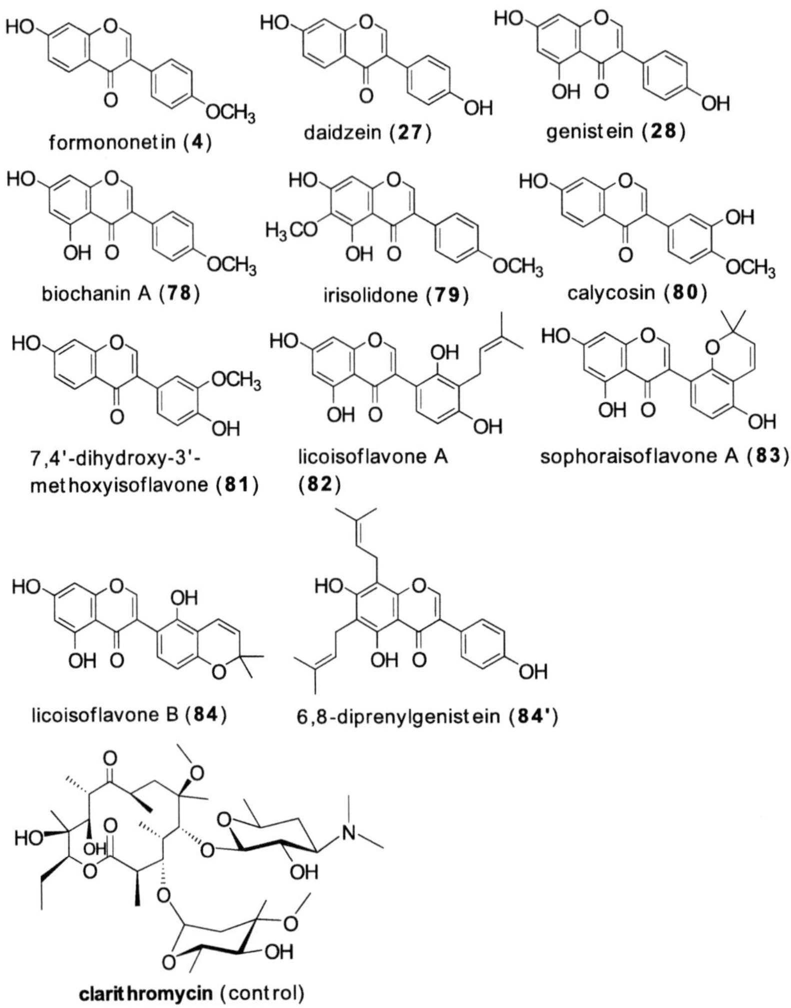

Fig. 16 Eleven isoflavones from *Sophora* species and a control

reaction system. This means that a radical-mediated oxidation mechanism might not be involved in the genistein (**28**)-induced cytotoxicity.

Moreover, licoisoflavone A (**82**) and sophoraisoflavone A (**83**) with prenyl groups produced higher anti-*Helicobacter pylori* activity. However, 6,8-diprenylgenistein (**84**′) with two prenyl groups had the highest cytotoxic activity but lower tumor-specificity. Because 6,8-diprenylgenistein (**84**) produced the highest amount of radicals and most efficiently scavenged the O_2^- radicals, 6,8-diprenylgenistein (**84**′) might induce cytotoxicity by the radical-mediated oxidation mechanism (Fig. 16) (Table 2) [15].

Table 2 Antitumor activity, radical generation, and O_2^- scavenging activity of isoflavones (**4**, **27**, **28**, **78–84**, **84′**)

Compd	Cytotoxic activity[1] (CC_{50}: µg/ml) (mM)				Anti-*H. pylori* activity	Radical generation	O_2^- scavenging activity
	Human oral tumor cell line		Human gingival fibroblast (HGF)	SI (HGF/HSC-2)	MIC_{50} (µg/ml)	(at pH 10.5)	SOD (U/1.5 mg)
	HSC-2	HSG					
27	114	177	625	5.5	> 100	< 0.03	0.13
4	596	584?	726	1.2	> 100	< 0.03	0.51
28	31	54	401	12.9	> 100	< 0.03	0.75
78	28	125	165	5.9	> 100	< 0.03	0.78
79	199	233	529	2.7	> 100	< 0.03	0.61
80	179	142	206	1.2	> 100	0.13	7.10
81	500?	524?	583	1.2	> 100	< 0.03	1.10
82	25	31	38	1.5	33	< 0.03	1.30
83	26	26	41	1.6	55	< 0.03	2.60
84	11	19	20	1.8	> 100	< 0.03	1.50
84′	10	13	15	1.5	> 100	0.30	14.60
clarithromycin					10		

[1] Near confluent HSC-2, HSG, and HGF cells were incubated for 24 h with various concentrations of **10**, **69–77**, and the relative viable cell number (A_{540}) was determined by MTT method. Each value represents mean from duplicate determinations. Control A_{540} values of HSC-2, HSG, and HGF cells were 0.97, 0.54, and 0.31, respectively.

Two isoflavones, daidzein (**27**) and genistein (**28**), and nine isoflavanones including sophoronol (**86**), isosophoranone (**87**), sophoraisoflavanone A (**17**), secundifloran (**88**), secundiflorol A (**89**), secundiflorol D (**90**), secundiflorol E (**91**), secundiflorol F (**92**), and tetrapterol G (**93**) were isolated from *Sophora tomentosa* L., and *Sophora moorcroftiana* Benth. ex Baker. These 12 isoflavones were examined for their cytotoxic activity against three normal human cells (gingival fibroblast, pulp cell, periodontal ligament fibroblast) and two human tumor cell lines (squamous cell carcinoma HSC-2, submandibular gland carcinoma HSG). Isoflavones with two isoprenyl groups (one in the A-ring and the other in the B-ring) such as tetrapterol G (**93**) and isosophoranone (**87**), and isoflavanones with a α,α-dimethylallyl group at C-5′ of the B-ring such as secundifloran (**88**), secundiflorol A (**89**), secundiflorol D (**90**) and secundiflorol E (**91**) had relatively higher cytotoxic activity. Secundiflorol A (**89**) had the highest tumor-specific cytotoxic action (tumor specificity index (TS) = 2.8), followed by genistein (TS = 2.4) (**28**), secundiflorol D (TS = 1.9) (**90**), secundiflorol E (TS = 1.9) (**91**), secundiflorol F (**YS30**, TS = 1.9) (**92**) > sophoraisoflavanone A (TS = 1.8) (**17**), sophoronol

(TS = 1.7) (**86**), secundifloran (TS = 1.7) (**88**) > tetrapterol G (TS = 1.6) (**93**) > isosophoranone (TS = 1.5) (**87**) > daidzein (TS = 1.1) (**27**) (Tables 3, 4).

When hydrophobicity was assessed by the octanol/water partition coefficient (log *P*), the maximum cytotoxic activity was observed at a log *P* value of around 4. Isoflavanoids with intermediate cytotoxic activity such as secundiflorol A (**89**), genistein (**28**), secundiflorol D (**90**), secundiflorol E (**91**), and secundiflorol F (**92**) showed relatively the higher tumor specificity.

Table 3 Antitumor activity of isoflavones (**27**, **28**) and isoflavanones (**17**, **86–93**) from *Sophora tomentosa* L., and *Sophora moorcroftiana* Benth. ex Baker

Compd	CC_{50} (μM)					TS
	Human oral tumor cell line		Human oral normal cell line			
	HSC-2	HSG	HGF	HPC	HPLF	
27	91	508	524	150	346	1.1
28	156	259	533	444	481	2.4
86	171	250	378	313	367	1.7
87	25	41	62	39	43	1.5
17	65	100	200	111	122	1.8
88	34	166	194	166	135	1.7
89	15	204	286	258	388	2.8
90	30	143	203	141	154	1.9
91	53	88	120	110	160	1.9
92	133	266	508	226	376	1.9
93	17	26	42	26	31	1.6

Table 4 Effects on NO production and radical scavenging activity of isoflavanones (**17**, **86–93**) from *Sophora tomentosa* L., and *Sophora moorcroftiana* Benth. ex Baker

Compd	NO production				NO scavenging activity	
	CC_{50} (μM)		IC_{50} (μM)	SI = CC_{50}/CC_{50}	Assay for NO scavenging activity	Assay for reduction activity
	LPS (−)	LPS (+)	LPS (+)			
86	129	126	32.0	4.0		
87	25	25	17.0	1.5		
17	58	51	8.4	6.0		
88	99	66	5.8	11.3	51	53
89	76	87	13.0	6.7		
90	66	64	3.1	20.8	51	51
91	63	83	17.0	4.9		
92	117	131	35.0	3.7		
93	28	24	12.0	1.9		
Control					100	100

All isoflavones [daidzein (**27**), genistein (**28**)] and isoflavanones (**17**, **86**–**93**) did not stimulate nitric oxide production by mouse macrophage-like Raw 264.7 cells, but almost completely inhibited NO production by the lipopolysaccharide (LPS)-activated Raw 264.7 cells. ESR spectroscopy showed that secundifloran (**88**) and secundiflorol D (**90**), which showed the most inhibitory effect for NO production, efficiently scavenged the superoxide anion and NO radicals. From these results, it is suggested that the inhibition of macrophage NO production by these isoflavanones might, at least in part, be explained by their radical scavenging or reduction activity (Fig. 17) (Tables 3, 4) [16].

Two new lavandulylated flavanones (2*S*)-2′-methoxykurarinone (**67**) and (–)-kurarinone (**68**) were isolated from the root of *Sophora flavescens*, together with two known lavandulyl flavanones sophoraflavanone G (**10**) and leachianone A (**66**), and two known isoflavonoids formononetin (**4**) and *l*-maackiain (**6**). The structures of (2*S*)-2′-methoxykurarinone (**67**) and (–)-kurarinone (**68**) were determined on the basis of optical rotation and spec-

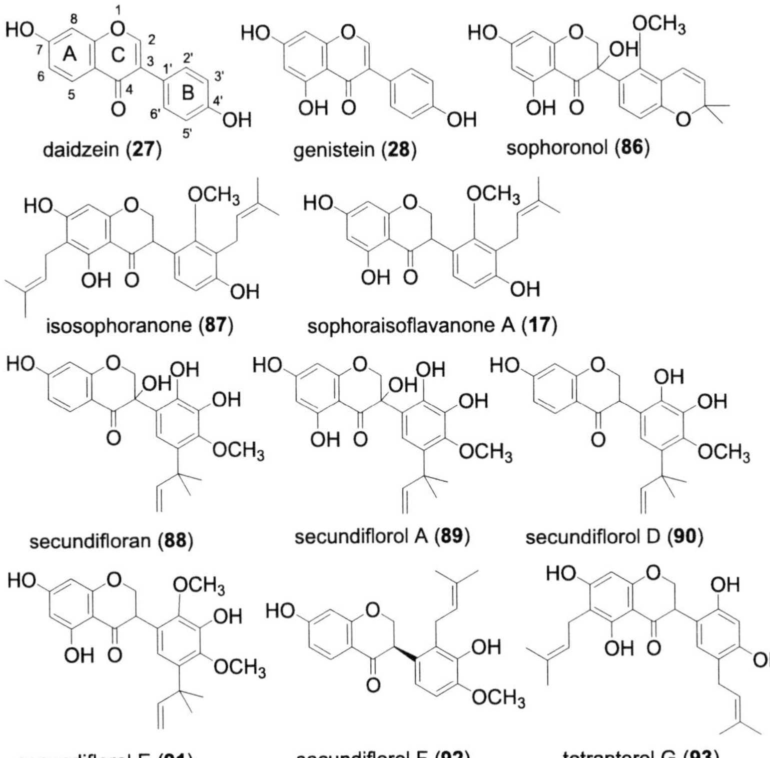

Fig. 17 Isoflavones (27, 28) and isoflavanones (17, 86–93) from *Sophora tomentosa* L., and *Sophora moorcroftiana* Benth. ex Baker

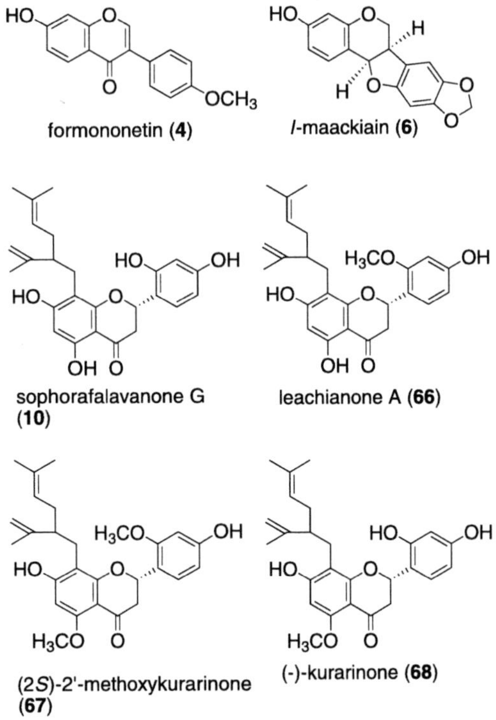

Fig. 18 Cytotoxic lavandulyl flavanones and two flavonoids from *Sophora flavescens*

tral evidence and by the comparison with known compounds. Four lavandulyl flavanones sophoraflavanone G (**10**), leachianone A (**66**), (2*S*)-2′-methoxykurarinone (**67**), and (–)-kurarinone (**68**) exhibited cytotoxic activity against human myeloid leukemia HL-60 cells (Fig. 18) [13, 17].

Kurarinone (**13**) of a lavandulyl flavanone was isolated from a polyphenolic extract of the roots of *Sophora flavescens*. First, kurarinone (**13**) had weak estrogenic activity both in the yeast screen and in the Ishikawa Var-I assay with EC_{50} values of 4.6 and 1.66 μM, respectively. Second, kurarinone (**13**) had the potent cytotoxic activity (IC_{50}) value = 22.2 μM) against human MCF-7/6 breast cancer cells in the sulforhodamine-B assay (Fig. 19) [18].

From the roots of *Sophora flavescens* collected in Taiwan, four new prenylflavonoids, sophoraflavanone K (**94**), sophoraflavanone L (**95**), 8-lavandulylkaempferol (**96**), and cyclokuraridin (**97**), a new arylbenzofuran, 2-(2,4-dihydroxy-5-prenylphenyl)-5,6-methylenedioxybenzofuran (**98**), and a new prenyldibenzoyl derivative of sophoradione (**99**), were isolated. The structures of these six compounds were determined by spectroscopic data analysis. The new compounds (**95–99**) had cytotoxic activity against the KB epidermoid carcinoma cell line (Fig. 20) [19].

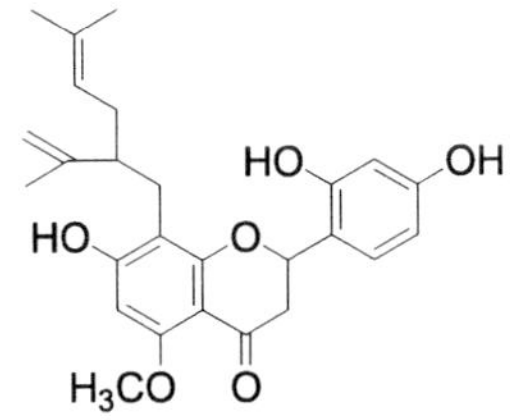

kurarinone (**13**)

Fig. 19 Estrogenic and antitumor activities of kurarinone (13) from the roots of *Sophora flavescens*

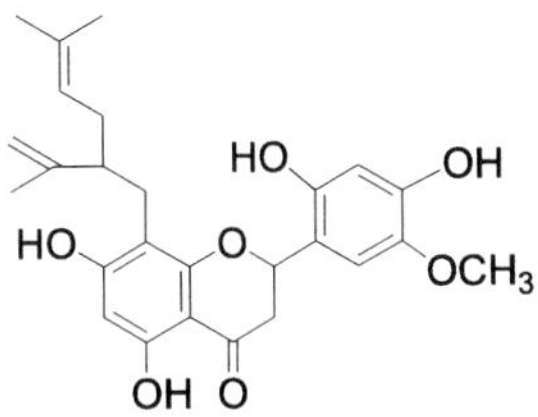

sophorafalavanone K (**94**)

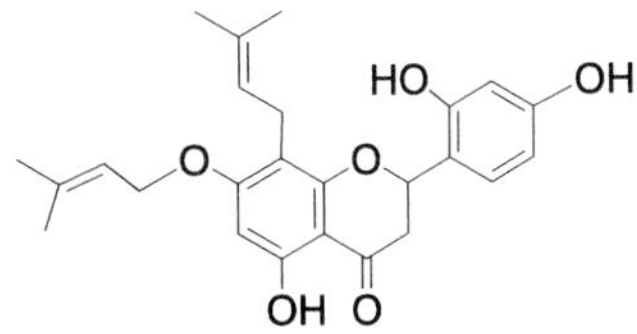

sophorafalavanone L (**95**)

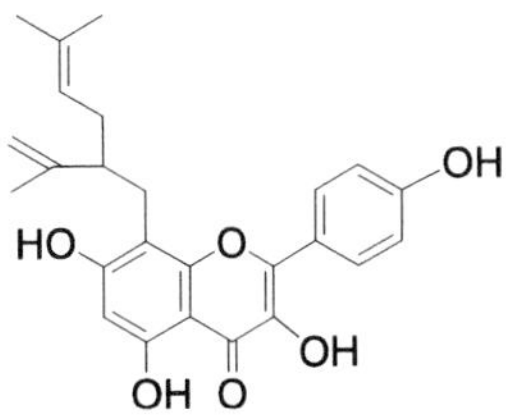

8-lavandulylkaempferol (**96**)

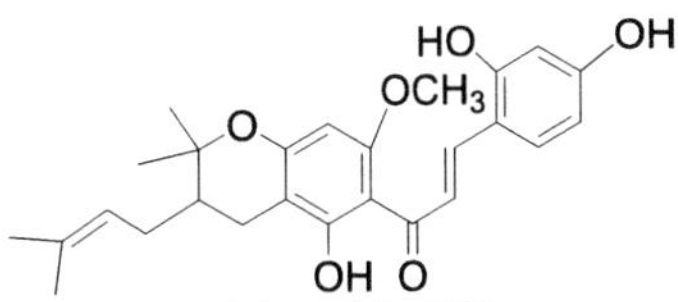

cyclokuraridin (**97**)

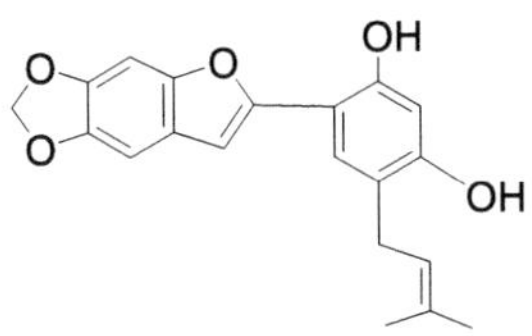

2-(2,4-dihydroxy-5-prenylphenyl)-5,6-methylenedioxybenzofuran (**98**)

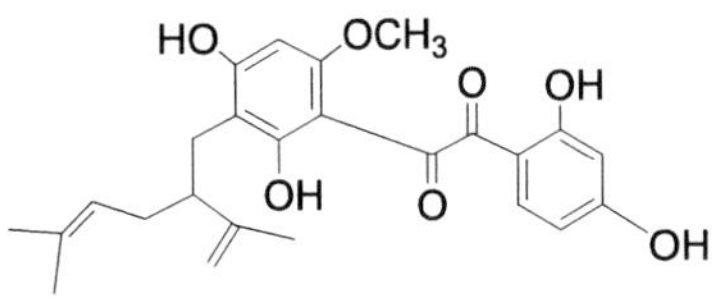

sophoradione (**99**)

Fig. 20 Phenolic compounds and their cytotoxic activity from *Sophora flavescens*

Six stilbenes sophorastilbene A (**100**), (+)-α-viniferin (**101**), (−)-ε-viniferin (**102**), resveratrol (**103**), piceatannol (**104**), and rhaponticin (**105**), and five flavonoids kaempferol (**106**), fisetin (**107**), quercetin (**43**), isoliquiritigenin (**108**), and butein (**109**) were examined for their tumor-specific cytotoxicity and apoptosis-inducing activity, using four human tumor cell lines including human squamous cell carcinoma HSC-2, HSC-3, submandibular gland carcinoma HSG, and promyelocytic leukemia HL-60, and three normal human oral cells including gingival fibroblast HGF, pulp cell HPC, and periodontal ligament fibroblast HPLF. Among the 11 compounds, five including sophorastilbene A (**100**), (+)-α-viniferin (**101**), piceatannol (**104**), quercetin (**43**), and isoliquiritigenin (**108**) showed significantly higher cytotoxicity against the tumor cell lines having tumor-specific indices of 3.6,

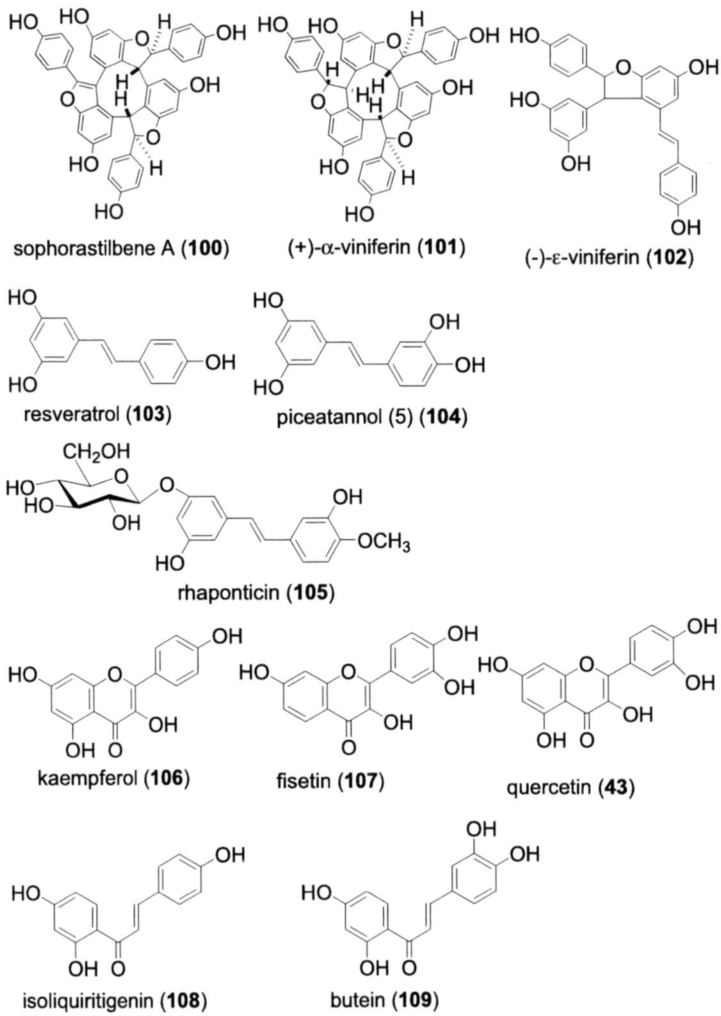

Fig. 21 Stilbenes (100–105) and flavonoids (43, 106–109)

4.7, > 3.5, > 3.3, and 4.0, respectively, when compared to the normal. Among the seven cell lines, HSC-2 and HL-60 cells were the most sensitive to the cytotoxic action of these compounds. Sophorastilbene A (**100**), piceatannol (**104**), quercetin (**43**), and isoliquiritigenin (**108**) induced dose-dependently internucleosomal DNA fragmentation and activation of caspase-3, caspase-8, and caspase-9 in HL-60 cells. (+)-α-Viniferin (**101**) had a similar activity at higher concentrations. All the compounds failed to induce DNA fragmentation and activated three caspases to a much lesser extent in HSC-2 cells. From Western blot analysis, sophorastilbene A (**100**), piceatannol (**104**), and quercetin (**43**) did not show induction on any consistent changes in their expression of pro-apoptotic proteins (Bax, Bad) and antiapoptotic protein (Bcl-2) in the HL-60 and HSC-2 cells. An undetectable expression of Bcl-2 protein in the control and the drug-treated HSC-2 cells could explain the relatively higher sensitivity of this cell line to stilbenes and flavonoids (Fig. 21) (Table 5) [20].

The structure–activity relationships of multidrug-resistance protein 1 (MRP1, ABCC1) inhibitors might aid in developing potent inhibitors against MRP1-mediated multidrug resistance. Six stilbenes sophorastilbene A (**100**), (+)-α-viniferin (**101**), (–)-ε-viniferin (**102**), resveratrol (**103**), piceatannol (**104**), and rhaponticin (**105**) were examined for their ability to inhibit the MRP1-mediated transport of 2′,7′-bis-(carboxypropyl)-5(6)-carboxyfluorescein (BCPCF) from human erythrocytes and into the inside-out erythrocyte membrane vesicles (IOVs). The concentrations of stilbenes de-

Table 5 Antitumor activity of stilbenes and flavonoids

Compd	CC_{50} (μM)							TS
	Tumor cell				Normal cell			
	HSC-2	HSC-3	HSG	HL-60	HGF	HPC	HPLF	
Stilbenes								
100	27	61	108	28	196	210	207	3.6
101	36	60	349	6	464	573	553	4.7
102	42	84	110	31	111	146	94	1.8
103	155	288	316	45	500	676	582	2.9
104	63	232	373	11	367	414	> 1,000	> 3.5
105	500	588	> 1,000	810	> 1,000	979	> 1,000	>< 1.4
Flavonoids								
106	743	> 1,000	> 1,000	29	> 1,000	> 1,000	> 1,000	>< 1.4
107	39	188	375	16	608	170	641	3.1
43	35	250	745	31	750	909	> 1,000	> 3.3
108	79	97	125	22	311	319	336	4.0
109	26	69	188	5	274	180	234	3.2

TS = $\left[\sum CC_{50}\ \text{(normal cells)}/\sum CC_{50}\ \text{(tumor cell lines)}\right] \times (4/3)$

creasing the BCPCF transport by 50% during 60 min of incubation at 37 °C (IC_{50}) were determined from their dose-response curves. Then, these stilbenes inhibited BCPCF transport in the cells in the following rank order: (+)-α-viniferin (IC_{50} = 0.8 μM) (**101**) > sophorastilbene A (IC_{50} = 3.1 μM) (**100**) > (−)-ε-viniferin (IC_{50} = 8.9 μM) (**102**) > piceatannol (IC_{50} = 57 μM) (**104**). Two stilbenes resveratrol (**103**) and rhaponticin (**105**) were ineffective. Three stilbenes (+)-α-viniferin (IC_{50} = 0.8 μM) (**101**), sophorastilbene A (IC_{50} = 3.7 μM) (**100**), and (−)-ε-viniferin (IC_{50} = 3.5 μM) (**102**) were also efficient as BCPCF transport inhibitors in the inside-out erythrocyte membrane vesicles (IOVs). From these results, the stilbenes might efficiently inhibit MRP1-mediated organic anion transport. The inhibitory potency of these stilbenes increased with oligomerization. The membrane might not be a strong barrier for the inhibitory activity of the trimeric stilbenes (Table 6) [21].

Table 6 Stilbene-induced inhibition of MRP1-mediated BCPCF transport in intact human erythrocytes and inside-out erythrocyte vesicles (IOVs)

Compd	Oligometric state	No. of hydroxyl group	IC_{50} (mM) for intact cells	IC_{50} (mM) for IOVs
Stilbenes				
100	Trimer	6	3.1 ± 0.1	3.7 ± 0.1
101	Trimer	6	0.8 ± 0.1	0.8 ± 0.2
102	Dimer	5	8.9 ± 0.6	3.5 ± 0.6
103	Monomer	3	–	
104	Monomer	4	57.0 ± 9.5	
105	Monomer	2 + glucose	–	
References				
Benzbromarone			4	
Indomethacin			10	
Euchrestaflavanone A (**75**)			3	
Sophoraflavanone H (**76**)			3	

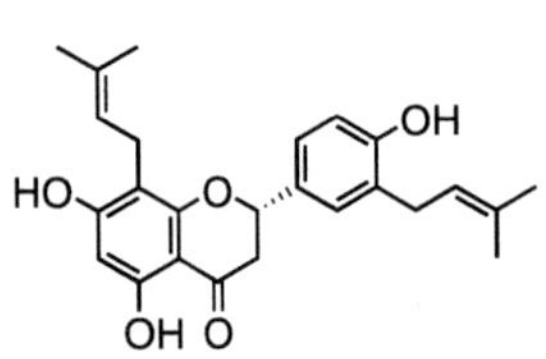

euchrestaflavanone A (**75**)

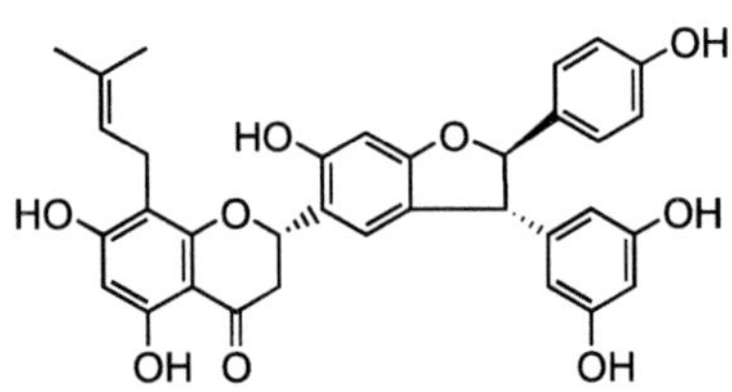

sophorafalavanone H (**76**)

2.4 Antimicrobial Activity

A new prenylated flavonol sophoflavescenol (**110**), together with five known flavonoids including kurarinol (**3**), kushenol K (**111**), kushenol H (**112**), trifolirhizin (**2**), and kuraridin (**8**), were isolated from the roots of *Sophora flavescens*. The structure of sophoflavescenol (**110**) was determined by spectroscopic analysis. Among the five known flavonoids kurarinol (**3**), kushenol K (**111**), and kushenol H (**112**) showed weak antiviral activity against *Herpes simplex* virus types I and II (Fig. 22) [22].

Ten prenylflavanones (**10**, **69–77**) of *Sophora tomentosa* L., and *Sophora moorcroftiana* Benth. ex Baker (Fig. 15) were examined for effects on bacteria such as *Staphylococcus aureus*, *S. typhimurium*, *Bacillus subtilis VBI*,

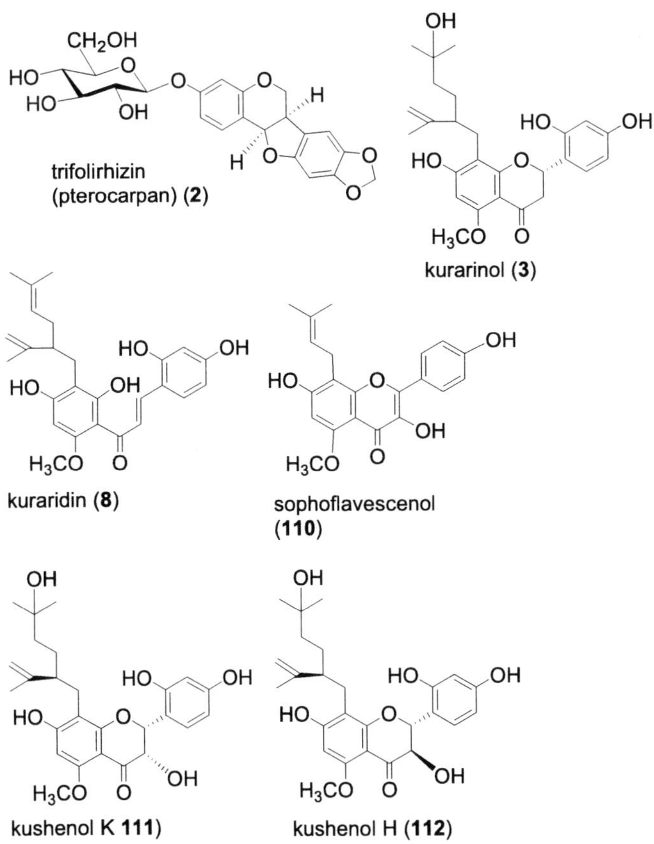

Fig. 22 Sophoflavescenol (**110**), flavonoids, and their anti-HIV activity from roots of *Sophora flavescens*

Table 7 Antimicrobial activity of prenylflavanones (**10**, **69–77**)

Compd	Gram-positive bacteria							Gram-negative bacteria							*Helicobacter pylori*
	Staphylococcus aureus			*Staphylococcus typhimurium*			*Bacillus subtilis*	*Shigella*		*Esherichia coli*		*Klebsiella* spp 14	*Providencia* spp 1	*Vibrio cholerae* 865	MIC_{50} (μg/ml) [1]
	6571	8530	8531	4	57	59	VB1	*dysenteriae* 1	*sonnei* 2	Row	R832				
69	> 100	> 100	> 100	> 100	> 100	> 100	> 100	> 100	> 100	> 100	> 100	> 100	> 100	> 100	> 100.0
70	25	100	100	> 100	> 100	50	> 100	> 100	> 100	50	> 100	> 100	50	100	> 100.0
71	25	50	50	> 100	> 100	> 100	> 100	50	> 100	> 100	25	> 100	> 100	25	> 100.0
72	50	25	25	> 100	> 100	> 100	> 100	50	> 100	> 100	12.5	> 100	> 100	25	68.0
73	25	25	25	> 100	> 100	> 100	> 100	25	> 100	> 100	25	> 100	> 100	50	> 100.0
10	12.5	12.5	12.5	> 100	> 100	> 100	> 100	12.5	> 100	> 100	12.5	> 100	> 100	12.5	5.8
74	25	12.5	> 100	> 100	> 100	> 100	> 100	25	> 100	> 100	> 100	> 100	> 100	100	21.0
75	12.5	12.5	> 100	> 100	> 100	> 100	> 100	25	> 100	> 100	> 100	> 100	> 100	> 100	> 100.0
76	25	12.5	> 100	> 100	> 100	> 100	> 100	50	> 100	> 100	> 100	> 100	> 100	> 100	2.6
77	12.5	25	> 100	> 100	> 100	> 100	> 100	12.5	> 100	> 100	> 100	> 100	> 100	> 100	2.1
Clarithromycin															10.0

[1] The MIC value (μg/ml) was determined for each compound by a broth microdilution method using *H. pylori* (ATCC43504) [26, 27]. Each **69** and **70**: each equivalent mixture of DL (*R*, *S*).

Shigella, *Escherichia coli*, *Klebsiella* spp. 14, *Providencia* spp. 1, and *Vibrio cholerae* 865.

Among these ten prenylflavanones (**10**, **69–77**), sophoraflavanone G (**10**) was highest for three Gram-positive bacteria including *Staphylococcus aureus* 6571, 8530, and 8531, followed by euchrestaflavanone A (**75**), sophoraflavanone H (**76**), and sophoraflavanone I (**77**). Sophoraflavanone G (**10**) also was effective against *Shigella dysenteria* 1, *Escherichia coli* R832, and *Vibrio cholerae* 865. Moreover, sophoraflavanone I (**77**) had the highest anti-*Helicobacter pylori*, followed by sophoraflavanone H (**76**) and sophoraflavanone G (**10**) (Table 7) [14].

Eleven isoflavonoids (**4**, **27**, **28**, **78–84**, **84′**) were examined for possible antimicrobial properties against 12 known Gram-positive and Gram-negative sensitive bacteria. Daidzein (**27**) and calycosin (**80**) failed to show antimicrobial activity and formononetin (**4**), genistein (**28**), biochanin A (**78**), irisolidone (**79**), 7,4′-dihydroxy-3′-methoxyisoflavone (**81**), licoisoflavone A (**82**), and licoisoflavone B (**84**) had moderate antimicrobial action. Sophoraisoflavone A (**83**) and 6,8-diprenylgenistein (**84**) had the highest antimicrobial properties. Sophoraisoflavone A (**83**) and 6,8-diprenylgenistein (**84**) were then tested in vitro against 214 strains of bacteria from one Gram-positive and six Gram-negative genera. The minimum inhibitory concentration (MIC) of sophoraisoflavone A (**83**) and 6,8-diprenylgenistein (**84**) was determined by the agar dilution method and ranged from 25 to 200 mg/l in most strains. At the concentrations of 30 and 60 μg/mouse, sophoraisoflavone A (**83**) and 6,8-diprenylgenistein (**84**) showed significant protection to mice challenged with 50 median lethal dose (MLD) of a virulent strain of *Salmonella typhimurium* (Fig. 6) (Table 8) [23].

Table 8 Antibacterial spectra of sophoraisoflavone A (**83**)

Bacterial strain	Number tested	Minimum inhibitory concentration (mg/l)				
		25	50	100	200	> 200
Staphylococcus aureus	39	39	39	39	39	39
Salmonella spp.	31	31	31	31	31	31
Shigella spp.	35	35	35	35	35	35
Klebsiella spp.	9	9	9	9	9	9
Pseudomonas spp.	11	11	11	11	11	11
Vibrio cholerae	69	69	69	69	69	69
Vibrio parahaemolyticus	20	20	20	20	20	20
Total	214	214	214	214	214	214

Table 8 (continued) Antibacterial spectra of 6,8-diprenylgenistein (**84′**)

Bacterial strain	Number tested	Minimum inhibitory concentration (mg/l)				
		25	50	100	200	> 200
Staphylococcus aureus	39	5	12	9	10	3
Salmonella spp.	31	4	5	6	12	4
Shigella spp.	35	6	8	5	10	6
Klebsiella spp.	9			2	5	2
Pseudomonas spp.	11		1	3	7	
Vibrio cholerae	69	6	8	24	24	7
Vibrio parahaemolyticus	20	2	5	5	8	
Total	214	23	39	54	76	22

Regarding the antibacterial activities of sophoraflavanone G (**10**) from *Sophora flavescens* in combination with two antimicrobial agents against oral bacteria, the combined effect of sophoraflavanone G (**10**) and the antimicrobial agents was estimated by the checkerboard method to obtain a fractional inhibitory concentration (FIC) index. The sophoraflavanone G (**10**) + ampicillin (**113**) combination was synergistic against *Streptococcus mutans, S. sanguinis, S. sobrinus, S. gordonii, Actinobacillus actinomycetemcomitans, Fusobacterium nucleatum, Prevotella intermedia*, and *Porphylomonas*

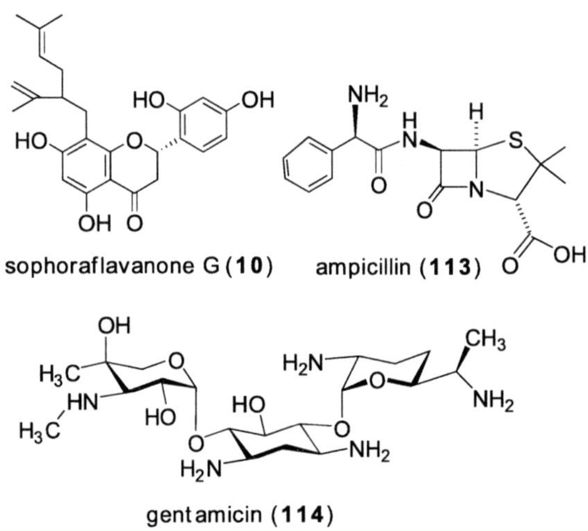

Fig. 23 The combination of sophoraflavanone G (10), ampicillin (113), and gentamicin (114) from the roots of *Sophora flavescens* for antibacterial activity

gingivalis, whereas the sophoraflavanone G (**10**) plus gentamicin (**114**) combination had a synergistic effect against *Streptococcus sanguinis*, *S. criceti*, *S. anginosus*, *Actinobacillus actinomycetemcomitans*, *Fusobacterium nucleatum*, *Prevotella intermedia*, and *P. gingivalis*. Neither combination exhibited any antagonistic interactions (FIC index > 4). In particular, the minimum inhibitory concentrations/minimum bactericidal concentrations (MICs/MBCs) for all the bacteria could be significantly reduced to 1/2–1/16 by the drug combinations. A synergistic interaction was also confirmed by time-kill studies for nine bacteria where the checkerboard suggested synergy. From these results, the strong bactericidal effect exerted by the drug combinations suggests that sophoraflavanone G (**10**) combined with other antibiotics might be microbiologically beneficial rather than antagonistic (Fig. 23) [24].

2.5 Anti-HIV Activity

Regarding anti-HIV activity ten prenylflavanones (**10**, **69–77**) in *Sophora tomentosa* L., and *Sophora moorcroftiana* Benth. ex Baker (Fig. 15), sophoraflavanone G (selectivity index (SI) = 5) (**10**) had selectively the highest anti-HIV activity, followed by euchrestaflavanone A (SI = 3) (**75**) and sophoraflavanone H (SI = 3) (**76**) (Table 9).

The prenylflavanones with either prenyl-, lavandulyl-, or geranyl groups on the A ring, two bioactive flavonostilbenes on the ring B, and stilbene

Table 9 Anti-HIV activity of prenylflavanones (**10**, **69–77**)

Compd	Anti-HIV activity		
	CC_{50} (μg/ml)	EC_{50} (μg/ml)	SI (CC_{50}/EC_{50})
69	> 200	> 200	>< 1
70	> 200	> 200	>< 1
71	125	> 200	< 1
72	95	> 200	< 1
73	21	> 200	< 1
10	23	4	5
74	113	> 200	< 1
75	69	22	3
76	66	23	3
77	114	> 200	< 1
Dextran sulfate (μg/ml)	> 1,000	0.44	2,273
AZT (μM)	277	0.04	6,925
ddC (μM)	753	0.26	2,896

Table 10 Anti-HIV activity of isoflavones (**4, 27, 28, 78–84, 84′**)

Compd	Anti-HIV activity		
	CC_{50} (μg/ml)	EC_{50} (μg/ml)	SI (CC_{50}/EC_{50})
27	> 200	> 200	>< 1
4	> 200	> 200	>< 1
28	9	> 40	< 1
78	24	> 40	< 1
79	> 200	> 200	>< 1
80	104	> 200	< 1
81	> 200	> 200	>< 1
82	39	> 40	< 1
83	41	> 200	< 1
84	5	8	< 1
84′	14	> 40	< 1
Dexatran sulfate	> 1,000	4.14	> 242
Curdlan sulfate	> 1,000	0.72	> 1,396
AZT (μM)	276	0.03	8,291
ddC (μM)	475	0.16	3,012

(resveratrol) had their anti-HIV activity, radical generation, and $O_2^{\cdot -}$ scavenging activity measured. A good relationship of anti-HIV activity with radical generation and $O_2^{\cdot -}$ scavenging activity in these prenylflavanones was observed [14].

Among these 11 isoflavones (**4, 27, 28, 78–84, 84′**), none induced anti-human immunodeficiency virus (HIV) activity (Fig. 16) (Table 10) [15].

2.6
Radical Generation and Radical O_2^- Scavenging Activity

First, regarding the radical generation of eight prenylflavanones (**10, 71–77**) in two plants of *Sophora tomentosa* L. and *Sophora moorcroftiana* Benth. ex Baker (Fig. 15), two sophoraflavanone H (**76**) and sophoraflavanone I (**77**) generated the highest electron-spin resonance signal of radicals when compared to the other six prenylflavanones. 6-Prenylnaringenin (**71**) was the second highest. However, for the other five prenylflavanones we could not find any detectable amount of radical signals (see Table 1 Antitumor activity, radical generation, and O_2^- scavenging activity of prenylflavanones and its related compounds (**10, 69–77**).

Second, on measuring the radical O_2^- scavenging activity of ten prenylflavanones (**10, 69–77**) in *Sophora tomentosa* L., and *Sophora moorcroftiana* Benth. ex Baker (Fig. 15), sophoraflavanone H (**76**) had the highest radical

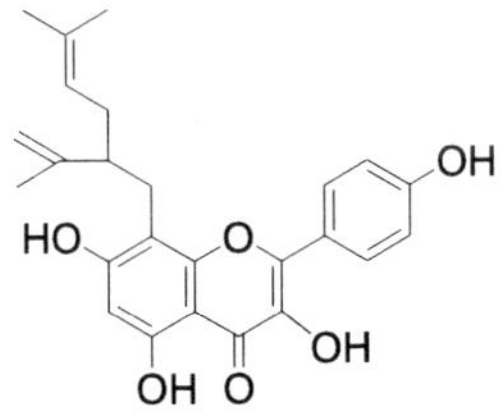

8-lavandulylkaempferol (**96**)

Fig. 24 8-Lavandulylkaempferol (**96**) from the root of *Sophora flavescensnes* AITON

O_2^- scavenging activity followed by sophoraflavanone I (**77**), sophoraflavanone D (**74**), euchrestaflavanone A (**75**), 6-prenylnaringenin (**71**), and sophoraflavanone G (**10**) (see Table 1 Antitumor activity, radical generation, and O_2^- scavenging activity of prenylflavanones (**10, 69–77**)).

The prenylflavanones with either prenyl-, lavandulyl-, or geranyl groups on the A ring, two bioactive flavonostilbenes on the ring B, and stilbene (resveratrol) had their radical generation and $O_2^{\bullet -}$ scavenging activity measured. A good relationship of biological activity with radical generation and O_2^- scavenging activity in these prenylflavanones was observed [14] [see Table 1 Antitumor activity, radical generation and O_2^- scavenging activity of prenylflavanones (**10, 69–77**)].

8-Lavandulylkaempferol (**96**) of a lavandulylated flavonoid was isolated from the roots of *Sophora flavescens* AITON. 8-Lavandulylkaempferol (**96**) might be a scavenger of both 1,1-diphenyl-2-picrylhydrazyl radicals and $ONOO^-$ (Fig. 24) [25].

From the roots of *Sophora flavescens*, five new compounds *trans*-hexadecyl ferulic acid (**115**), *cis*-octadecyl ferulic acid (**116**), *trans*-hexadecyl sinapic acid (**117**), (–)-4-hydroxy-3-methoxy-(6a*R*,11a*R*)-8,9-methylenedioxypterocarpan (**118**), and desmethylanhydroicaritin (**119**), along with nine known inactive compounds including (–)-maackiain (**6**), xanthohumol (**11**), formononetin (**4**), (2*S*)-2′-methoxykurarinone (**67**), (2*S*)-3β,7,4′-trihydroxy-5-methoxy-8-(γ,γ-dimethylallyl)-flavanone (**120**), (2*S*)-7,4′-dihydroxy-5-methoxy-8-(γ,γ-dimethylallyl)-flavanone (**121**), umbelliferone (**122**), kuraridin (**8**), and trifolirhizin (**2**) were found. The five compounds *trans*-hexadecyl ferulic acid (**115**), *cis*-octadecyl ferulic acid (**116**), *trans*-hexadecyl sinapic acid (**117**), (–)-4-hydroxy-3-methoxy-(6a*R*,11a*R*)-8,9-methylenedioxypterocarpan (**118**), and desmethylanhydroicaritin (**119**) exhibited 1,1-diphenyl-2-picrylhydrazyl (DPPH) free radical scavenging effects at IC_{50} values of 33.01 ± 0.20, 57.06 ± 0.16, 39.84 ± 0.36, 35.83 ± 0.47, and 18.11 ± 0.04 μM, respectively. When L-ascorbic acid was used as a positive control, L-ascorbic acid exhibited an IC_{50} value of 7.39 ± 0.01 μM. The new compounds *trans*-hexadecyl ferulic acid (**115**), *cis*-octadecyl ferulic acid (**116**), *trans*-hexadecyl sinapic acid (**117**), (–)-4-hydroxy-3-methoxy-(6a*R*,11a*R*)-8,9-methylenedioxypterocarpan

(**118**), and desmethylanhydroicaritin (**119**) also appeared to exert significant scavenging effects on authentic $ONOO^-$, with IC_{50} values of 5.76 ± 1.19, 15.06 ± 1.64, 8.17 ± 4.97, 1.95 ± 0.29, and 4.06 ± 2.41 μM, respectively. Penicillamine ($IC_{50} = 2.36 \pm 0.79$ μM) (**123**) was used as a positive control. In addition, the five compounds xanthohumol (**11**), *cis*-octadecyl ferulic acid (**116**), (–)-4-hydroxy-3-methoxy-(6a*R*,11a*R*)-8,9-methylenedioxypterocarpan (**118**), desmethylanhydroicaritin (**119**), and (2*S*)-3β,7,4′-trihydroxy-5-methoxy-8-(γ,γ-dimethylallyl)-flavanone (**120**) were isolated from this plant for the first time (Fig. 25) [26].

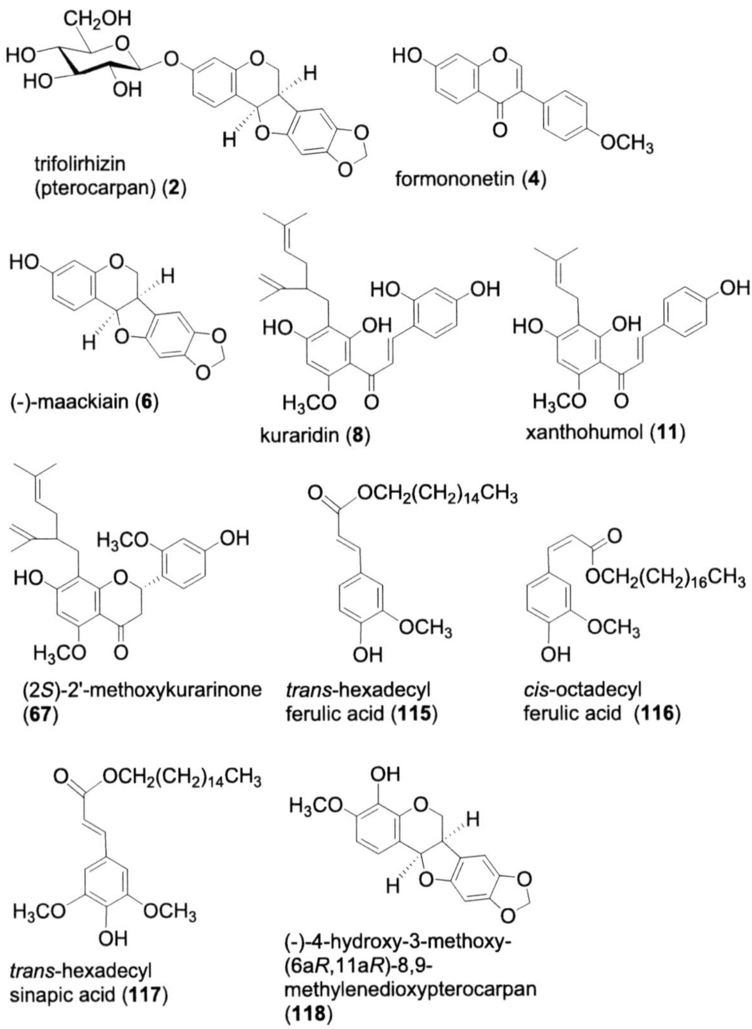

Fig. 25 Fourteen flavonoids from roots of *Sophora flavescens*

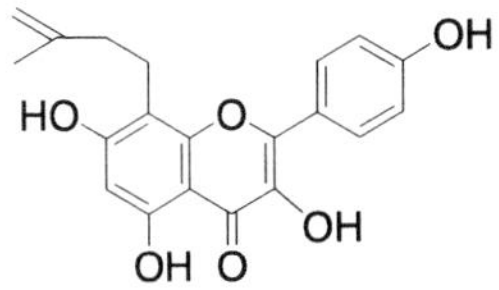

desmethylanhydroicaritin
(**119**)

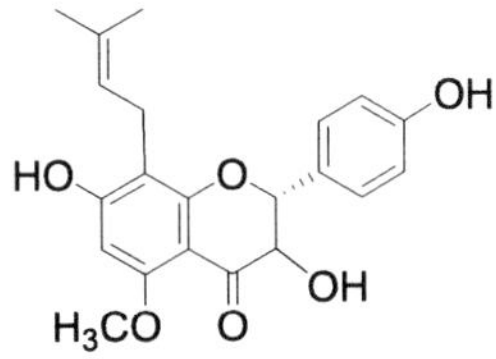

(2*S*)-3β,7,4'-trihydroxy-5-methoxy-8-(γ,γ-dimethylallyl)-flavanone
(**120**)

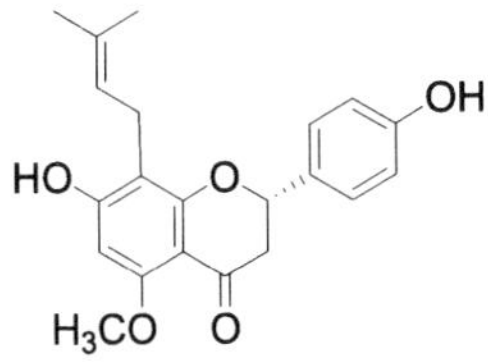

(2*S*)-7,4'-dihydroxy-5-methoxy-8-(γ,γ-dimethylallyl)-flavanone
(**121**)

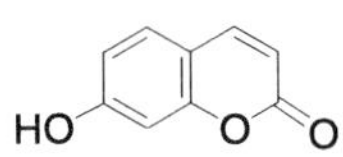

umbelliferone
(**122**)

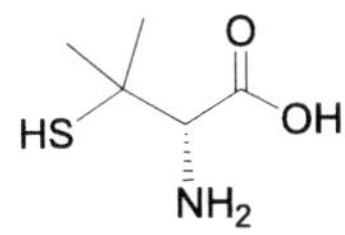

penicillamine
(positive control)(**123**)

Fig. 25 (continued)

Regarding oxidative damage by the radical generator 2,2′-azobis(2-amidinopropane)dihydrochloride (AAPH) in renal epithelial LLC-PK(1) cells, the MeOH extract and fractions from *Sophora flavescens* showed dose-dependently 1,1-diphenyl-2-picrylhydrazyl (DPPH) radical-scavenging effects, whereas only the BuOH and CH_2Cl_2 fractions showed dose-dependently protective effects against LLC-PK(1) cellular damage induced by AAPH. Notably, the BuOH fraction had the most effective antioxidative capacity. By a bioassay-linked HPLC/MS method, the active constituents from the BuOH fraction of *Sophora flavescens* were found to be sophoraflavanone G (**10**) and kurarinone (**13**) with potent antioxidant effects against the DPPH radical, with IC_{50} values of 5.26 and 7.73 μg/ml, respectively. Moreover, sophoraflavanone G (**10**) and kurarinone (**13**) dose-dependently could recover cell viability which was decreased by the AAPH treatment, suggesting their protective roles against cellular oxidative damage. The results suggest that *Sophora flavescens* with sophoraflavanone G (**10**) and kurarinone (**13**) could have antioxidative and kidney-protective potential (Fig. 26) [27].

In addition to measuring the antioxidant effects of prenylated flavonoids from *Sophora flavescens* by using in vitro 1,1-diphenyl-2-picrylhydrazyl (DPPH), 2,2′-azino-bis-3-ethylbenzothiazoline-6-sulfonic acid (ABTS), peroxynitrite ($ONOO^-$), and total reactive oxygen species (ROS) assays, the examination of the inhibition of tert-butylhydroperoxide (*t*-BHP)-induced intracellular ROS generation and *t*-BHP-induced activation of nuclear factor-κB (NF-κB) was also added as a further examination of kuraridinol (**9**),

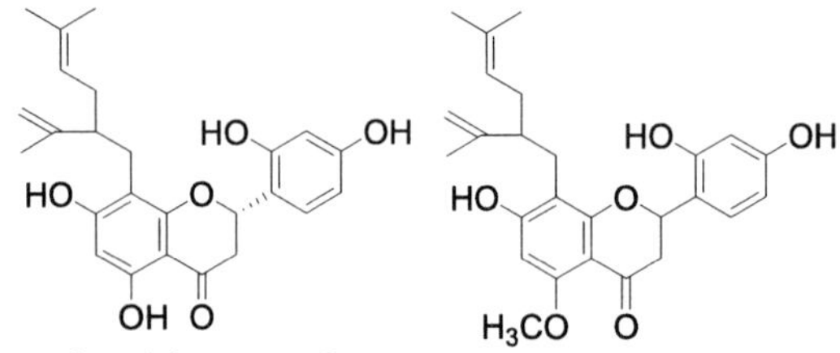

sophorafalavanone G (**10**)

kurarinone (**13**)

Fig. 26 Sophoraflavanone G (**10**) and kurarinone (**13**) with antioxidative and kidney-protective effects from *Sophora flavescens*

kurarinol (**3**), and kurarinone (**13**) from *Sophora flavescens*. The ethyl acetate (EtOAc) soluble fraction of *Sophora flavescens* with the two major prenylated chalcones kuraridin (**8**) and kuraridinol (**9**), along with a minor prenylated flavonol of kushenol C (**22**) scavenged highly for the DPPH.

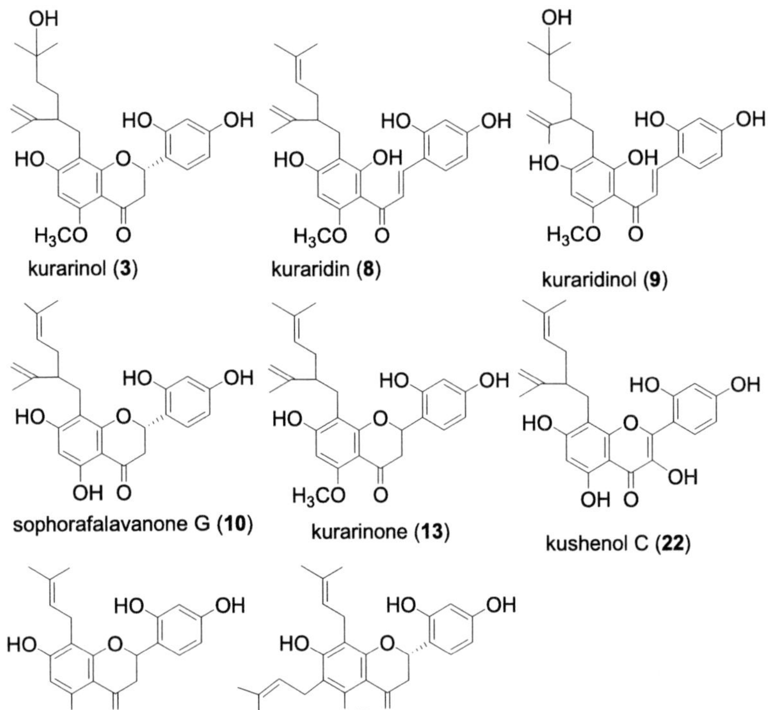

Fig. 27 Antioxidant prenylated flavonoids from roots of *Sophora flavescens*

Interestingly, this result was in contrast to the prenylated flavanones sophoraflavanone G (**10**) and kurarinone (**13**) because these were also isolated from the methylene chloride fraction of the same *Sophora flavescens*. The five flavanones kushenol E (**124**), leachianone G (**125**), kurarinol (**3**), sophoraflavanone G (**10**), and kurarinone (**13**) exhibited significant antioxidant potential against the ABTS, $ONOO^-$, and total ROS assays. However, the prenylated chalcones and prenylated flavonol had more potent scavenging and inhibitory activities when compared to the prenylated flavanones. Then, from these facts, it could be suggested that the prenylated chalcones and prenylated flavonol might play a more important role in the marked antioxidant capacities of *Sophora flavescens* when compared to the prenylated flavanones. Furthermore, kuraridinol (**8**), kurarinol (**3**), and kurarinone (**13**) had significantly inhibitory activities against the intracellular ROS levels as well as NF-κB activation by the *t*-BHP. Additionally, the results showed that *Sophora flavescens* and its prenylated flavonoids could have good anti-inflammatory activity with also significant antioxidant activity (Fig. 27) [28].

2.7
Enzyme Inhibitory Activity

During research of the cyclic guanosine monophosphate (cGMP)-specific phosphodiesterase type 5 (PDE5) inhibitors from *Sophora flavescens*, the extracts showed potent inhibitory activity against cyclic guanosine monophosphate-specific phosphodiesterase type 5 (cGMP-PDE5) prepared from rat diaphragm. Then, the inhibitory activities of five flavonoids kurarinol (**3**), kuraridin (**8**), sophoflavescenol (**110**), kushenol K (**111**), and kushenol H (**112**) from *S. flavescens* were measured against the cGMP-PDE5 to identify potent cGMP PDE5 inhibitory constituents. Among the five flavonoids, sophoflavescenol (**110**) with a C-8 prenylated flavonol showed the most potent inhibitory activity ($IC_{50} = 0.013\ \mu M$) against cGMP-PDE5 with 31.5- and 196.2-fold selectivity over two enzymes of PDE3 and PDE4, respectively (Fig. 28) [22, 29, 30].

Among 12 prenylated flavonoids with the resorcinol moiety from the roots of *Sophora flavescens* studied for tyrosinase inhibitor activity by activity-guided fractionation, the three flavonoids kuraridin (**8**), kurarinone (**13**), and norkurarinol (**18**) showed stronger inhibitory activity ($IC_{50} = 1.1$, 1.3, and 2.1 μM, respectively) when compared to that of kojic acid **126**): $IC_{50} = 11.3\ \mu M$) as a control of tyrosinase inhibition. The substitution of a lavandulyl or hydroxylavandulyl group at the C-8 position and a methoxy or hydroxy group at the C-5 position might be essential for their inhibitory effect (Fig. 29) [31].

For the development of a skin-whitening agent, *Sophora flavescens* was examined for tyrosinase inhibitory activity. The ethanol extract and dichloromethane fraction from *S. flavescens* showed significantly their inhibi-

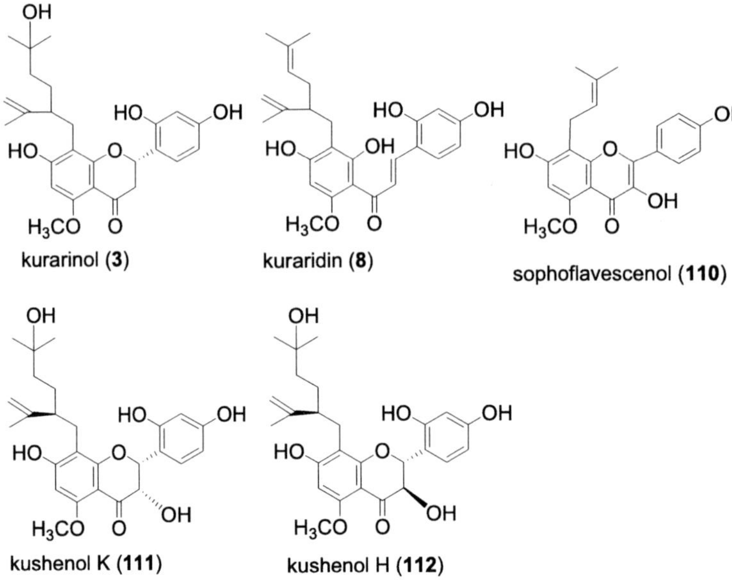

Fig. 28 Inhibitory cyclic guanosine monophosphate (cGMP)-specific phosphodiesterase type 5 (PDE5) flavonoids from *Sophora flavescens*

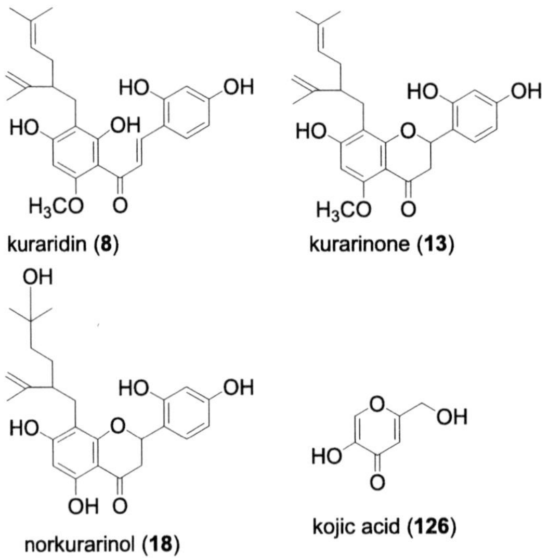

Fig. 29 Prenylated flavonoids with tyrosinase inhibitory activity from *Sophora flavescens* and kojic acid (a control) (126)

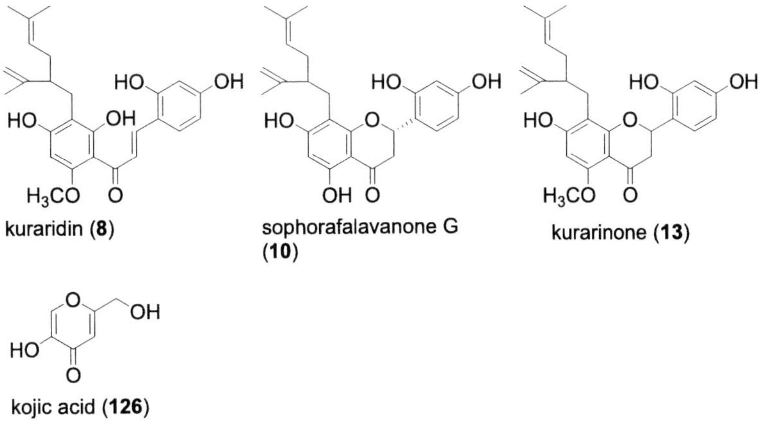

Fig. 30 Tyrosine inhibitory prenylated flavonoids from *Sophora flavescens* and kojic acid (126) as a control of tyrosine inhibition

tion of mushroom tyrosinase. From the dichloromethane fraction, the three known prenylated flavonoids sophoraflavanone G (**10**), kuraridin (**8**), and kurarinone (**13**) were found. The three prenylated flavonoids kuraridin (**8**), sophoraflavanone G (**10**), and kurarinone (**13**) showed significantly the higher tyrosinase inhibitory activity when compared to kojic acid (IC_{50} = 20.5 μM) (**126**). The IC_{50} values were 6.6, 0.6, and 6.2 μM for sophoraflavanone G (**10**), kuraridin (**8**), and kurarinone (**13**), respectively (Figs. 15, 30) [32].

The five prenylflavonoid derivatives kurarinone (**13**), kuraridin (**8**) of a chalcone of kurarinone (**13**), kurarinol (**3**), kushenol H (**112**) and kushenol K (**111**) from the roots of *Sophora flavescens* were examined for their inhibitory effects on diacylglycerol acyltransferase (DGAT), an enzyme that participates in bile acid biosynthesis. These flavonoids inhibited dose-dependently DGAT activity with their IC_{50} values of 10.9 μM kurarinone (**13**), 9.8 μM kuraridin (**8**), 8.6 μM kurarinol (**3**), 142.0 μM kushenol H (**112**), and 250 μM kushenol K (**111**), respectively. Three prenylflavonoids without C3-OH, kurarinone (**13**), kuraridin (**8**), and kurarinol (**3**), had higher DGAT inhibition when compared to the two flavanoids kushenol H (**112**) and kushenol K (**111**) with C3-OH.

On the other hand, flavonoids without side chains such as quercetin (**43**), naringenin (**69**), hesperetin (**70**), and kaempferol (**106**) did not inhibit the enzyme activity at a final concentration of 800 μM. These results suggest that the lavandulyl side chain and the position of the hydroxy group might be important for the high DGAT inhibitory activity. Kurarinone (**13**) also inhibited de novo synthesis of triacylglycerol (TG) in Raji cells (Fig. 31) [22, 33].

Monoamine oxidase (MAO) inhibitors for Parkinson's syndrome from the methanol extract of the roots of *Sophora flavescens* showed an inhibitory effect on mouse brain monoamine oxidase. The two flavonoids formononetin (**4**) and kushenol F (**127**) were highly inhibitory. Three oxyma-

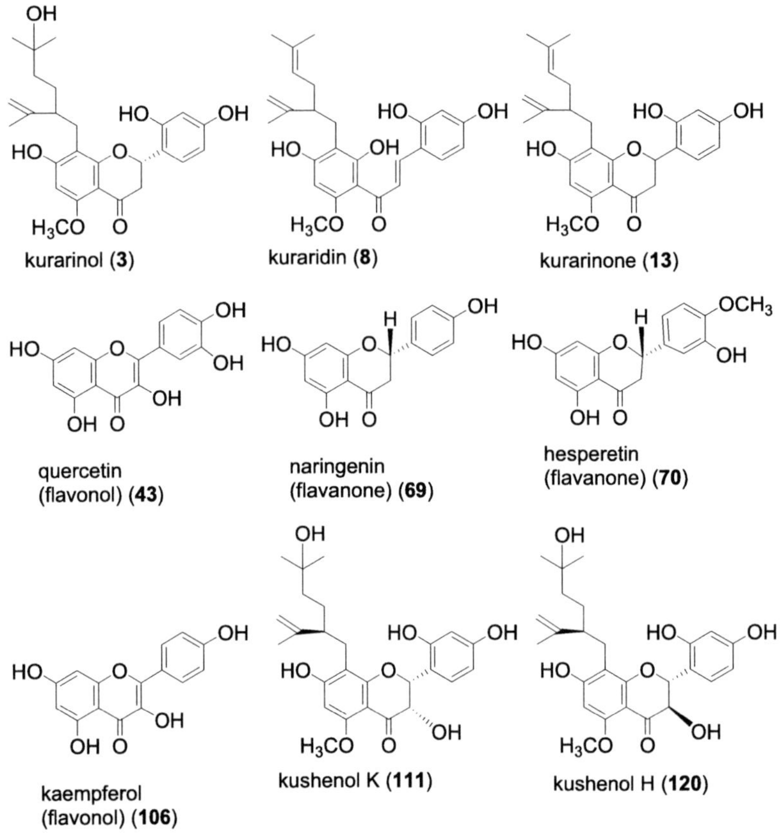

Fig. 31 Prenylflavonoids with anti-diacylglycerol acyltransferase (DGAT) from the roots of *Sophora flavescens*, and two flavanones and two flavonols without anti-DGAT

trine (ammothaminine) (**128**) of quinolizidine alkaloids, trifolirhizin (**2**) and β-sitosterol (**129**) were inactive. Two active MAO inhibitors, formononetin (**4**) and kushenol F (**127**), showed dose-dependently remarkable inhibitory effects on MAO with IC_{50} values of 13.2 and 69.9 μM, respectively. Formononetin (**4**) had a slightly more potent inhibitory effect against MAO-B ($IC_{50} = 11.0$ μM) when compared to that of MAO-A ($IC_{50} = 21.2$ μM). Kushenol F (**127**) also preferentially inhibited MAO-B activity compared to MAO-A activity with IC_{50} values of 63.1 and 103.7 μM, respectively (Fig. 32) [34].

The methanol extract of *Sophora flavescens* showed potent glycosidase inhibitory activity. Active components were identified as well-known flavonoid antioxidants including kurarinol (**3**), maackian (**6**), isoxanthohumol (**7**), kuraridin (**8**), sophoraflavanone G (**10**), 2′-methoxykurarinone (**23**), kushenol A (**26**), 8-prenylkaempferol (**63**), and (–)-kurarinone (**68**). All of the flavonoids were effective inhibitors of α-glucosidase and β-amylase. Interestingly, la-

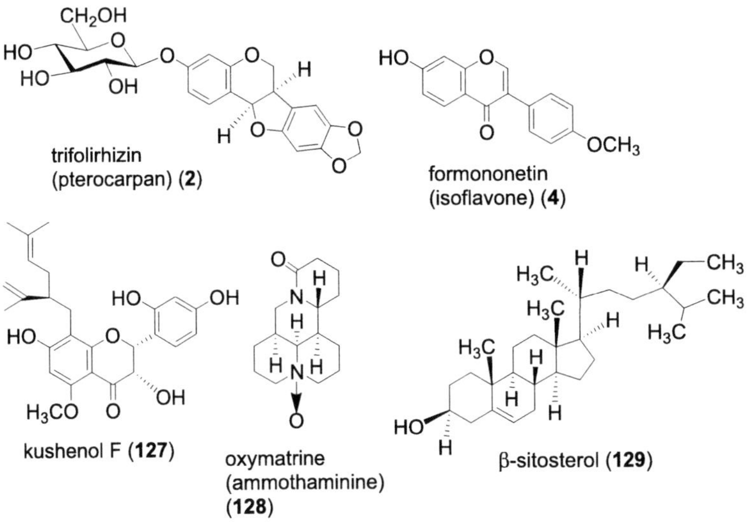

Fig. 32 Monoamine oxidase inhibitory components from the roots of *Sophora flavescens*

vandulylated flavanones 1–5 had strong α-glucosidase inhibitory activities, with IC_{50} values of 45, 68, 37, 155, and 179 μM, respectively. Kushenol A (**26**) which does not bear a 4′-hydroxy group showed selective α-glucosidase inhibitory activity. The lavandulylated chalcone of kuraridin (**8**) exhibited an IC_{50} value of 57 μM against β-glucosidase, which is the first report of a chalcone displaying glycosidase inhibition. From these results, it could be suggested that an 8-lavandulyl group in the B-ring was a key factor for the glycosidase inhibitory activities. The inhibition pattern was noncompetitive for α-glucosidase, whereas mixed inhibition was observed for β-amylase (Fig. 33) [35].

The methanol radix extract of *Sophora flavescens* of a Chinese folklore medicine had potent Na^+-glucose cotransporter (SGLT) inhibitory activity. The flavonoids with active SGLT inhibitory activity were kurarinone (**13**), sophoraflavanone G (**10**), kushenol K (**111**), and kushenol N (**21**) (Fig. 34) [36].

The three flavanones sophoraflavanone G (**10**), kurarinone (**13**), and kurarinol (**3**) from the root of *Sophora flavescens* have significantly high tyrosinase inhibitory activity. Regarding the principal pharmacological features of kurarinol (**3**), kurarinol (**3**) could lead to the inhibition of the oxidation of L-tyrosine to melanin by mushroom tyrosinase (IC_{50} of 100 nM). The inhibition kinetic analyses showed that the tyrosinase inhibitory activity of the two flavonoids sophoraflavanone G (**10**) and kurarinone (**13**) might be noncompetitive.

However, on the other hand, similar analysis shows kurarinol (**3**) might be a competitive inhibitor. The two flavonoids sophoraflavanone G (**10**) and ku-

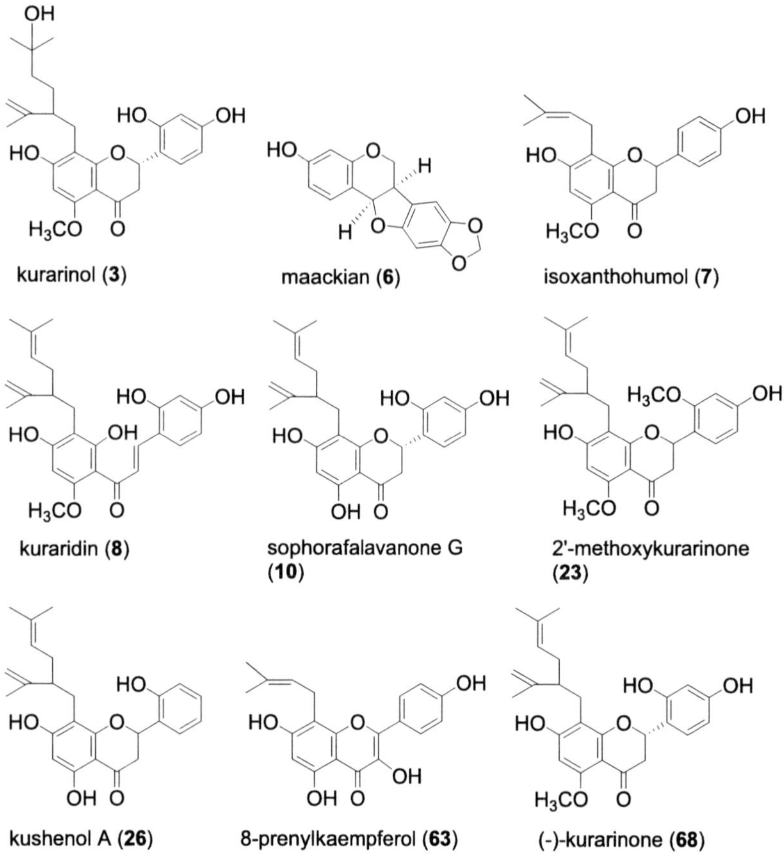

Fig. 33 Glycosidase inhibitory flavonoids from *Sophora flavescens*

rarinone (**13**) exhibited potent antibacterial activity with 10 μg/disk against Gram-positive bacteria, whereas kurarinol (**3**) did not present any antibacterial activity. Interestingly, kurarinol (**3**) could inhibit production of melanin in *S. bikiniensis* without affecting the growth of the microorganism. Kurarinol (**3**) is thus apparently different from the other tyrosinase inhibitors such as sophoraflavanone G (**10**) and kurarinone (**13**). In addition, kurarinol (**3**) showed relatively low cytotoxic activity ($EC_{50} > 30$ μM) when compared to the two flavonoids sophoraflavanone G (**10**) and kurarinone (**13**). It might be suggested that there is an interaction of the lavandulyl group within kurarinol (**3**) with the tyrosinase. Selectively, the terminal hydroxy function within the lavandulyl group of kurarinol (**3**) might be the most important for optimal binding of the two components (Fig. 35) [37].

In previous study, the prenylated flavonoids kuraridin (**8**), sophoraflavanone G (**10**), kurarinone (**13**), and kushenol F (**127**) from *Sophora flavescens* showed tyrosinase inhibitor activity. Interestingly, based on the level of these

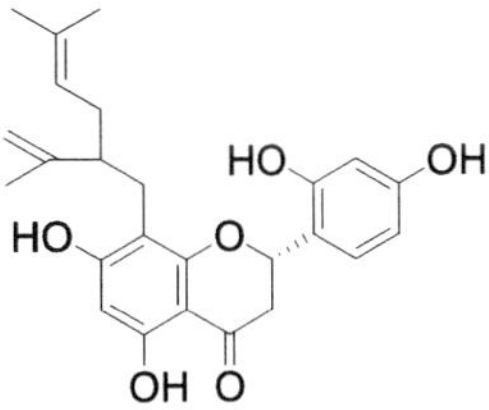

sophorafalavanone G (**10**)

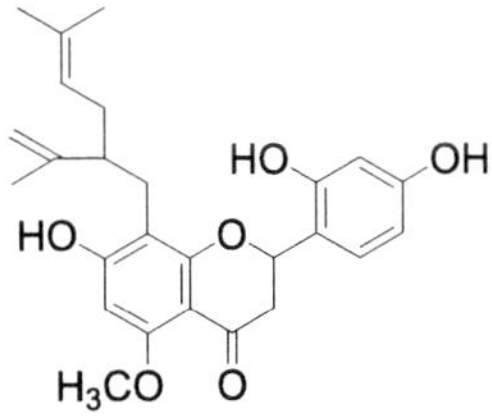

kurarinone (**13**)

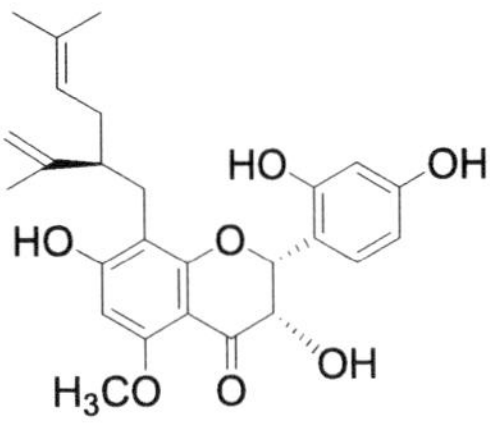

kushenol N (**21**)

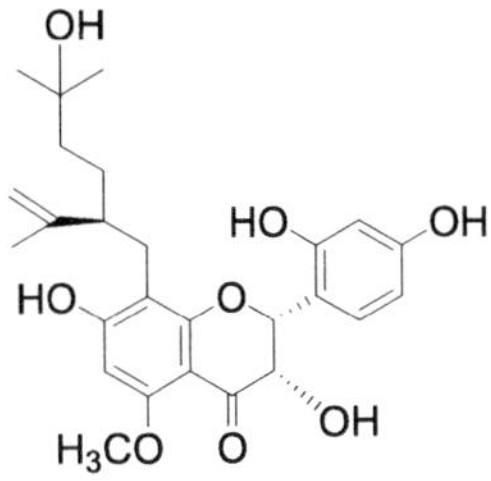

kushenol K (**111**)

Fig. 34 Na^{+}-glucose cotransporter (SGLT) inhibitory flavonoids from radix of *Sophora flavescens*

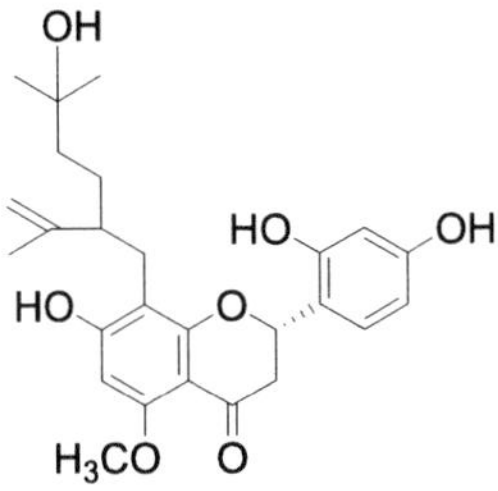

kurarinol (**3**)

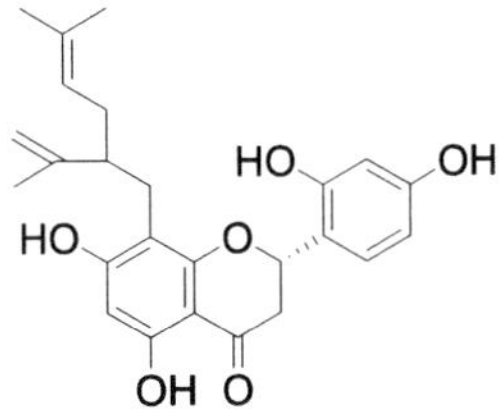

sophorafalavanone G (**10**)

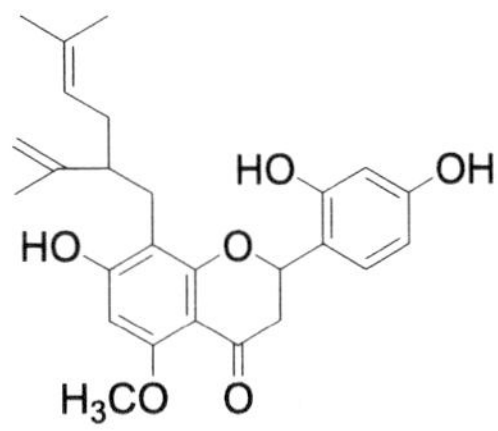

kurarinone (**13**)

Fig. 35 Typical tyrosinase inhibitory flavonoids from roots of *Sophora flavescens*

inhibitors in the extract, its inhibitory effect on tyrosinase activity was higher than expected. Further studies were carried out to determine any other possible constituents that might contribute to the extract's stronger inhibitory activity. Kurarinol (**3**), kuraridinol (**9**), and trifolirhizin (**2**) from *Sophora* root extract could inhibit tyrosinase activity. When compared to kojic acid ($16.22 \pm 1.71\ \mu M$) (**126**), the three flavonoids kurarinol (**3**), kuraridinol (**8**), and trifolirhizin (**2**) had potent tyrosinase inhibitory activity with IC_{50} values of 8.60 ± 0.51, 0.88 ± 0.06, and $506.77 \pm 4.94\ \mu M$, respectively. When the three compounds were further tested for their inhibitory effects on melanogenesis in cultured B16 melanoma cells, $50\ \mu M$ of kurarinol (**3**), kuraridinol (**8**), and trifolirhizin (**2**) markedly inhibited below around 50% melanin synthesis. From these results, it is suggested that kurarinol (**3**), kuraridinol (**9**), and trifolirhizin (**2**) could exert varying degrees of inhibition on tyrosinase-dependent melanin biosynthesis, and therefore, are candidates for skin-whitening agents (Fig. 36) [38].

For the prevention and treatment of diabetic complications it is essential to intake aldose reductase inhibitors (ARIs) and inhibitors of advanced glycation endproduct (AGE) formation in our daily diet and these are important targets. The inhibitory activities of prenylated flavonoids from *Sophora flavescens* were examined on rat-lens aldose reductase (RLAR), human recombinant aldose reductase (HRAR), and in the formation of advanced glycation endproducts. Among these flavonoids, the two prenylated chalcones desmethylanhydroicaritin (**119**) and 8-lavandulylkaempferol (**96**) along with the five prenylated flavanones kurarinol (**3**), kurarinone (**13**), (2*S*)-2′-methoxykurarinone (**67**), (2*S*)-3β,7,4′-trihydroxy-5-methoxy-8-(γ,γ-dimethylally)-flavanone (**120**), and kushenol E (**124**) were potent inhibitors of RLAR, with IC_{50} values of 0.95, 3.80, 2.13, 2.99, 3.77, 3.63, and 7.74 μM,

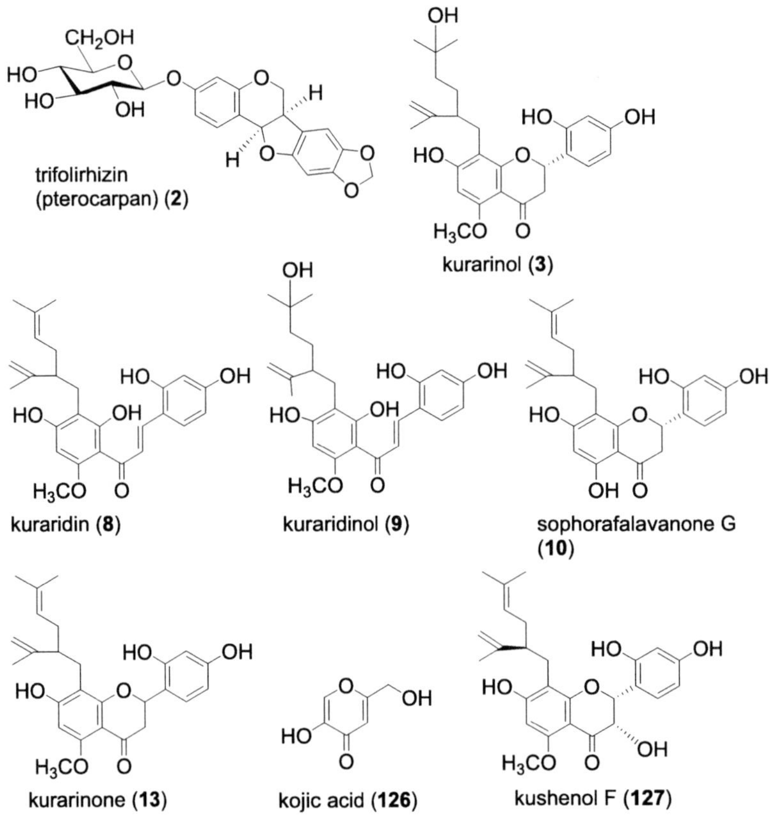

Fig. 36 Inhibitory effects of trifolirhizin (2), kurarinol (3), and kuraridinol (9) from *Sophora flavescens* on tyrosinase and melanin synthesis

respectively, when compared to quercetin (IC_{50} 7.73 μM) (**43**). In the HRAR assay, most of these prenylated flavonoids showed significantly their inhibitory activity when compared to quercetin (IC_{50} 2.54 μM) (**43**). Especially, the three prenylated flavonols desmethylanhydroicaritin (IC_{50} 0.45 μM) (**119**), 8-lavandulylkaempferol (IC_{50} 0.79 μM) (**96**), and kushenol C (IC_{50} 0.85 μM) (**22**), as well as a prenylated chalcone, kuraridin (IC_{50} 0.27 μM) (**8**), and a prenylated flavanone, (2*S*)-7,4′-dihydroxy-5-methoxy-8-(γ,γ-dimethylally)-flavanone (IC_{50} 0.37 μM) (**121**) showed significant inhibitory activities when compared to a control of the potent aldose reductase (AR) inhibitor epalrestat (IC_{50} 0.28 μM) (**130**). Interestingly, prenylated flavonoids such as desmethylanhydroicaritin (IC_{50} 104.3 μg/ml) (**119**), 8-lavandulylkaempferol (IC_{50} 132.1 μg/ml) (**96**), kushenol C (IC_{50} 84.6 μg/ml) (**22**), and (2*S*)-3β,7,4′-trihydroxy-5-methoxy-8-(γ,γ-dimethylally)-flavanone (IC_{50} 261.0 μg/ml) (**120**), which harbor a 3-hydroxyl group, also had good inhibitory activity

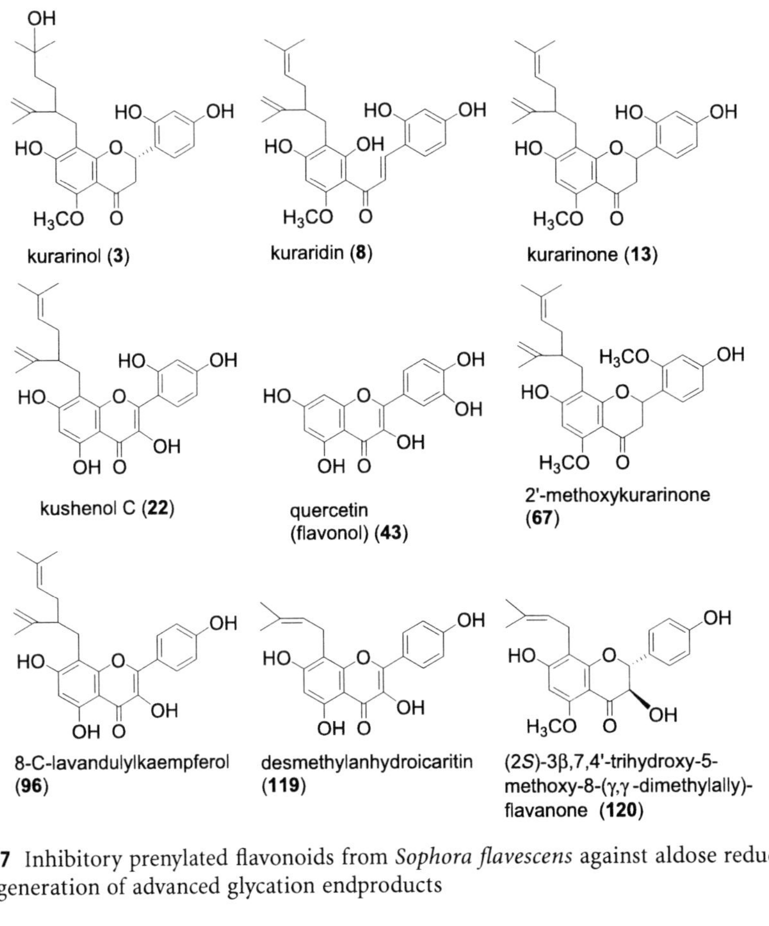

Fig. 37 Inhibitory prenylated flavonoids from *Sophora flavescens* against aldose reductase and generation of advanced glycation endproducts

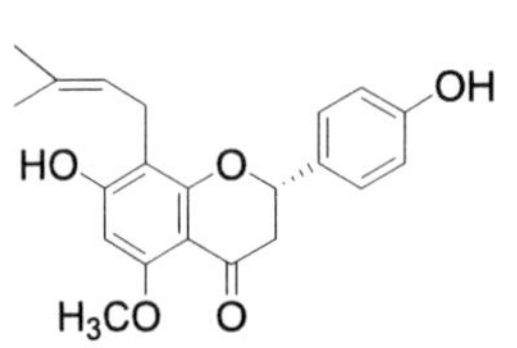

(2*S*)-7,4'-dihydroxy-5-methoxy-8-(γ,γ-dimethylally)-flavanone (**121**)

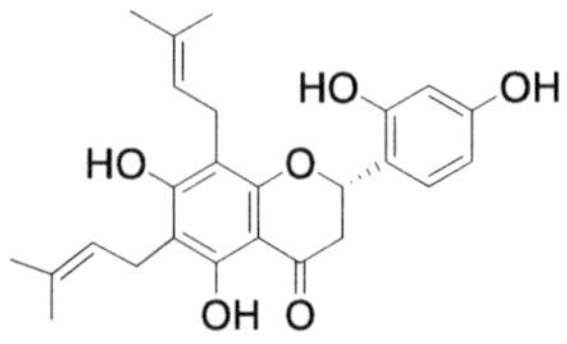

kushenol E (**124**)

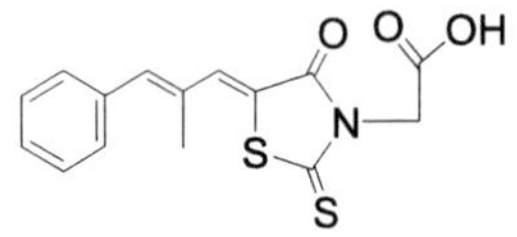

epalrestat (**130**)

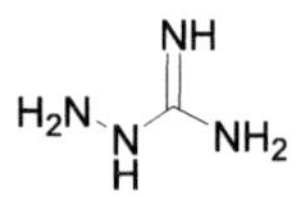

aminoguanidine (**131**)

Fig. 37 (continued)

against advanced glycation endproduct formation when compared to a positive control of aminoguanidine (IC_{50} 115.7 μg/ml) (**131**). From these results, because the prenylated flavonoids of *S. flavescens* could inhibit the processes that underlie diabetic complications and age-related diseases, it is suggested that the prenylated flavonoids might be a therapeutic agent for treatment and prevention of diabetic complications (Fig. 37) [39].

Eight known prenylated flavonoids from the root of *Sophora flavescens* were examined for in vitro anti-allergic activity (Fig. 38). Among these eight flavonoids, kushenol N (**21**), sophoraflavanone G (**10**), and leachianone A (**66**) demonstrated significantly their inhibition of the release of *β*-hexosaminidase from the cultured RBL-2H3 cells with IC_{50} values ranging from 15 to 30 μM (Fig. 38) (Table 11) [17].

Among 11 novel chalcones studied for inhibition of interleukin-5 (IL-5), first, 1-(6-benzyloxy-2-hydroxyphenyl)-3-(4-hydroxyphenyl)propenone (**132**) with 78.8% inhibition at 50 μM, $IC_{50} = 25.3$ μM) was shown as a potent inhibitor of IL-5. When this activity is compared to that of budenoside (**142**) or sophoricoside (**44**), the benzyloxy group appears to be critical for the enhancement of the IL-5 inhibitory activity. For identification of the role of this hydrophobic moiety, cyclohexyloxy (**135**), cyclohexylmethoxy (**134**), cyclohexylethoxy (**136**), cyclohexylpropoxy (**137**), 2-methylpropoxy (**138**), 3-methylbutoxy (**139**), 4-methylpentoxy (**140**), and 2-ethylbutoxy (**141**) analogs were compared for their effects on IL-5 bioactivity. Cyclohexylmethoxy($IC_{50} = 12.6$ μM) (**134**), cyclohexyloxy ($IC_{50} = 12.2$ μM) (**135**), and 2-ethylbutoxy ($IC_{50} = 16.1$ μM) (**141**) exhibited the most potent activity. Considering the $c \log P$ values of chalcones, the alkoxy group might contribute to the cell permeability of chalcones for the enhancement of activity, rather than playing a role in the ligand motif binding to the receptor. The

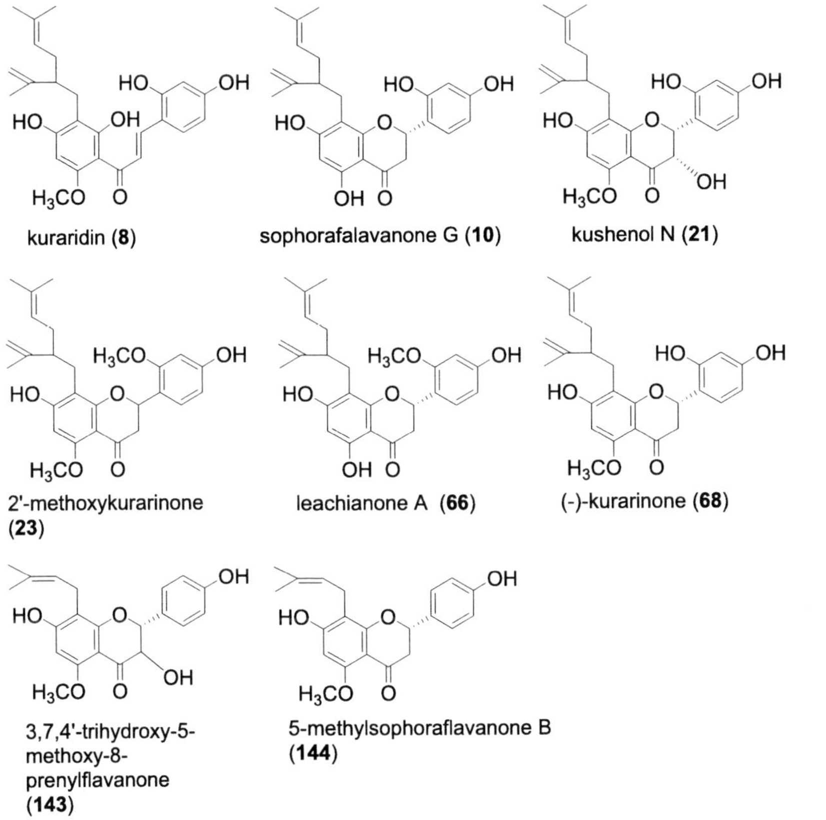

Fig. 38 Prenylated flavonoids with anti-allergic activity from *Sophora flavescens*

Table 11 Prenylated flavonoids from *Sophora flavescens* for the inhibition of mast cell degranulation

Compound	IC_{50} (mM)
10	20.0 ± 3.2
21	27.1 ± 2.1
23	> 100
66	16.5 ± 1.1
68	> 100
143	> 100
144	> 100

optimum alkoxy group in ring A of 2′ or 6′ might be one which provides the $c \log P$ of chalcones in the range of 4.22 to 4.67 (Table 12) [40].

Table 12 Inhibition of interleukin-5 by chalcones

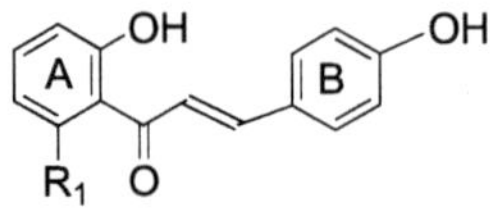

Chalcone	Substituent (R_1)	Inhibition (%) at 50 μM	IC_{50} (μM)	$c\log P$
132	benzyloxy	78.8	25.3	4.23
133	H	50.0	48.2	2.81
134	cyclohexylmethoxy	99.0	12.6	4.66
135	cyclohexyloxy	91.5	12.2	4.23
136	cyclohexylethoxy	71.8	29.8	5.00
137	cyclohexylpropoxy	91.3	27.2	5.42
138	2-methylpropoxy	76.6	23.2	3.90
139	3-methylbutoxy	78.0	16.8	4.25
140	4-methylpentoxy	99.0	12.6	4.67
141	2-ethylbutoxy	82.5	16.1	4.74
budenoside (**142**)		70.3	26.8	

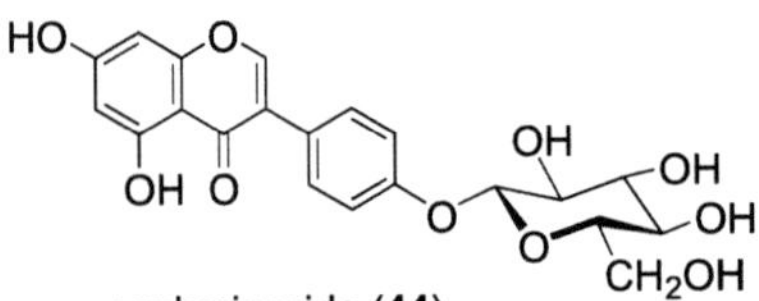

sophoricoside (**44**)

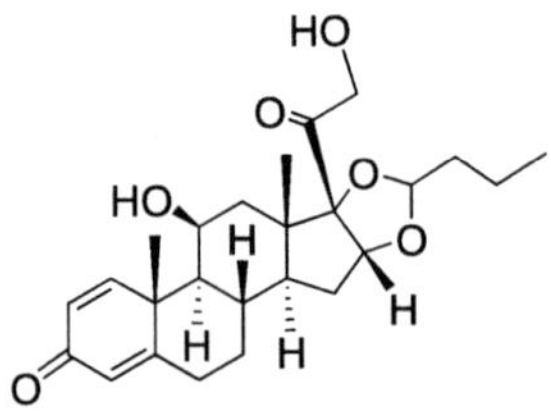

budenoside (**142**)

References

1. Zhang L, Xu L, Xiao SS, Liao QF, Li Q, Liang J, Chen XH, Bi KS (2007) J Pharm Biomed Anal 44:1019
2. Song T, Barua K, Buseman G, Murphy PA (1998) Am J Clin Nutr 68(6 Suppl):1474S
3. Setchell KD, Zimmer-Nechemias L, Cai J, Heubi JE (1997) Lancet 350(9070):23
4. Setchell KD, Zimmer-Nechemias L, Cai J, Heubi JE (1998) Am J Clin Nutr 68(6 Suppl):1453S
5. Song WO, Chun OK, Hwang I, Shin HS, Kim BG, Kim KS, Lee SY, Shin D, Lee SG (2007) J Med Food 10:571
6. Liu IM, Sheu SJ (1989) Am J Chin Med 17:179
7. Min B, Oh SR, Lee HK, Takatsu K, Chang IM, Min KR, Kim Y (1999) Planta Med 65:408
8. Tang Y, Lou F, Wang J, Zhuang S (2001) J Nat Prod 64:1107
9. Park JA, Kim HJ, Jin C, Lee KT, Lee YS (2003) Arch Pharm Res 26:1009
10. Hillerns PI, Wink M (2005) Planta Med 71:1065
11. Joo SS, Kang HC, Lee MW, Choi YW, Lee DI (2003) Arch Pharm Res 26:1029
12. Shirataki Y, Yokoe I, Noguchi M, Tomimori T, Komatsu M (1988) Chem Pharm Bull 36:2220

13. Kang TH, Jeong SJ, Ko WG, Kim NY, Lee BH (2000) J Nat Prod 63:680
14. Shirataki Y, Motohashi N, Tani S, Sakagami H, Satoh K, Nakashima H, Mahapatra SK, Ganguly K, Dastidar SG, Chakrabarty AN (2001) Anticancer Res 21:275
15. Shirataki Y, Tani S, Sakagami H, Satoh K, Nakashima H, Gotoh K, Motohashi N (2001) Anticancer Res 21:2643
16. Shirataki Y, Wakae M, Yamamoto Y, Hashimoto K, Satoh K, Ishihara M, Kikuchi H, Nishikawa H, Minagawa K, Motohashi N, Sakagami H (2004) Anticancer Res 24:1481
17. Quan W, Lee HJ, Kim CY, Noh CW, Um BH, Oak MH, Kim KM (2008) Planta Med 74:168
18. De Naeyer A, Vanden Berghe W, Pocock V, Milligan S, Haegeman G, De Keukeleire D (2004) J Nat Prod 67:1829
19. Shen CC, Lin TW, Huang YL, Wan ST, Shien BJ, Chen CC (2006) J Nat Prod 69:1237
20. Chowdhury SA, Kishino K, Satoh R, Hashimoto K, Kikuchi H, Nishikawa H, Shirataki Y, Sakagami H (2005) Anticancer Res 25:2055
21. Bobrowska-Hägerstrand M, Lillås M, Mrówczyñska L, Wróbel A, Shirataki Y, Motohashi N, Hägerstrand H (2006) Anticancer Res 26:2081
22. Woo ER, Kwak JH, Kim HJ, Park H (1998) J Nat Prod 61:1552
23. Dastidar SG, Manna A, Kumar KA, Mazumdar K, Dutta NK, Chakrabarty AN, Motohashi N, Shirataki Y (2004) Int J Antimicrob Agents 23:99
24. Cha JD, Jeong MR, Jeong SI, Lee KY (2007) J Microbiol Biotechnol 17:858
25. Jung HJ, Kang SS, Woo JJ, Choi JS (2005) Arch Pharm Res 28:1333
26. Jung HJ, Kang SS, Hyun SK, Choi JS (2005) Arch Pharm Res 28:534
27. Piao XL, Piao XS, Kim SW, Park JH, Kim HY, Cai SQ (2006) Biol Pharm Bull 29:1911
28. Jung HA, Jeong DM, Chung HY, Lim HA, Kim JY, Yoon NY, Choi JS (2008) Biol Pharm Bull 31:908
29. Shin HJ, Kim HJ, Kwak JH, Chun HO, Kim JH, Park H, Kim DH, Lee YS (2002) Bioorg Med Chem Lett 12:2313
30. Lee NK, Son KH, Chang HW, Kang SS, Park H, Heo MY, Kim HP (2004) Arch Pharm Res 27:1132 (structure of sophoflavescenol is not correct)
31. Son JK, Park JS, Kim JA, Kim Y, Chung SR, Lee SH (2003) Planta Med 69:559
32. Kim SJ, Son KH, Chang HW, Kang SS, Kim HP (2003) Biol Pharm Bull 26:1348
33. Chung MY, Rho MC, Ko JS, Ryu SY, Jeune KH, Kim K, Lee HS, Kim YK (2004) Planta Med 70:258
34. Hwang JS, Lee SA, Hong SS, Lee KS, Lee MK, Hwang BY, Ro JS (2005) Arch Pharm Res 28:190
35. Kim JH, Ryu YB, Kang NS, Lee BW, Heo JS, Jeong IY, Park KH (2006) Biol Pharm Bull 29:302
36. Sato S, Takeo J, Aoyama C, Kawahara H (2007) Bioorg Med Chem 15:3445
37. Ryu YB, Westwood IM, Kang NS, Kim HY, Kim JH, Moon YH, Park KH (2008) Phytomedicine 15:612
38. Hyun SK, Lee WH, Jeong da M, Kim Y, Choi JS (2008) Biol Pharm Bull 31:154
39. Jung HA, Yoon NY, Kang SS, Kim YS, Choi JS (2008) J Pharm Pharmacol 60:1227
40. Yang HM, Shin HR, Cho SH, Song GY, Lee IJ, Kim MK, Lee SH, Ryu JC, Kim Y, Jung SH (2006) Arch Pharm Res 29:969

Top Heterocycl Chem (2009) 16: 93–133
DOI 10.1007/7081_2007_100

Published online: 9 January 2008

Quantitative Structure–Cytotoxicity Relationship of Bioactive Heterocycles by the Semi-empirical Molecular Orbital Method with the Concept of Absolute Hardness

Mariko Ishihara[1] (✉) · Hiroshi Sakagami[2] · Masami Kawase[3] · Noboru Motohashi[4]

[1]Division of Basic Chemistry, Department of Oral Biology and Tissue Engineering, Meikai University School of Dentistry, Sakado, 350-0283 Saitama, Japan
mariko@dent.meikai.ac.jp

[2]Division of Pharmacology, Meikai University School of Dentistry, Sakado, 350-0283 Saitama, Japan

[3]Faculty of Pharmaceutical Sciences, Matsuyama University, Matsuyama, 790-8578 Ehime, Japan

[4]Meiji Pharmaceutical University, Kiyose, 204-8588 Tokyo, Japan

Abstract The relationship between the cytotoxicity of *N*-heterocycles (13 4-trifluoromethylimidazole, 15 phenoxazine and 12 5-trifluoromethyloxazole derivatives), *O*-heterocycles (11 3-formylchromone and 20 coumarin derivatives) and seven vitamin K_2 derivatives against eight tumor cell lines (HSC-2, HSC-3, HSC-4, T98G, HSG, HepG2, HL-60, MT-4) and a maximum of 15 chemical descriptors was investigated using CAChe Worksystem 4.9 project reader. After determination of the conformation of these compounds and approximation to the molecular form present in vivo (biomimetic) by CONFLEX5, the most stable structure was determined by CAChe Worksystem 4.9 MOPAC (PM3). The present study demonstrates the best relationship between the cytotoxic activity and molecular shape or molecular weight of these compounds. Their biological activities can be estimated by hardness and softness, and by using η–χ activity diagrams.

Keywords CONFLEX · Cytotoxicity · Descriptor · Heterocycles · QSAR · Semi-empirical molecular orbital

1 Structure–Activity Relationship by the Semi-empirical Molecular Orbital Method with the Concept of Absolute Hardness

1.1 Quantitative Structure–Activity Relationship

It is empirically known that compounds with similar structures display similar biological activity [1–3]. Quantitative structure–activity relationship (QSAR) between physico-chemical properties and biological activity has been utilized in the field of medicine and pharmaceutical sciences for a long time. This method is a popular means for searching the pharmacologically active molecular structures by correlating the pharmacological activity to chemical structure factors (so-called descriptors). The numerical expression of the QSAR dates back to the report of Hansch–Fujita in 1964 [4, 5]. This method, called Hansch analysis or the Hansch approach, is one that analyzes the relationship between the chemical structure and biological activity.

The toxicity of chemical materials is generated by the intestinal absorption, the transport to the binding components (such as receptors), and the subsequent chemical reaction with the target molecules. Chemical compounds are transported through the outer cell membrane and reach the active sites. Since the cell membrane is mainly composed of lipids, materials with higher hydrophobicity are expected to be more easily transported within the cells. On the other hand, the reaction between the active sites and chemical compounds, in most cases, depends on the electron characteristics and three-dimensional configuration. Therefore, the octanol–water distribution coefficient (log*P*) is usually used to investigate the QSAR. This descriptor is an indicator of the molecule-to-molecule interaction and the binding strength between the drug and its receptor. The classical QSAR (Hansch–Fujita method) is useful in various fields of medical science.

With the progress in the chemical calculation of the molecule, analyses with HOMO/LUMO (highest occupied molecular orbital/lowest unoccupied molecular orbital) energy, absolute hardness/absolute electron negativity, and research into these new and old descriptors have been reported recently. Furthermore, new methods with neural network computers, as well as multiregression analysis, cluster analysis, and major component analysis have been applied to investigate the relationship between the property and function of the molecule and of each descriptor.

1.2 Calculation of Molecular Orbital

Mathematical calculation, based on quantum mechanics, was established in the 1920s and developed with the rapid progress and spread of computers. Mathematical calculation is the field in which the approximated equation is solved by computers.

To carry out the molecular orbital (MO) method program, the following capabilities are needed: rapid calculation time, high precision (both quantitative and qualitative), easy manipulation and calculation of various molecular characteristics. Lots of software must fulfil these requirements. Generally, the MO methods are separated into four groups: (i) empirical MO method, (ii) semi-empirical MO method, (iii) non-empirical MO method (Hartree–Fock approximated level), (iv) non-empirical MO method, with consideration of the electron relationship. The semi-empirical method, with the experimental data included in the software, makes it possible to solve the Schrödinger equation more easily. On the other hand, the non-experimental method utilizes the standard numerical approximation, requiring a high-speed computer due to the large quantity of calculations. Alternatively, there are "quantitative" and "qualitative" MO methods. Despite its high precision, the fourth method described above is time-consuming and is only applicable to low molecular weight material. The authors utilized the semi-empirical MO method, which has a broad application range and short calculation time.

The MOPAC (molecular orbital package), which has been explored by Dr. James J.P. Stewart, is the method by which the electron characteristics of a molecule is calculated. The MOPAC determines both an optimum geometry and the electron properties of molecule by solving the Schrödinger equation, using the MINDO/3 [6], MNDO [7], or AM1 [8] semi-empirical Hamiltonians developed by M.J.S. Dewar, the MIND-d Hamiltonian developed by W. Thiel, or the PM3 [9] or PM5 semi-empirical Hamiltonians developed by J.J.P. Stewart.

Recent dramatic progress with computers has made it possible to perform the quantum mechanics calculation by the MO method (ab initio calculation method) and density functional theory (DFT) [10, 11]. Although these two methods are frequently used now, they are time-consuming. On the other hand, MOPAC (PM3 method) requires much shorter calculation time, and therefore seems to be more useful. In order to calculate the most stable structure of the test sample with the PM3 method, however, it is recommendable to search for the configuration due to the instability of the initially input structure [12]. Recently, CONFLEX (Conflex, Tokyo, Japan) has been explored for estimating the configuration of the soft compounds with higher degree of freedom. The authors determined the most stable configuration of dipalmitoil phosphatidylcholine with CONFLEX. We found that PM3 method 3 alone produced an inaccurate V-shaped configuration of dipalmitoil phosphatidylcholine. On the other hand, the PM3 method combined with CONFLEX revealed more accurately the bending structure of acyl-groups in the side chain of dipalmitoil phosphatidylcholine [13]. These data suggest that the CONFLEX/PM3 method is more useful than the MP3 method alone to estimate the biological activity of physiologically active compounds. Although the DFT method is the mainstream of computer chemistry, the COFLEX/PM3 method produces similar estimation values. This article summarizes our recent extensive study of QSAR of bioactive heterocycles (4-trifluoromethylimidazole, phenoxazine, 5-trifluoromethyloxazole, 3-formylchromone and coumarin derivatives) and vitamin K_2 derivatives by the semi-empirical molecular orbital method, with the concept of absolute hardness.

1.2.0.1 Calculation

The most stable conformation was calculated by CONFLEX 5 (Confluex, Tokyo). The optimization of the structure was done by the semi-empirical MO method (PM3), using CAChe Worksystem 4.9 (Fujitsu, Tokyo) PM3 method, in the presence (COSMO) or in the absence (non-COSMO) of water. The descriptors studied were: heat of formation (COSMO, non-COSMO)(Kcal/mol), stability of hydration (= COSMO – non-COSMO) (ΔH) (Kcal/mol), dipole moment (D), electron affinity (eV), ionization potential (eV), octanol–water distribution coefficient (log$P$) (determined by ACD/Log P DB6.0, Fujitsu) as an index of hydrophobicity, highest occupied molecular orbital energy (E_{HOMO}) (eV), lowest unoccupied molecular orbital energy (E_{LUMO}) (eV), maximum length of molecule (Å), absolute hardness (η) (eV), absolute electron negativity (χ) (eV), reactivity index (ω) (eV) [14–18], and hardness/absolute electronegativity diagram (η–χ diagram).

η, χ, and ω were determined by the following equations:

$$\eta = (\varepsilon_{\mathrm{LUMO}} - \varepsilon_{\mathrm{HOMO}})/2$$
$$\chi = -(\varepsilon_{\mathrm{LUMO}} + \varepsilon_{\mathrm{HOMO}})/2$$
$$\omega = \chi^2/2\eta\,.$$

Their QSAR between CC_{50} and each descriptor delineated from the molecular structure was investigated by CAChe Worksystem 4.9 project reader.

1.2.0.2 Assay for Cytotoxic Activity

Near confluent human promyleocytic leukemia (HL-60), human T-cell leukemia (MT-4) and human squamous cell carcinoma cell lines (HSC-2, HSC-3, HSC-4), human submandibular gland carcinoma (HSG), and human glioblastoma cells (T98G) were cultured for 24 h in RPMI1640 (only for HL-60 and MT-4 cells) or DMEM (for all other cells) supplemented with 10% fetal bovine serum containing various concentrations of each 5-trifluoromethyloxazole derivative, as described previously [19]. The 50% cytotoxic concentration (CC_{50}) of each compound was determined from the dose–response curve. Their QSAR between CC_{50} and each descriptor delineated from the molecular structure was investigated by CAChe.

2
N-Heterocycles

2.1
4-Trifluoromethylimidazole

Imidazole derivatives are well known to inhibit the drug metabolizing enzymes via direct binding to the heme-Fe, and thus they enhance the pharmacological actions of the concomitantly administered drugs [20]. They are also known as inhibitors of p38 MAP kinase, anti-inflammatory agents, angiotensin II receptor antagonists, fungicides, and herbicides [21]. Some imidazole derivatives have been reported to induce apoptosis in Ehrlich ascites tumor cells, via a mechanism involving the activation of a pro-apoptotic protein BAK and a caspase-activated DNase [22]. On the other hand, we have recently found that 4-trifluoromethyl-1,2-diphenylimidazole (**5**), 1-benzyl-4-trifluoromethyl-2-phenylimidazole (**7**), and 5-[1-ethoxy-2,2,2-trifluoro-1-(trifluoromethyl)ethyl]-1-methyl-2-phenyl-1*H*-imidazole (**12**), which showed higher cytotoxicity against human tumor cell lines (oral squamous cell carcinoma HSC-2, HSC-3, HSC-4, promyelocytic leukaemia HL-60) as compared to human normal oral cells (gingival fibroblast, pulp cell, periodontal ligament fibroblast), induced autophagic cell death characterized by the secondary lysosome in tumor cell lines [23]. We investigated here the QSAR of 13 4-trifluoromethylimidazoles derivatives (**1–13** in (Fig. 1)) by conventional and recent techniques of computation chemistry, such as the concept of absolute hardness [16–18].

Twelve 4-trifluoromethylimidazoles (**1–12**) were synthesized by the method of Kawase et al. [24, 25]. Compound **13** was purchased from Aldrich, USA. Cytotoxicity assays and determination of 50% cytotoxic concentration (CC_{50}) against human promyelocytic leukaemia HL-60 and human oral squamous cell carcinoma HSC-3 cell lines were performed as described elsewhere [23].

By determining the most stable conformation of 13 imidazole derivatives, their structure was approximated to the molecular form present in vivo (biomimetic). The most stable structure was next determined by CAChe Worksystem 4.9 MOPAC (PM3) (Fig. 2). The CC_{50} value (determined by experiments) against HL-60 and HSC-3 cells, and chemical descriptors (determined by calculations) such as heat of formation (COSMO), stability of

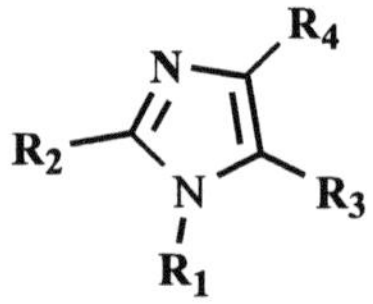

[1-12]

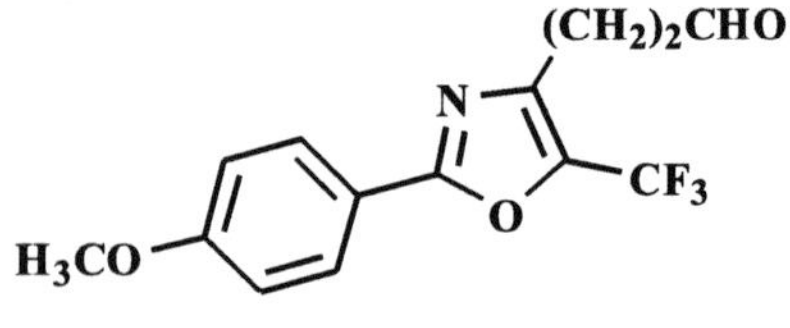

[13]

Compd.	R_1	R_2	R_3	R_4
[1]	CH_3	C_6H_5	H	CF_3
[2]	CH_3	$C_6H_5CH_2$	H	CF_3
[3]	CH_3	t-Bu	H	CF_3
[4]	C_6H_5	CH_3	H	CF_3
[5]	C_6H_5	C_6H_5	H	CF_3
[6]	$C_6H_5CH_2$	CH_3	H	CF_3
[7]	$C_6H_5CH_2$	C_6H_5	H	CF_3
[8]	$C_6H_5CH_2$	OH	$C_6H_5CH_2$	CF_3
[9]	CH_3	C_6H_5	H	C_2F_5
[10]	CH_3	C_6H_5	H	C_3F_7
[11]	CH_3	C_6H_5	CHO	H
[12]	CH_3	C_6H_5	$C(OEt)(CF_3)_2$	H
[13]	oxazole derivative			

Fig. 1 Structures of 4-trifluoromethylimidazole derivatives used in this study

hydration, dipole moment, ionization potential, electron affinity, molecule length, molecular weight (M.W.), log*P*, E_{HOMO}, E_{LUMO}, η, χ, and ω are listed in Table 1. Using these values, we investigated whether there was any relationship between the CC_{50} and each of these descriptors (Fig. 3).

In HL-60 cells, there was no relationship between CC_{50} and heat of formation (Fig. 3a), stability of hydration (Fig. 3b), dipole moment (Fig. 3c), ionization potential (Fig. 3d), E_{HOMO} (Fig. 3e), molecular weight (Fig. 3f), or η (that expresses the extent of elicitation) (Fig. 3g). On the other hand, there was correlation between CC_{50} and electron affinity (Fig. 3h), molecule length (Fig. 3i) or E_{LUMO} (Fig. 3j) ($r^2 = 0.375\sim0.435$). When CC_{50} was plotted vs. log*P*, a parabolic curve was produced, with a maximum cytotoxicity (least CC_{50} value) at log*P* of 4.4 (Fig. 3k), slightly higher than the optimal log*P* values reported for derivatives of prenylalcohol, vitamin K_2 [26], gallic acid [27], and coumarin [28] (log*P* of 2–3). There was also a good relationship between CC_{50} and absolute electron negativity (χ) ($r^2 = 0.704$) (Fig. 3l), and slight, but apparent relationship between CC_{50} and ω ($r^2 = 0.479$) (Fig. 3m). Since ω reflects the extent of reactivity, these results suggest the relationship between the reactivity and cytotoxicity of imidazole compounds.

The relationship between the electron structure and the cytotoxicity of imidazole derivatives was next investigated using the absolute hardness (η) vs. absolute electron negativity (χ) (Fig. 4) [16–18]. By applying this η–χ activity

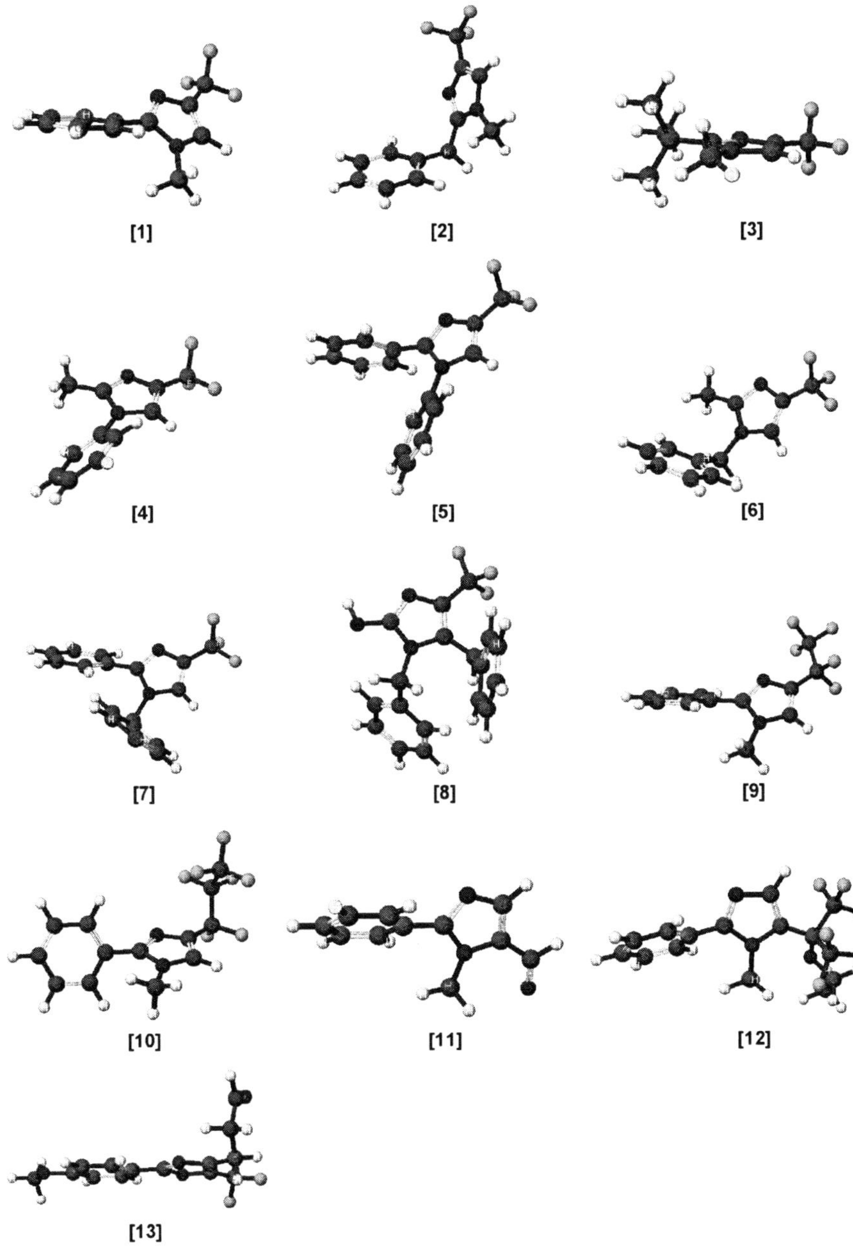

Fig. 2 The most stable conformation of each 4-trifluoromethylimidazole derivative used

diagram, we found that imidazole derivatives can be separated into two groups. Compounds with higher cytotoxicity (lower CC_{50}) distributed within the area surrounded by the box. Their cytotoxicity strongly depended on the χ value,

Table 1 CC_{50} and chemical descriptors for imidazoles

Compd.	HL-60 CC_{50} (μM)	HSC-3 CC_{50} (μM)	Heat of formation (Kcal/mol)	ΔH	Dipole moment (D)	Electron affinity (eV)	Ionization potential (eV)	log P	E_{HOMO} (eV)	E_{LUMO} (eV)	η (eV)	χ (eV)	ω (eV)	M.W.	Length (Å)
1	255.0	400.0	−119.034	14.358	9.474	0.531	9.620	2.070	−9.620	−0.531	4.544	5.075	2.834	226.20	9.872
2	149.0	400.0	−125.760	14.559	9.338	0.107	9.913	2.220	−9.913	−0.107	4.903	5.010	2.560	240.23	9.610
3	354.0	400.0	−165.771	12.209	9.349	−0.130	9.963	1.670	−9.963	0.130	5.046	4.916	2.395	206.21	7.939
4	31.0	294.0	−121.529	13.324	9.995	0.600	9.788	2.810	−9.788	−0.600	4.594	5.194	2.936	226.20	9.642
5	3.0	38.0	−87.642	14.438	10.056	0.639	9.841	4.100	−9.841	−0.639	4.601	5.240	2.984	288.27	9.646
6	31.0	399.0	−127.267	14.693	9.547	0.258	10.001	2.220	−10.001	−0.258	4.871	5.130	2.701	240.23	9.803
7	8.1	86.0	−94.046	14.777	9.505	0.544	9.735	3.850	−9.735	−0.544	4.596	5.140	2.874	302.29	9.557
8	145.0	262.0	−145.019	15.855	8.081	0.287	9.664	4.610	−9.664	−0.287	4.689	4.976	2.640	332.33	8.923
9	102.0	320.0	−214.115	12.585	9.544	0.510	9.591	2.820	−9.591	−0.510	4.540	5.051	2.809	276.21	9.623
10	27.0	215.0	−311.384	13.385	9.455	0.766	9.619	3.910	−9.619	−0.766	4.426	5.193	3.046	326.22	10.335
11	122.0	400.0	0.394	17.452	2.624	0.629	9.585	1.780	−9.585	−0.629	4.478	5.107	2.912	186.20	9.724
12	10.8	83.0	−309.112	14.063	8.150	0.686	9.711	4.010	−9.711	−0.686	4.513	5.199	2.995	352.30	11.190
13	29.0	230.0	−228.517	16.608	5.053	1.122	9.348	2.920	−9.348	−1.122	4.113	5.235	3.332	299.25	12.116

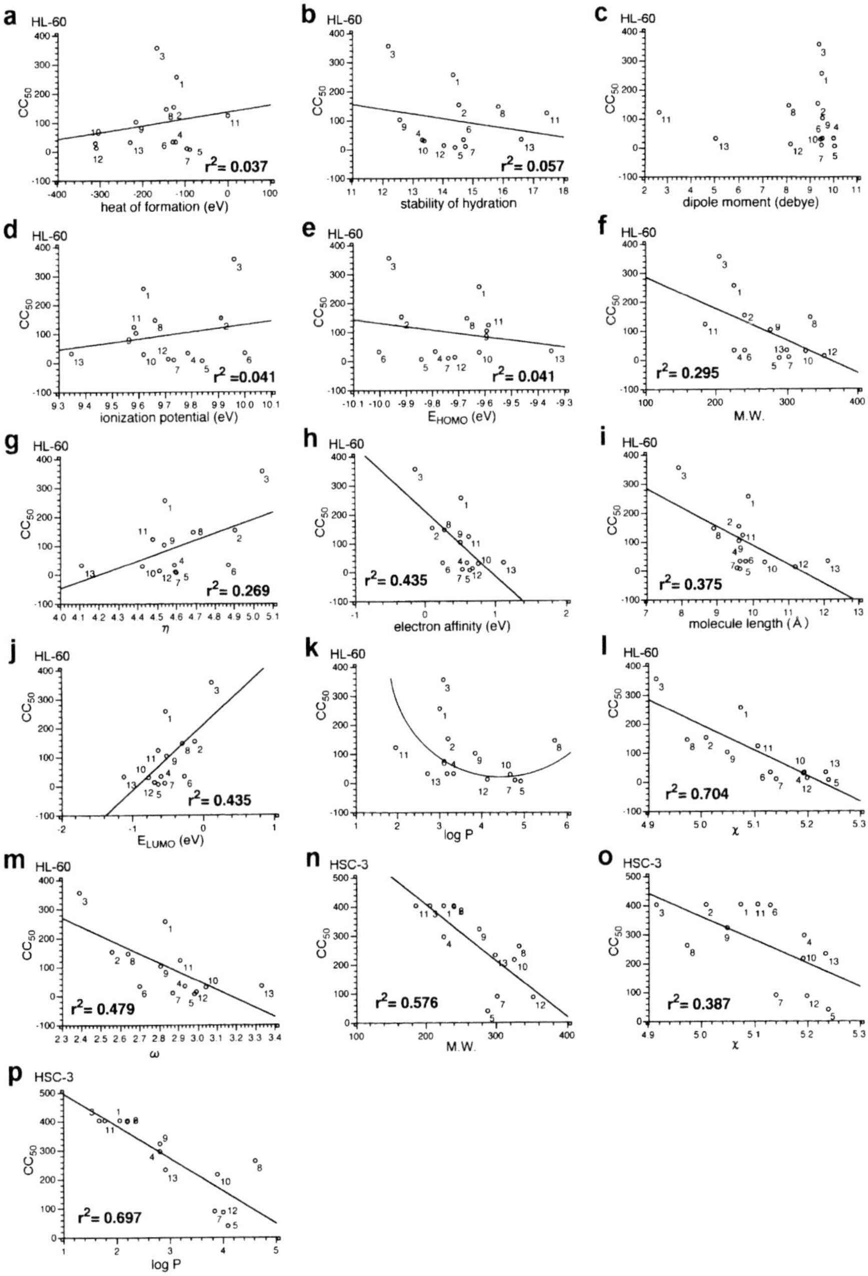

Fig. 3 Relationship between CC_{50} and each descriptor of 4-trifluoromethylimidazole derivatives against HL-60 (**a–m**) and HSC-3 cells (**n–p**)

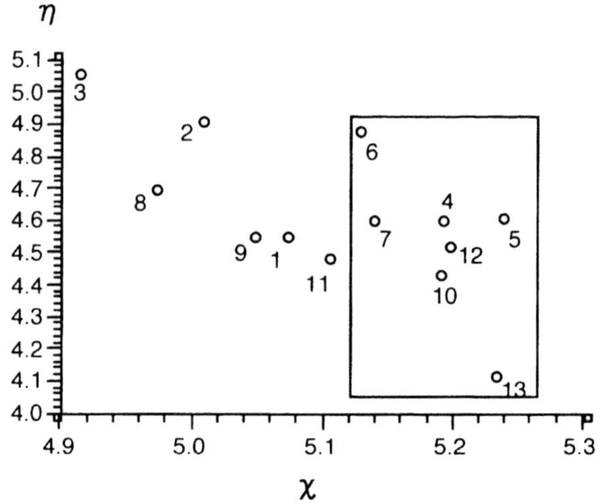

Fig. 4 η–χ activity diagram of 4-trifluoromethylimidazole derivatives against HL-60 cells. The *box* surrounds compounds with higher cytotoxicity (lower CC_{50})

but not on the η value. When the χ value was above 5.12, the cytotoxic activity goes up (the CC_{50} value goes down). On the other hand, when the χ value was below 5.12, the cytotoxic activity goes down (the CC_{50} value goes up). The value of χ determined by this calculation method may be useful for estimating the cytotoxic activity of newly synthesized imidazole derivatives.

In HSC-3 cells, relationship was found between the CC_{50} and the molecular weight ($r^2 = 0.576$) (Fig. 3n), χ ($r^2 = 0.387$) (Fig. 3o), or log$P$ ($r^2 = 0.697$) (Fig. 3p). It should be noted that a straight line, instead of the usual parabolic curve, was delineated when CC_{50} value was plotted vs. log$P$ in HSC-3 cells. However, there was no relationship between the CC_{50} and other descriptors (Table 2).

We investigated the QSAR of newly synthesized imidazole derivatives in HL-60 and HSC-3 cells, based on the concept of chemical hardness. A good correlation was found between the CC_{50} and χ in HL-60 cells. The biological

Table 2 Relationship coefficients between CC_{50} against HSC-3 cells and each chemical descriptor of imidazoles

Cells	Heat of formation (Kcal/mol)	ΔH	Dipole moment (D)	Electron affinity (eV)	Ionization potential (eV)	log P	E_{HOMO} (eV)	E_{LUMO} (eV)	η (eV)	χ (eV)	ω (eV)	M.W.	Length (Å)
HSC-3	0.089	0.000	0.028	0.250	0.029	C	0.029	0.250	0.160	0.387	0.290	0.576	0.129

C correlated

activity of imidazole derivatives in HL-60 cells can be estimated by the η–χ activity diagram.

2.2 Phenoxazine Derivatives

Phenoxazines are a group of *N*-heterocycles having three six-ring structures with nitrogen and oxygen atoms [29]. Phenoxazines have shown antitumor [30], antimicrobial [31], antiviral [32], anti-inflammatory [33] and multidrug resistant reversal activity [34], prevented human amyloid disorders [35], and protected neuronal cell death from an oxidative stress [36]. However, the mechanism for the induction of antitumor activity, and the type of cell death induced by phenoxazines have not been well understood.

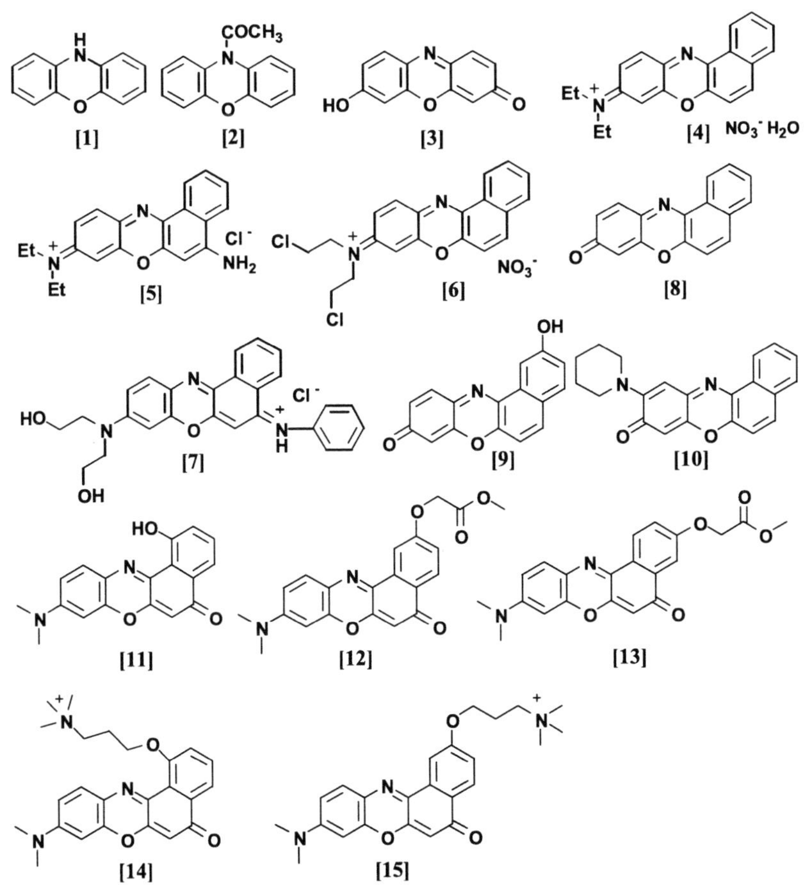

Fig. 5 Structures of phenoxazine derivatives used in this study

We have recently found that among 24 phenoxazine derivatives, **7** and **8** showed the highest tumor-specificity indexes of 4.3 and 4.8, respectively [37]. **7** and **8** did not apparently induce intranucleosomal DNA fragmentation, nor activate caspase-3 in HSC-2, HSC-4, and human glioblastoma T98G cells. We investigated the QSAR of 15 phenoxazine derivatives (Fig. 5) [38], using conventional and recent techniques of computation chemistry, such as the concept of chemical hardness [16–18].

By determining the most stable conformation of 15 phenoxazine derivatives, their structure was approximated to the molecular form present in vivo (biomimetic). The most stable structure was next determined by CAChe Worksystem 4.9 MOPAC (PM3) (Fig. 6). The CC_{50} value (determined by experiments) of phenoxazines against HSC-2, HSC-3, HSC-4, and HL-60 cells, and chemical descriptors (determined by calculations): heat of formation

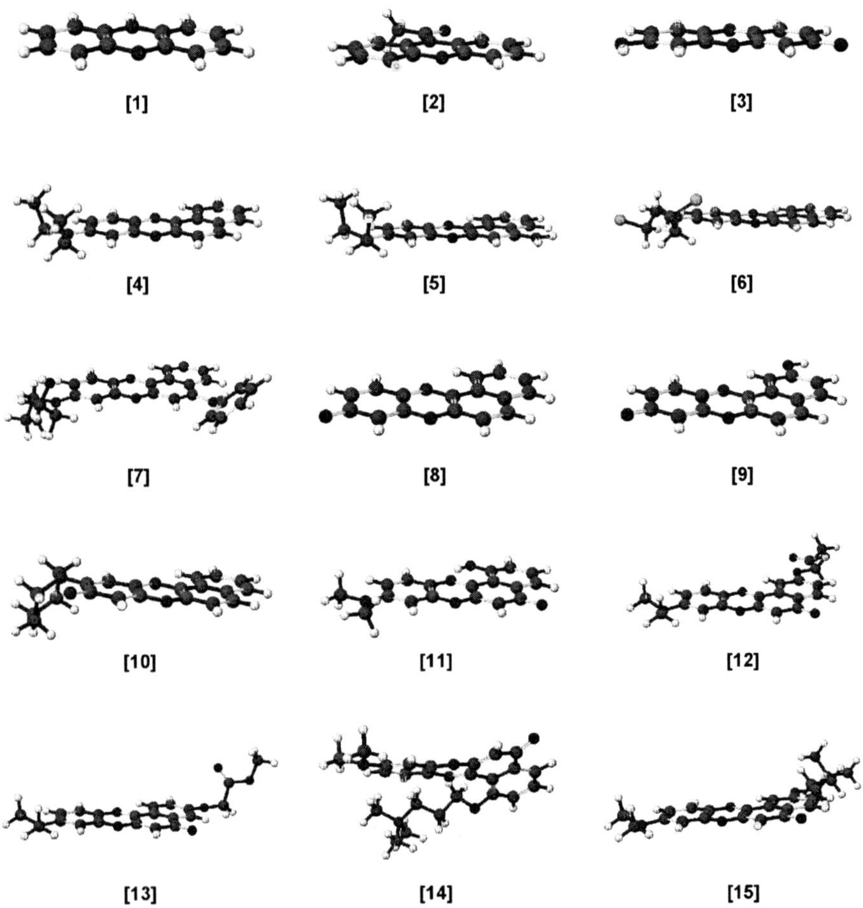

Fig. 6 The most stable conformation of each phenoxazine derivative used

(COSMO), ionization potential, electron affinity, log$P$, E_{HOMO}, E_{LUMO}, η, χ, ω, and molecular weight (not determined by calculation) are listed in Table 3. Using these values, whether there are any relationships between CC_{50} and each of these descriptors was investigated (Fig. 7).

In HSC-2 and HSC-4 cells, there was correlation between CC_{50} and electron affinity ($r^2 = 0.580$, 0.515, respectively) (Fig. 7a), E_{LUMO} ($r^2 = 0.580$, 0.515, respectively) (Fig. 7b), absolute hardness, which indicates the extent of

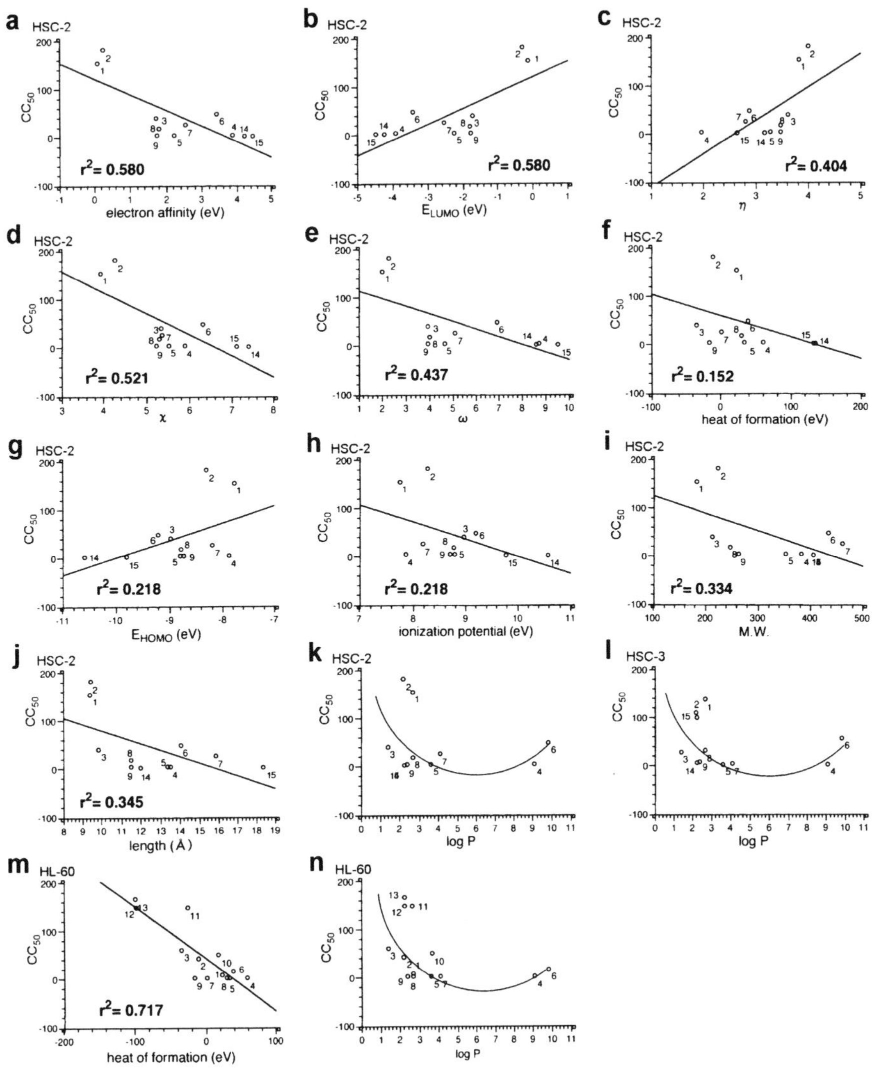

Fig. 7 Relationship between CC_{50} and each chemical descriptor of phenoxazine derivatives against HSC-2 (**a–k**), HSC-3 (**l**), and HL-60 cells (**m, n**)

Table 3 CC_{50} and chemical descriptors for phenoxazines

Compd.	HL-60 CC_{50} (μM)	HSC-2 CC_{50} (μM)	HSC-3 CC_{50} (μM)	HSC-4 CC_{50} (μM)	Heat of formation (Kcal/mol)	Electron affinity (eV)	Ionization potential (eV)	log *P*	E_{HOMO} (eV)	E_{LUMO} (eV)	η (eV)	χ (eV)	ω (eV)	M.W.	Length (Å)
[1]	6.9	154.0	137.0	164.0	22.957	0.104	7.757	2.681	−7.757	−0.104	3.827	3.930	2.018	183.21	9.432
[2]	42.0	181.0	109.0	157.0	−10.092	0.264	8.284	2.206	−8.284	−0.264	4.010	4.274	2.278	225.24	9.444
[3]	60.0	40.0	28.0	115.0	−33.951	1.721	8.963	1.396	−8.963	−1.721	3.621	5.342	3.940	213.19	9.864
[4]	1.6	3.5	2.6	4.4	60.007	3.900	7.869	9.118	−7.869	−3.900	1.984	5.884	8.726	383.39	13.544
[5]	1.6	4.5	2.0	4.5	34.226	2.239	8.788	3.645	−8.788	−2.239	3.275	5.513	4.641	353.85	13.392
[6]	16.0	48.0	56.0	78.0	39.007	3.437	9.202	9.840	−9.202	−3.437	2.883	6.319	6.926	434.27	14.055
[7]	1.6	25.0	3.8	28.0	1.653	2.540	8.183	4.108	−8.183	−2.540	2.821	5.361	5.094	461.94	15.873
[8]	1.6	18.0	31.0	17.0	29.457	1.802	8.767	2.683	−8.767	−1.802	3.483	5.285	4.010	247.24	11.514
[9]	1.6	3.2	6.9	8.0	−15.490	1.760	8.710	2.398	−8.710	−1.760	8.475	5.235	3.943	263.24	11.531
[10]	49.0	–	–	–	18.102	1.679	8.294	3.669	−8.294	−1.679	3.307	4.986	3.758	330.37	12.691
[11]	147.0	–	–	–	−24.755	1.553	8.249	2.663	−8.249	−1.553	3.348	4.901	3.588	306.00	13.675
[12]	148.0	–	–	–	−96.408	1.510	8.520	2.268	−8.520	−1.510	3.505	5.015	3.588	378.00	17.092
[13]	165.0	–	–	–	−98.723	1.509	8.462	2.268	−8.462	−1.509	3.477	4.985	3.575	378.00	17.986
[14]	–	1.6	6.2	3.1	136.03	4.225	10.580	2.251	−10.580	−4.225	3.177	7.402	8.623	406.00	11.996
[15]	–	1.6	100.0	19.0	132.172	4.465	9.779	2.251	−9.779	−4.465	2.657	7.122	9.545	406.00	18.394

excitation (η) ($r^2 = 0.404$, 0.40, respectively) (Fig. 7c), absolute electron negativity (χ) ($r^2 = 0.521$, 0.436, respectively) (Fig. 7d), or reactivity index (ω) ($r^2 = 0.437$, 0.405, respectively) (Fig. 7e). However, there was no relationship between CC_{50} and heat of formation (Fig. 7f), E_{HOMO} (Fig. 7g), ionization potential (Fig. 7h), molecular weight (Fig. 7i) or molecular length (Fig. 7j) ($r^2 = 0.152 \sim 0.345$). Generally, the interaction between cells and drugs are generated by the shape of the molecule. The molecular structure of phenoxazine derivatives determined by calculation was extended and planar. The structure of compound **14** bent at the center and overlapped with each other (see Fig. 6). The cytotoxic activity of phenoxazine derivatives became maximum at the log*P* = 5.9 (Fig. 7k), slightly higher than the optimal log*P* values reported for derivatives of prenylalcohol, vitamin K_2 [26], gallic acid [27] and coumarin [28] (log*P* of 2–3).

The relationship between the electron structure and the cytotoxicity of phenoxazine derivatives was next investigated, using the absolute hardness/absolute electron negativity (η–χ) activity diagram (Fig. 8). In HSC-2 cells, compounds with higher cytotoxicity (lower CC_{50}) are distributed within the area surrounded by box. Their cytotoxicity strongly depended on the χ value, but not on the η value. Compounds with relatively higher cytotoxicity (lower CC_{50}) showed higher χ values ($\chi > 5.28$). Compounds with relatively lower cytotoxicity (higher CC_{50}) showed lower χ values ($\chi < 4.27$). The value of χ determined by this method may be useful for estimation of the cytotoxic activity of newly synthesized phenoxazine derivatives.

In HSC-3 cells, relationship was found only between the CC_{50} and log*P* (Fig. 7l). In HL-60 cells, CC_{50} was correlated only to heat of formation ($r^2 = 0.717$) (Fig. 7m) or log*P* (Fig. 7n), but not to other descriptors (Table 4).

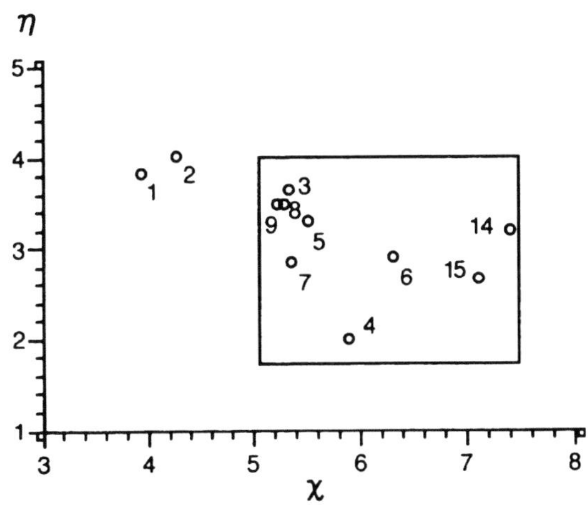

Fig. 8 η–χ activity diagram between the electron structure and the cytotoxicity of phenoxazine derivatives. The *box* surrounds compounds with higher cytotoxicity (lower CC_{50})

Table 4 Relationship coefficients between CC_{50} against HL-60, HSC-3, and HSC-4 cells and each chemical descriptor of phenoxazines

Cells	Heat of formation (Kcal/mol)	Electron affinity (eV)	Ionization potential (eV)	log *P*	E_{HOMO} (eV)	E_{LUMO} (eV)	η (eV)	χ (eV)	ω (eV)	M.W.	Length (Å)
HL-60	0.171	0.073	0.000	C	0.000	0.073	0.074	0.055	0.105	0.012	0.288
HSC-3	0.001	0.180	0.031	C	0.031	0.180	0.171	0.137	0.089	0.178	0.031
HSC-4	0.219	0.515	0.152	C	0.152	0.515	0.403	0.436	0.405	0.383	0.360

C correlated

We investigated the QSAR of newly synthesized phenoxazine derivatives, based on the concept of chemical hardness. The cytotoxic activity of phenoxazines against HSC-2 and HSC-4 cells can be estimated from the following chemical descriptors: electron affinity, η, χ, and log*P*. However, only log*P* was correlated to CC_{50} in HSC-3 cells, whereas only heat of formation and log*P* were corrected to CC_{50} in HL-60 cells. This similarity observed between HSC-2 and HSC-4 cells, and the difference from HSC-3 cells, may be due to difference in the expression of outer cell membrane proteins, since we found abundance of two P-glycoproteins, MDR and MRP, in HSC-2 and HSC-4 cells, as compared with HSC-3 cells (Hashimoto et al., in preparation).

2.3 5-Trifluoromethyloxazole Derivatives

1,3-Oxazole, also known simply as oxazole, is one of the most prominent heterocycles, and the oxazole motif occurs within the framework of many important pharmacophores and natural products [39]. The benefit of introducing the trifluoromethyl group into organic molecules, especially in the area of drugs or pesticides, is now well documented [40]. Therefore, trifluoromethylated nitrogen heterocycles are an attractive class of compounds because of their potential biological activities, and many five-membered aromatic nitrogen heterocycles have been prepared [41]. Outstanding applications of such molecules were found in the pharmaceutical field, as illustrated by Celecoxib (antiarthritic) bearing a trifluoromethyl substituent on the pyrazole-ring, a recent drug used in the treatment of human diseases [42]. As an extension of the search of highly selected antitumor agents against human promyleocytic leukemia (HL-60) and human squamous cell carcinoma cell lines (HSC-2, HSC-3, HSC-4), and human glioblastoma cells (T98G), we investigated the QSAR of newly synthesized 5-trifluoromethyloxazole derivatives [43–45] (Fig. 9) using conventional and recent techniques of computation chemistry, such as the concept of chemical hardness [16–18].

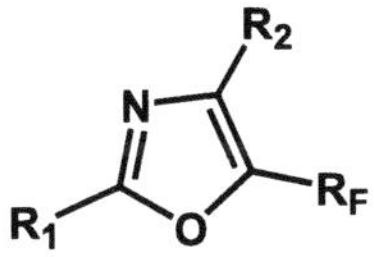

[1]-[12]

Compd.	R_1	R_2	R_F
[1]	C_6H_5	C_6H_5	CF_3
[2]	2-Thienyl	C_6H_5	CF_3
[3]	C_6H_5	$CH_2C_6H_5$	CF_3
[4]	tert-Butyl	$CH_2C_6H_5$	CF_3
[5]	$C_6H_5CH{=}C(CH_3)$	$CH_2C_6H_5$	CF_3
[6]	C_6H_5	$(CH_2)_3OH$	CF_3
[7]	4-ClC_6H_4	$(CH_2)_3OH$	CF_3
[8]	4-$CH_3OC_6H_4$	$(CH_2)_3OH$	CF_3
[9]	2-$CH_3OC_6H_4$	$(CH_2)_3OH$	CF_3
[10]	$C_6H_5CH{=}C(CH_3)$	$(CH_2)_3OH$	CF_3
[11]	$C_6H_5CH{=}C(CH_3)$	$(CH_2)_2CHO$	CF_3
[12]	H	$(CH_2)_3NHCOC_6H_5$	CF_3

Fig. 9 Structures of 5-tifluoromethyloxazol derivatives used in this study

The optimization of the structure was done by the semi-empirical molecular-orbital method, using CAChe Worksystem 4.9 PM3 method, in the presence (COSMO) or in the absence (non-COSMO) of water (Fig. 10). The CC_{50} values of 12 5-trifluoromethyloxazole derivatives against HL-60, HSC-2, HSC-3, HSC-4, and T98G cells (experimental values), and chemical descriptors such as heat of formation, stability of hydration (= COSMO – non-COSMO) (ΔH), dipole moment (D), electron affinity, ionization potential, log$P$, E_{HOMO}, E_{LUMO}, maximum length of molecule, molecular weight, η, χ, and ω (determined by calculation) are listed in Table 5 (COSMO or non-COSMO). The relationship between the CC_{50} and each descriptor was then investigated (Fig. 11). The relationship coefficient between CC_{50} and each chemical descriptor (non-COSMO or COSMO) is listed in Table 6. There was no relationship between the CC_{50} and heat of formation, χ, or molecular weight (regardless of whether non-COSMO or COSMO) in all five cells. There was only minor relationship between CC_{50} and ΔH, dipole moment, or the maximum length of the molecule except for T98G cells (ΔH) (Fig. 11, C-1), T98G (dipole moment) (COSMO), HL-60 (maximum length) (COSMO, non-COSMO).

On the other hand, relationship was found between the CC_{50} and electron affinity in all cells, especially in HL-60 cells ($r^2 = 0.857$) (Fig. 11, A-1), except for HSC-3 cells in COSMO. In non-COSMO, the relationship between these parameters was found only in HL-60 and HSC-2 cells in non-COSMO (Table 6). Similarly, relationship was found between the CC_{50} and ionization

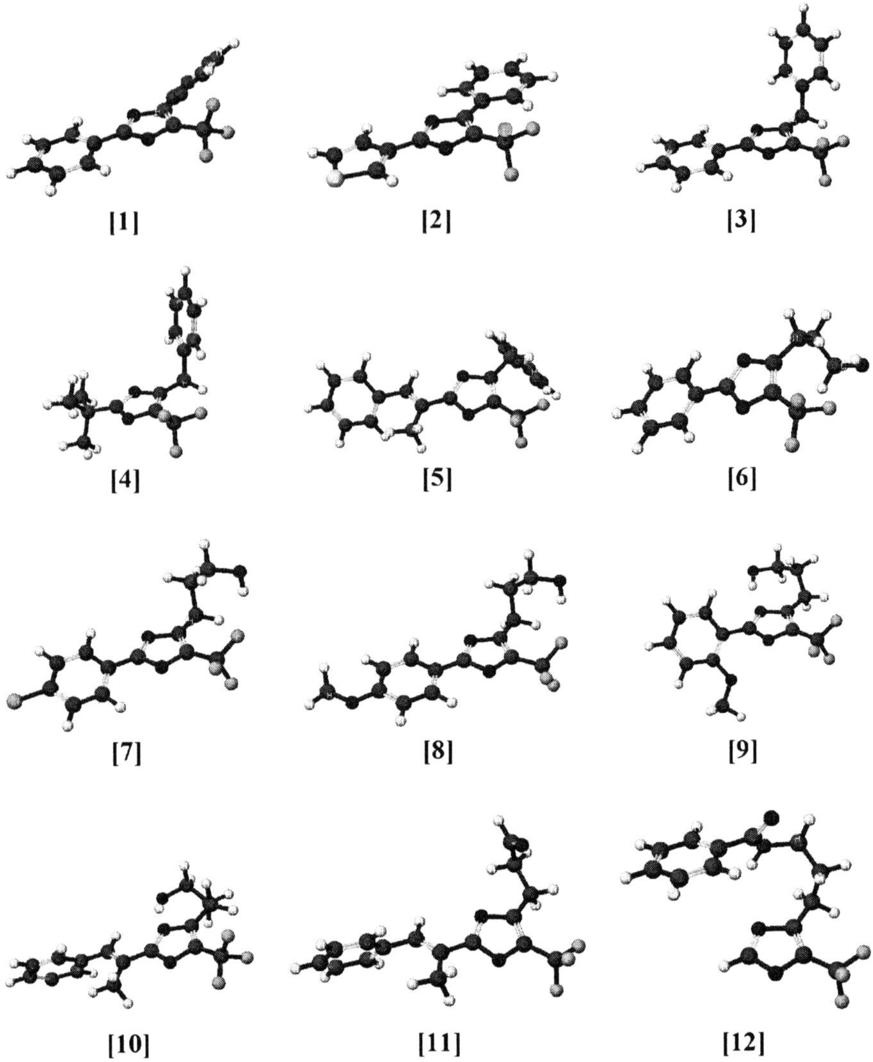

Fig. 10 The most stable conformation of each 5-tifluoromethyloxazol derivative used

potential in HL-60 ($r^2 = 0.750$) (Fig. 11, A-2), HSC-4 ($r^2 = 0.544$) in COSMO, and in HSC-4 ($r^2 = 0.422$) in non-COSMO. E_{HOMO} (Fig. 11, A-3 and B-1) and ionization potential gave identical values, whereas E_{LUMO} (Fig. 11, A-4, B-2, and C-2) and electron affinity were the same. Good relationship was found between the CC_{50} and η (Fig. 11, B-3 and C-3), both in non-COSMO (except for HSC-3 cells) and COSMO, especially in HL-60 cells ($r^2 = 0.903$) (Fig. 11, A-5), but not in HSC-2 and HSC-3 cells. Good relationship was found between CC_{50} and ω (Fig. 11, C-4) especially in HL-60 cells ($r^2 = 0.761$) (Fig. 11, A-6), but not in HSC-3 (COSMO) and HSC-3, HSC-4, T98G cells (non-COSMO).

Table 5 CC_{50} and chemical descriptors for 5-trifluoromethyloxazole derivatives

COSMO

Compd.	HL-60 CC_{50} (μM)	HSC-2 CC_{50} (μM)	HSC-3 CC_{50} (μM)	HSC-4 CC_{50} (μM)	T98G CC_{50} (μM)	Heat of formation (Kcal/mol)	ΔH	Dipole moment (D)	Electron affinity (eV)	Ionization potential (eV)	E_{HOMO}	E_{LUMO}	η (eV)	χ (eV)	ω (eV)	M.W.	Length (Å)	log P
1	43.0	177.0	138.0	140.0	150.0	−112.707	−8.392	3.120	1.228	9.622	9.622	−1.228	4.197	5.425	3.506	289.0	12.075	5.184
2	45.0	195.0	196.0	137.0	188.0	−104.602	−8.218	2.443	1.228	9.483	9.483	−1.228	4.098	5.386	3.540	295.0	11.230	3.782
3	52.0	152.0	90.0	130.0	180.0	−119.426	−8.705	3.315	1.167	9.739	−9.739	−1.167	4.286	5.453	3.469	303.0	11.341	5.436
4	120.0	363.0	257.0	231.0	296.0	−167.015	−7.751	2.543	0.689	9.943	−9.943	−0.689	4.627	5.316	3.054	283.0	10.681	5.514
5	36.0	162.0	88.0	82.0	151.0	−112.777	−9.301	3.258	1.130	9.491	−9.491	−1.130	4.181	5.310	3.373	343.0	12.788	6.256
6	52.0	250.0	246.0	255.0	279.0	−200.871	−11.896	5.161	1.170	9.748	−9.748	−1.170	4.289	5.459	3.474	271.0	12.314	3.468
7	35.0	141.0	140.0	123.0	190.0	−207.606	−12.261	3.943	1.262	9.553	−9.553	−1.262	4.146	5.407	3.527	306.0	11.371	3.986
8	35.0	195.0	261.0	150.0	225.0	−241.844	−14.202	6.082	1.154	9.337	−9.377	−1.154	4.111	5.266	3.372	301.0	12.878	3.215
9	44.0	280.0	221.0	200.0	267.0	−237.769	−12.784	3.959	1.023	9.550	−9.550	−1.023	4.263	5.287	3.278	301.0	10.227	3.215
10	29.0	128.0	120.0	111.0	168.0	−193.299	−11.256	2.078	1.122	9.532	−9.532	−1.122	4.205	5.327	3.374	311.0	12.078	4.288
11	74.0	330.0	249.0	211.0	344.0	−181.013	−15.037	5.731	1.051	9.547	−9.547	−1.051	4.248	5.299	3.305	401.0	12.194	3.791
12	142.0	275.0	260.0	320.0	330.0	−196.773	−18.597	6.378	0.539	10.016	−10.016	−0.539	4.738	5.278	2.939	298.0	9.466	2.664

Table 5 (continued)

Non-COSMO

Compd.	HL-60 CC_{50} (μM)	HSC-2 CC_{50} (μM)	HSC-3 CC_{50} (μM)	HSC-4 CC_{50} (μM)	T98G CC_{50} (μM)	Heat of formation (Kcal/mol)	Dipole moment (D)	Electron affinity (eV)	Ionization potential (eV)	log P	E_{HOMO}	E_{LUMO}	η	χ	ω	M.W.	Length (Å)
1	43.0	177.0	138.0	140.0	150.0	−104.315	2.533	1.118	9.466	5.184	−9.466	−1.118	4.174	5.292	3.355	289.0	12.075
2	45.0	195.0	196.0	137.0	188.0	−96.384	2.184	1.253	9.365	3.782	−9.365	−1.253	4.056	5.309	3.475	295.0	11.230
3	52.0	152.0	90.0	130.0	180.0	−110.721	2.599	1.062	9.631	5.436	−9.631	−1.062	4.285	5.346	3.336	303.0	11.341
4	120.0	363.0	257.0	231.0	296.0	−159.264	2.198	0.549	9.696	5.514	−9.696	−0.549	4.574	5.122	2.868	283.0	10.681
5	36.0	162.0	88.0	82.0	151.0	−103.476	2.73	0.979	9.023	6.256	−9.450	0.979	4.235	5.215	3.210	343.0	12.788
6	52.0	250.0	246.0	255.0	279.0	−188.975	3.817	1.173	9.742	3.468	−9.742	−1.173	4.285	5.457	3.476	271.0	12.314
7	35.0	141.0	140.0	123.0	190.0	−195.345	2.972	1.327	9.561	3.986	−9.561	−1.327	4.117	5.444	3.599	306.0	11.371
8	35.0	195.0	261.0	150.0	225.0	−227.642	4.715	1.100	9.318	3.215	−9.318	−1.100	4.109	5.209	3.301	301.0	12.878
9	44.0	280.0	221.0	200.0	267.0	−224.985	3.478	1.014	9.534	3.215	−9.534	−1.014	4.260	5.274	3.265	301.0	10.227
10	29.0	128.0	120.0	111.0	168.0	−182.043	2.163	1.189	9.577	4.288	−9.577	−1.189	4.194	5.383	3.455	311.0	12.078
11	74.0	330.0	249.0	211.0	344.0	−165.976	4.441	0.944	9.522	3.791	−9.522	−0.944	4.304	5.248	3.200	401.0	12.194
12	142.0	275.0	260.0	320.0	330.0	−178.176	4.035	0.858	9.739	2.664	−9.739	−0.858	4.441	5.298	3.161	298.0	9.466

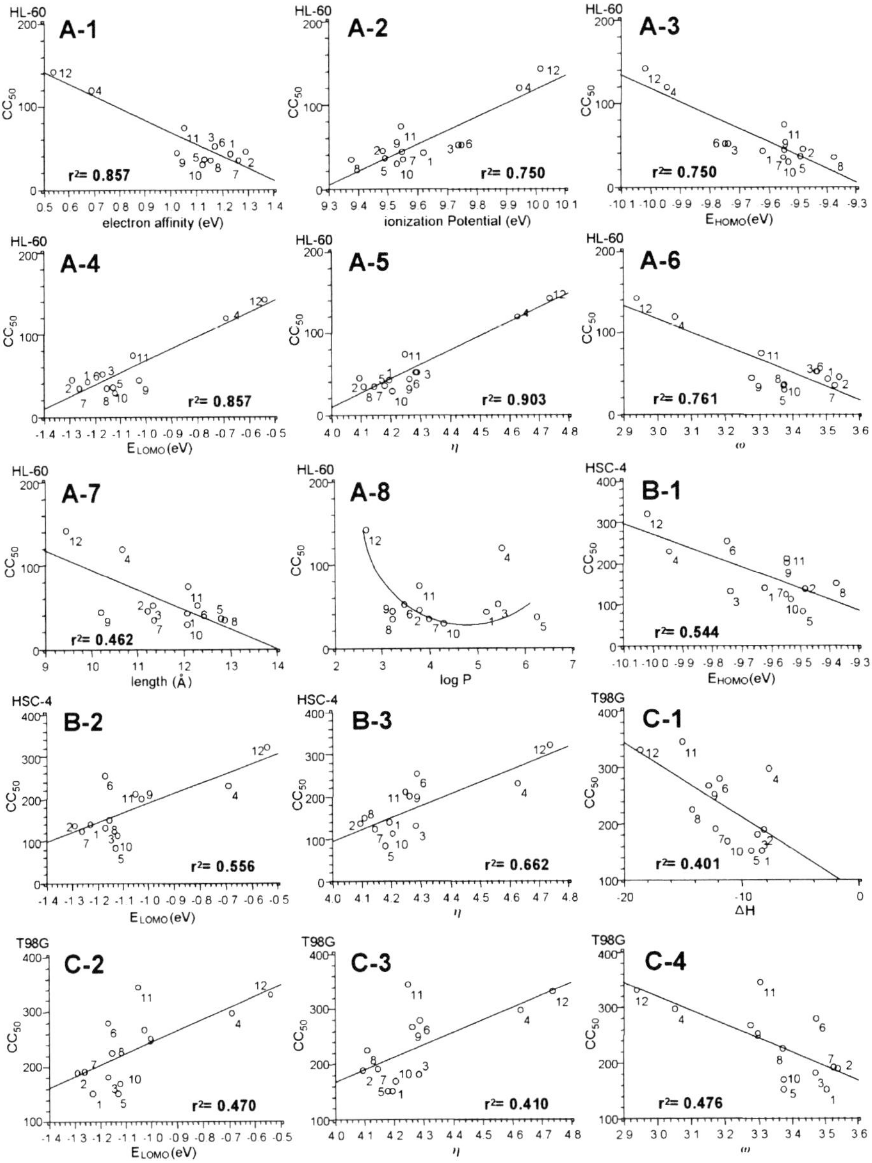

Fig. 11 Relationship between CC_{50} and each chemical descriptor of 5-tifluoromethyl-oxazol derivatives against HL-60 (**A-1** to **A-8**), HSC-4 (**B-1** to **B-3**), and T98G cells (**C-1** to **C-4**)

The cytotoxic activity of 5-trifluoromethyloxazole derivatives became maximum at log $P = 4.6$ (Fig. 11, A-8), slightly higher than the optimal log$P$ values reported for derivatives of prenylalcohol, vitamin K_2 [26], gallic acid [27] and coumarin [28] (log$P$ of 2–3).

Table 6 Relationship coefficients between CC_{50} against the indicated cells and each chemical descriptor of 5-trifluoromethyloxazole derivatives

Cells	Heat of formation (Kcal/mol)	ΔH	Dipole moment (D)	Electron affinity (eV)	Ionization potential (eV)	log P	E_{HOMO} (eV)	E_{LUMO} (eV)	η (eV)	χ (eV)	ω (eV)	M.W. (Å)	Length
COSMO													
HL-60	0.002	0.136	0.109	0.857	0.750	C	0.750	0.857	0.903	0.100	0.761	0.003	0.462
HSC-2	0.071	0.073	0.123	0.448	0.249	C	0.249	0.448	0.391	0.161	0.459	0.020	0.182
HSC-3	0.321	0.283	0.383	0.267	0.086	C	0.086	0.267	0.192	0.189	0.290	0.006	0.092
HSC-4	0.161	0.345	0.363	0.556	0.544	C	0.544	0.556	0.662	0.045	0.496	0.029	0.357
T98G	0.247	0.401	0.413	0.470	0.261	C	0.261	0.470	0.410	0.169	0.476	0.042	0.210
Non-COSMO													
HL-60	0.001	0.136	0.026	0.598	0.387	C	0.387	0.598	0.711	0.142	0.499	0.003	0.453
HSC-2	0.067	0.073	0.125	0.577	0.164	C	0.164	0.577	0.534	0.265	0.534	0.020	0.170
HSC-3	0.309	0.283	0.394	0.168	0.157	C	0.157	0.168	0.145	0.088	0.157	0.006	0.087
HSC-4	0.143	0.345	0.235	0.255	0.442	C	0.442	0.255	0.453	0.004	0.179	0.029	0.348
T98G	0.226	0.401	0.375	0.315	0.273	C	0.273	0.315	0.417	0.050	0.258	0.042	0.195

C correlated

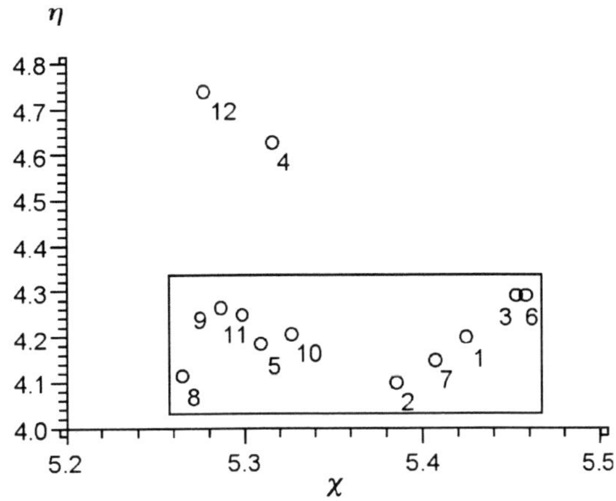

Fig. 12 η–χ diagram of 5-tifluoromethyloxazol derivatives. The *box* surrounds compounds with higher cytotoxicity (lower CC_{50})

The relationship between the electron structure and the cytotoxicity of 5-trifluoromethyloxazole derivatives was next investigated, using the η–χ activity diagram (Fig. 12). In HL-60 cells, compounds with higher cytotoxicity (lower CC_{50}) distributed within the area surrounded by box. Their cytotoxicity strongly depended on the η value, but not on the χ value. Compounds with relatively higher cytotoxicity (lower CC_{50}) showed lower η value ($\eta < 4.29$). Compounds with relatively lower cytotoxicity (higher CC_{50}) showed higher η value ($\eta > 4.62$). The value of η determined by this method may be useful for estimation of the cytotoxic activity of newly synthesized 5-trifluoromethyloxazole derivatives.

We performed the QSAR analysis of the CC_{50} values of 5-trifluoromethyloxazole derivatives. Higher relationship coefficients were found in COSMO than in non-COSMO. This suggests that calculation should be performed under biomimetic conditions, although longer calculation time is required. The concept of absolute hardness is applicable to estimate the cytotoxicity of 5-trifluoromethyloxazoles using chemical descriptors η, ω, or the η–χ activity diagram.

3
O-Heterocycles

3.1
3-Formylchromone Derivatives

The chromones (4*H*-1-benzopyran-4-ones) have been reported to exhibit antifungal, antiviral, antimicrobial, antiallergic, antitublin, and antitumor activities [46–49]. In addition, many flavonoids are based on the chromone structure and have been found to possess therapeutically interesting biological activities [50]. We have recently reported some tumor-specific cytotoxic activity and anti-*H. pylori* activity of 11 3-formylchromones (Fig. 13) and four related compounds [51]. We investigated the QSAR of these 3-formylchromones using conventional and recent techniques of computation chemistry, such as the concept of chemical hardness.

By determining the most stable conformation of 11 3-formylchromone derivatives, their structure was approximated to the molecular form present in vivo (biomimetic). The most stable structure was next determined by CAChe Worksystem 4.9 MOPAC (PM3) (Fig. 14). The most stable conformation of all these compounds displayed a planar structure. Compounds **2**, **3**, **4**, and **9** had additionally protruding branches from the coplanar surface (see Fig. 14). The CC_{50} value (determined by experiment) of 3-formylchromone derivatives against HSG, HL-60, HSC-3, HSC-2, MT-4 cells, and chemical descriptors (determined by calculations): heat of formation (COSMO, non-COSMO), ΔH, dipole moment, electron affinity, ionization potential, log*P*, E_{HOMO}, E_{LUMO}, η, χ, ω, M.W., length of molecule, surface area, and surface volume are listed in Table 7. Using these values, the relationship between CC_{50} and each of the descriptors was investigated (Table 8), and the regression lines were drawn (Fig. 15).

In HSG cells, the best relationship coefficient was observed between CC_{50} and ΔH ($r^2 = 0.719$) (Fig. 15, A-1), followed by electron affinity ($r^2 = 0.565$)

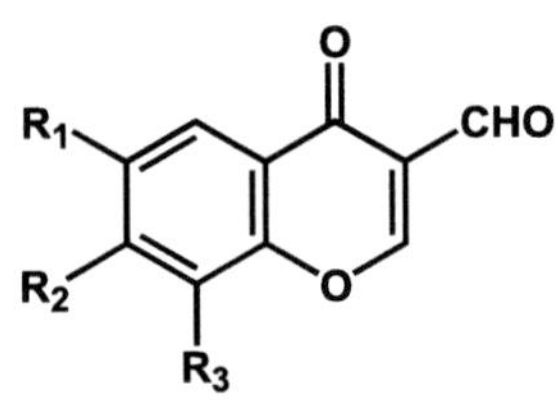

[1] - [11]

Compd	R_1	R_2	R_3
[1]	H	H	H
[2]	CH_3	H	H
[3]	i-Pr	H	H
[4]	CH_3O	H	H
[5]	NO_2	H	H
[6]	F	H	H
[7]	Cl	H	H
[8]	Br	H	H
[9]	Cl	CH_3	H
[10]	Cl	H	Cl
[11]	Br	H	Br

Fig. 13 Structures of 3-formylchromone derivatives used in this study

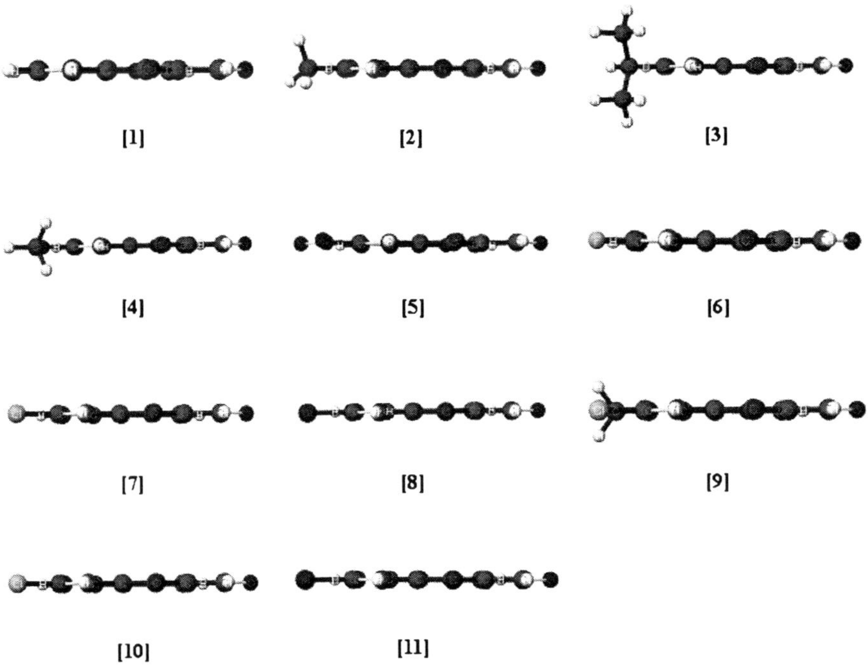

Fig. 14 The most stable conformation of each 3-formylchromone derivative used

(Fig. 15, A-2), E_{LUMO} ($r^2 = 0.566$) (Fig. 15, A-3), E_{HOMO} ($r^2 = 0.470$) (Fig. 15, A-4), ionization potential ($r^2 = 0.470$) (Fig. 15, A-5), χ ($r^2 = 0.616$) (Fig. 15, A-6) and ω ($r^2 = 0.636$) (Fig. 15, A-7), (Table 8). When the value for **1** for ΔH, which was off from the regression line, was omitted, higher relationship coefficients were produced ($r^2 = 0.906$) (data not shown). The omission of **1** also elevated the relationship coefficients for electron affinity, E_{LUMO}, χ, and ω to 0.872, 0.872, 0.711, and 0.835, respectively (data not shown). When CC_{50} was plotted vs. log*P*, a parabolic curve was produced, under the condition that the data for **5** was omitted (Fig. 15, A-8).

In HL-60 cells, moderate relationship was found between CC_{50} and five descriptors: ΔH ($r^2 = 0.439$) (Fig. 15, B-1), electron affinity ($r^2 = 0.463$) (Fig. 15, B-2), E_{LUMO} ($r^2 = 0.463$) (Fig. 15, B-3), χ ($r^2 = 0.433$) (Fig. 15, B-4), and ω ($r^2 = 0.486$) (Fig. 15, B-5) (Table 8). When the values for **1** and **6**, which was off the regression line, were omitted, the relationship coefficients for ΔH, electron affinity, E_{LUMO}, χ, and ω were elevated up to 0.595, 0.777, 0.777, 0.520, and 0.700, respectively (Table 8).

In HSC-3 cells, there was moderate relationship between CC_{50} and dipole moment ($r^2 = 0.449$), but not with other descriptors (Table 8).

In HSC-2 and MT-4 cells, there was no clear-cut relationship between CC_{50} and any descriptors (Table 8).

Table 7 CC_{50} and chemical descriptors for 3-formylchromone derivatives

Compd.	HSG CC_{50} (μM)	HL-60 CC_{50} (μM)	HSC-3 CC_{50} (μM)	HSC-2 CC_{50} (μM)	MT-4 CC_{50} (μM)	Heat of formation (Kcal/mol)	ΔH	Dipole moment (D)	Electron affinity (eV)	Ionization potential (eV)	log *P*	E_{HOMO}	E_{LUMO}	η	χ	ω	M.W.	Length (Å)	Surface area (Å^2)	Volume range (Å^3)
1	332.0	59.0	225.0	89.0	177.0	−80.357	−16.644	6.801	0.847	9.727	0.350	−9.727	−0.847	4.440	5.287	3.148	174.150	8.395	136.514	81.970
2	128.0	13.0	184.0	47.0	211.0	−89.659	−16.538	6.925	0.849	9.542	0.860	−9.542	−0.849	4.346	5.195	3.105	188.180	9.837	154.120	91.880
3	95.0	20.0	172.0	42.0	49.0	−98.909	−16.379	6.881	0.850	9.612	1.760	−9.612	−0.850	4.381	5.231	3.123	216.230	10.066	189.085	110.620
4	166.0	30.0	99.0	80.0	474.0	−119.702	−18.311	5.461	0.896	9.365	0.560	−9.365	−0.896	4.234	5.131	3.108	204.180	9.968	162.113	95.480
5	546.0	78.0	262.0	84.0	162.0	−107.417	−36.827	7.883	1.315	10.047	0.760	−10.047	−1.315	4.366	5.681	3.697	219.150	9.622	156.472	91.240
6	91.0	8.0	165.0	22.0	88.5	−123.644	−17.66	6.27	0.943	9.715	0.600	−9.715	−0.943	4.386	5.329	3.238	192.140	9.468	140.085	83.380
7	92.0	17.0	73.0	46.0	85.8	−86.508	−16.84	6.327	0.923	9.510	1.170	−9.510	−0.923	4.293	5.216	3.169	208.600	8.738	149.383	89.540
8	169.0	41.0	81.0	64.0	111.0	−78.185	−17.948	6.259	0.940	9.770	1.320	−9.770	−0.940	4.415	5.355	3.247	253.050	8.923	154.171	93.710
9	115.0	30.0	102.0	42.0	86.2	−95.402	−16.667	6.753	0.943	9.489	1.670	−9.489	−0.943	4.573	5.216	3.184	222.620	9.345	166.587	99.570
10	215.0	52.0	117.0	56.0	361.0	−90.687	−16.631	5.334	1.042	9.443	1.910	−9.443	−1.042	4.200	5.242	3.271	243.040	8.785	162.340	97.270
11	216.0	40.0	131.0	75.0	278.0	−63.874	−18.767	5.244	1.068	9.823	2.210	−9.823	−1.068	4.378	5.445	3.387	331.950	8.933	171.680	105.290

Table 8 Relationship coefficients between CC_{50} against the indicated cells and each chemical descriptor of 3-formylchromone derivatives

Cells	Heat of formation (Kcal/mol)	ΔH	Dipole moment (D)	Electron affinity (eV)	Ionization potential (eV)	log P	E_{HOMO} (eV)	E_{LUMO} (eV)	η (eV)	χ (eV)	ω (eV)	M.W.	Length (Å)	Surface area (Å^2)	Volume range (Å^3)
HSG	0.001	0.719	0.184	0.565	0.470	C	0.470	0.566	0.711	0.616	0.636	0.001	0.001	0.037	0.059
HL-60	0.048	0.439	0.049	0.463	0.292	NC	0.292	0.463	0.433	0.486	0.486	0.032	0.002	0.007	0.013
HSC-3	0.038	0.345	0.449	0.118	0.382	C	0.382	0.118	0.311	0.311	0.208	0.109	0.004	0.031	0.054
HSC-2	0.090	0.189	0.000	0.121	0.115	C	0.115	0.121	0.134	0.141	0.141	0.037	0.009	0.007	0.007
MT4	0.009	0.000	0.384	0.011	0.129	C	0.129	0.011	0.035	0.035	0.000	0.037	0.028	0.003	0.003

C correlated, *NC* not correlated

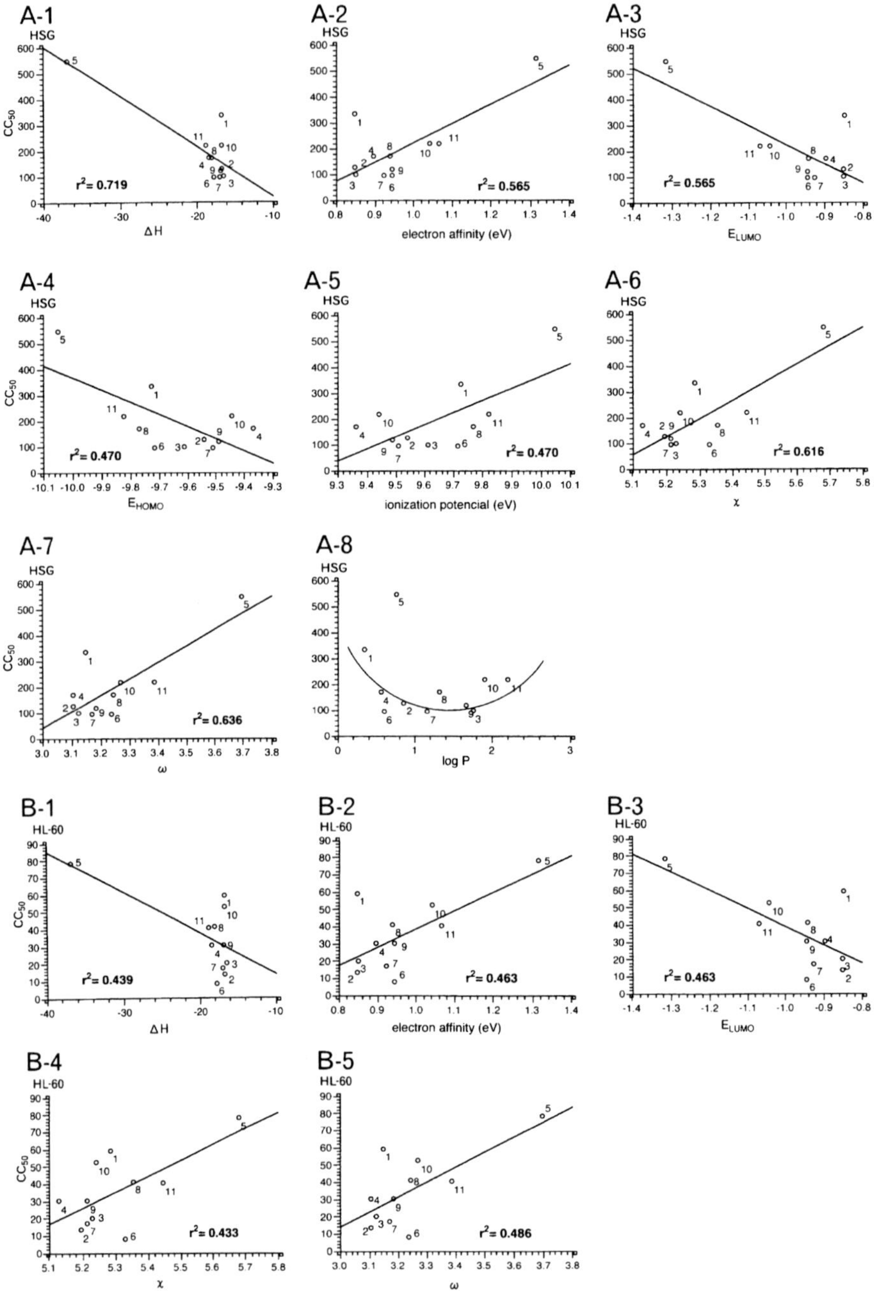

Fig. 15 Relationship between CC_{50} and each chemical descriptor of 3-formylchromone derivatives against HSG (**A-1** to **A-8**) and HL-60 cells (**B-1** to **B-5**)

In conclusion, there was the best relationship between CC_{50} and descriptors in HSG cells. This cell line may provide the opportunity to search more active 3-formylchromone derivatives using QSAR with the concept of absolute hardness.

3.2 Coumarin and its Derivatives

Coumarin (2*H*-pyran-2-one) and its derivatives (Fig. 16) are widely distributed in nature and exhibit a broad pharmacological profile, including anticancer activity [52] and scavenging activity of superoxide anions generated by activated neutrophils [53]. Introduction of substituents at C-2, C-4, and C-7 of the heterocyclic ring of coumarin induced various biological activities. These coumarin derivatives are known to induce apoptosis in human leukemia cells by increasing the cytosolic translocation of cytochrome c and activation of cysteine protease 32 kDa proenzyme [54, 55]. However, few efforts have been made to establish the relationship between the structure and conventional cytotoxic activity of coumarins. In this study, we investigated the relationship between the 50% cytotoxic concentration (CC_{50}) of 20 coumarin derivatives (Fig. 16) [56] and 12 physical parameters (descriptors) calculated with the CONFLEX/PM3 method. The following descriptors were used: ① heat of formation, ② stability of hydration, ③ dipole moment, ④ electron affinity, ⑤ ionization potential, ⑥ hydrophobicity (log*P*), ⑦ highest occupied molecular orbital (HOMO) energy, ⑧ lowest unoccupied molecular orbital (LUMO) energy, ⑨ $\Delta\varepsilon(\varepsilon_{HOMO} - \varepsilon_{LUMO})$, ⑩ absolute hardness ($\eta$), ⑪ absolute electron negativity (χ), and ⑫ η–χ; where ε_{HOMO} is the electron energy of HOMO; ε_{LUMO} is the electron energy of LUMO; $\eta = (\varepsilon_{LUMO} - \varepsilon_{HOMO})/2$; and $\chi =- (\varepsilon_{LUMO} + \varepsilon_{HOMO})/2$.

The CC_{50} and 12 descriptors for each compound are shown in Table 9. The relationship between CC_{50} and each descriptor is shown in Fig. 17, ① to ⑫. There was no relationship between CC_{50} and ①, ②, ③, ④, nor ⑧ ($r^2 = 0.13{\sim}0.36$). When CC_{50} was plotted vs. log*P*, a parabolic curve was produced, with a maximum cytotoxicity (least CC_{50}) at log*P* of 2.5 (Fig. 17, ⑥). There was highly significant relationship between CC_{50} and ⑤, ⑦, ⑨, or ⑩ ($r^2 = 0.756{\sim}0.802$), suggesting that these four descriptors can be utilized for the estimation of the cytotoxic activity. HOMO and LUMO energy values can be used as markers for electron-donating and electron-accepting capabilities, respectively. With the increase in HOMO energy, the molecule would have an increasing electron-donating property. On the other hand, the decrease in the LUMO energy leads to the increase in the electron-accepting property. HOMO energy reflects the ionization potential. The relationship between the CC_{50} and ionization potential ($r^2 = 0.756$) indicates that the cytotoxicity of coumarin derivatives is produced from this electron-donating property.

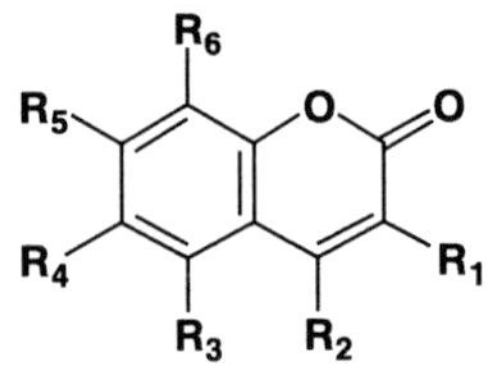

Compd.	R_1	R_2	R_3	R_4	R_5	R_6
1	H	H	H	H	H	H
2	H	H	H	H	OH	H
3	H	H	H	OH	OH	H
4	H	H	H	OCH_3	OH	H
5	H	H	OCH_3	OCH_3	OH	H
6	H	CH_3	H	H	OH	H
7	H	CH_3	H	OH	H	H
8	H	CH_3	H	OH	OH	H
9	H	CH_3	OH	H	OH	H
10	H	CH_3	H	OH	OCH_3	H
11	H	CH_3	H	OCH_3	OH	H
12	CH_3	H	H	H	OH	H
13	CH_3	CH_3	H	H	OH	H
14	CH_3	CH_3	H	OH	OH	H
15	CH_3	CH_3	H	OH	OCH_3	H
16	CH_3	CH_3	H	OCH_3	OH	H
17	-$(CH_2)_3$-		H	OH	OH	H
18	-$(CH_2)_3$-		OH	H	OH	H
19	-$(CH_2)_3$-		H	OH	OCH_3	H
20	-$(CH_2)_3$-		H	OCH_3	OH	H

Fig. 16 Chemical structure of coumarin derivatives

Recently, the hazardous effects of endocrine disruptors (environmental hormones) such as bisphenol A and nonylphenol on the human body have been reported, and led to the initiation of many studies concerning the detection and structural determination of these compounds present in tiny amounts [57]. As one of the QSAR analyses of environmental hormones, the relationship between their biological activity and chemical hardness has been reported [16]. By applying these analytical methods, the molecular toxicity and estrogen-like activity of environmental hormones have been found

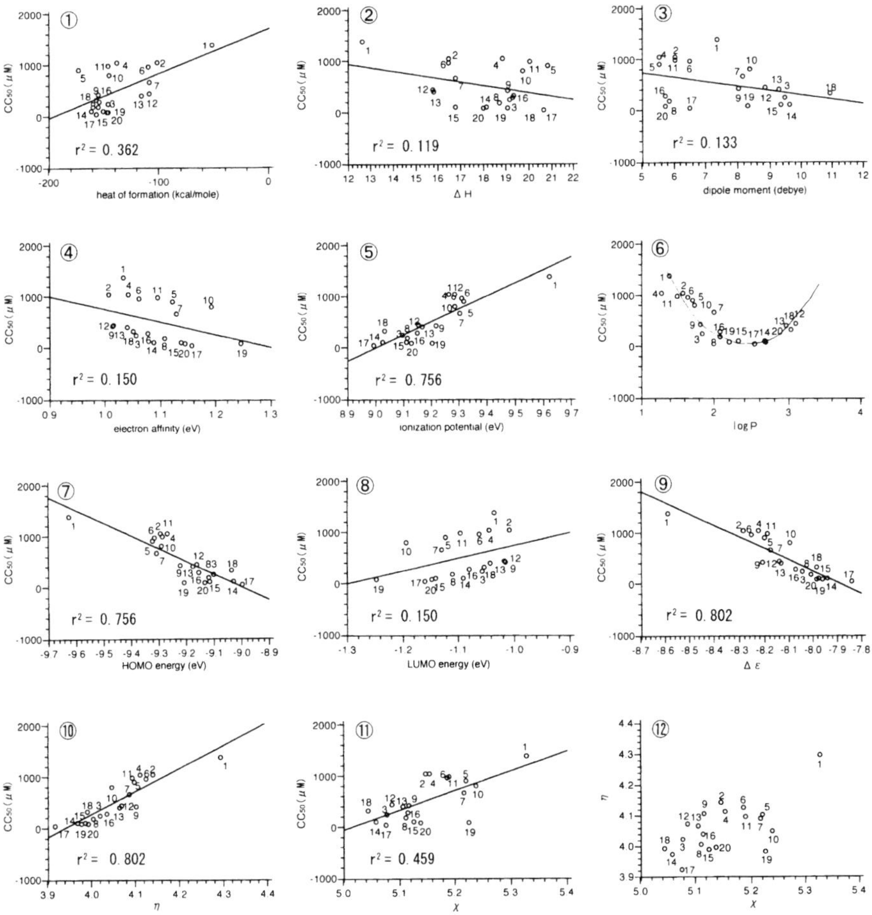

Fig. 17 Relationship between CC_{50} of coumarin derivatives against HSC-2 cells and each descriptor

to strongly correlate with the η value. We have used this method to investigate the relationship between CC_{50} and η values. We found that the cytotoxic activity of coumarin derivatives against HSC-2 cells vs. η produced a well-fitted straight line ($r^2 = 0.802$, Fig. 17, ⑩). The η–χ diagram was next determined, as one of the methods for investigating the electron state of the molecule. By applying this η–χ diagram to coumarin derivatives, we found that their cytotoxicity depended on the the η value (Fig. 17, ⑩), but not on the χ (Fig. 17, ⑪) nor the η–χ value (Fig. 17, ⑫). Compounds with relatively higher cytotoxicity showed lower η values (<4.04). Compounds with relatively lower cytotoxicity (except for compound 18) showed higher η values (>4.04). This η–χ activity diagram may be useful for the study of cytotoxic activity.

Table 9 CC_{50} value and 12 descriptors for coumarin derivatives

Compd.	HSC-2 CC_{50} (μM)	Heat of formation (Kcal/mol)	ΔH	Dipole moment (D)	Electron affinity (eV)	Ionization potential (eV)	log P	ε_{HOMO} (eV)	ε_{LUMO} (eV)	$\Delta\varepsilon$	η (eV)	χ (eV)
1	1370	−150.799	12.671	7.400	1.035	9.623	1.39	−9.623	−1.035	−8.588	4.294	5.329
2	1025	−100.588	16.491	6.046	1.007	9.290	1.58	−9.290	−1.007	−8.282	4.141	5.148
3	236	−145.390	19.172	9.478	1.056	9.099	1.84	−9.099	−1.056	−8.043	4.021	5.078
4	1042	−137.096	18.892	5.568	1.043	9.266	1.28	−9.266	−1.043	−8.223	4.111	5.154
5	901	−172.396	20.858	5.564	1.123	9.320	1.71	−9.320	−1.123	−8.198	4.099	5.221
6	950	−109.326	16.482	6.525	1.062	9.313	1.64	−9.313	−1.062	−8.251	4.125	5.187
7	664	−107.911	16.818	8.175	1.130	9.350	2.00	−9.305	−1.030	−8.175	4.087	5.217
8	172	−154.099	18.757	5.874	1.109	9.117	2.08	−9.117	−1.109	−8.008	4.004	5.113
9	417	−154.078	19.078	8.063	1.015	9.221	1.81	−9.221	−1.015	−8.206	4.103	5.118
10	791	−144.433	19.752	8.393	1.193	9.287	1.73	−9.287	−1.193	−8.094	4.047	5.240
11	971	−145.733	20.062	6.055	1.096	9.285	1.49	−9.285	−1.096	−8.189	4.095	5.190
12	443	−108.071	15.812	8.851	1.017	9.158	3.10	−9.158	−1.017	−8.140	4.070	5.087
13	405	−115.471	15.849	9.290	1.041	9.171	2.96	−9.171	−1.041	−8.130	4.065	5.106
14	92	−160.324	18.154	9.641	1.089	9.028	2.67	−9.028	−1.089	−7.939	3.970	5.059
15	105	−150.052	16.792	9.358	1.039	9.114	2.32	−9.114	−1.139	−7.974	3.987	5.126
16	282	−153.639	19.305	5.762	1.079	9.151	2.08	−9.151	−1.079	−8.072	4.036	5.115
17	41	−155.885	20.685	6.531	1.059	8.996	2.54	−8.996	−1.159	−7.838	3.919	5.178
18	325	−156.667	19.367	10.964	1.052	9.035	3.04	−9.035	−1.052	−7.983	3.992	5.044
19	73	−145.465	19.088	8.332	1.247	9.204	2.20	−9.204	−1.247	−7.958	3.979	5.226
20	82	−146.697	18.028	5.757	1.046	9.131	2.68	−9.131	−1.146	−7.985	3.993	5.139

In conclusion, the present study demonstrates that the cytotoxic activity of coumarin derivatives show a strong linear relationship with the absolute hardness, η. Hardness and softness, apart from the electron accepting and donating properties of the molecule, are important factors for estimation of their cytotoxic activity. From the η value, the CC_{50} value of the novel coumarin compounds can be estimated. CONFLEX is useful for calculating the hardness and softness of the molecule using the PM3 method.

4 Vitamin K_2 Derivatives

Vitamin K_2 (menaquinone) represents a series of compounds in which the phytyl side chain of phytonadione has been replaced by a side chain built up of 1–14 isoprenyl units. It has been reported that vitamin K_2 with four isoprenyl units (**4**) induced the monocytic differentiation of human myeloid leukemia cell lines [58], or apoptosis in isolated osteoclast [59] and human ovary cancer cells [60]. We have recently found that vitamin K_2 derivatives (**1**–**3**) induced some tumor-specific cytotoxicity and non-apoptotic cell death in oral carcinoma [61]. As an extension of the search for tumor-specific substances targeted against human oral squamous, hepatocellular carcinoma, and promyelocytic leukemia cell lines, we investigated the QSAR of seven vitamin K_2 derivatives (**1**–**7**) (Fig. 18), by conventional and recent techniques of computation chemistry, such as the concept of absolute hardness [16–18].

By determining the conformation of seven vitamin K_2 derivatives with CONFLEX5, their structure was approximated to the molecular form present in vivo (biomimetic). The most stable structure was next determined by CAChe Worksystem 4.9 MOPAC (PM3) (Fig. 19). The three-dimensional structure of these seven compounds could be roughly separated into two groups: more extended form (**1**–**3**) and spherical form (**4**–**7**). The molecular

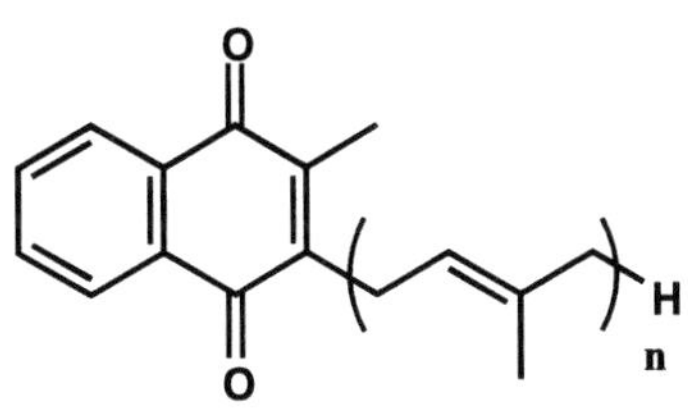

Vitamin K_2

[1] n =1 [2] n =2 [3] n =3

[4] n =4 [5] n =5 [6] n =6

[7] n =7

Fig. 18 Structures of vitamin K_2 derivatives used in this study

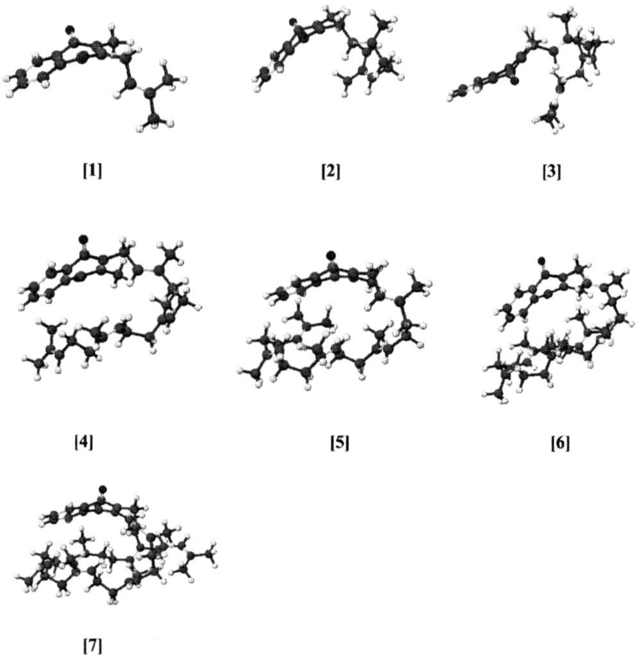

Fig. 19 The most stable conformation of each vitamin K_2 derivative used

length of **4**–**7** was approximately 10 Å, although the isoprenyl unit in the side chain increased from **4** to **7**.

The CC_{50} value (determined by experiment) against HSC-2, HSC-3, HSC-4, $HepG_2$, and HL-60 cells and chemical descriptors (determined by calculation) such as heat of formation (Fig. 20, A-1, B-1, C-1, D-1, E-1), electron affinity (A-2, B-2), ionization potential (A-3, B-3), log*P* (A-11, B-11), E_{HOMO} (A-4, B-4), E_{LUMO} (A-5, B-5), η (A-6, B-6), χ (A-7, B-7), ω (A-8, B-8, C-2), molecular length (A-10, B-10, C-4, D-3, E-3) and molecular weight (A-9, B-9, C-3, D-2, E-2) are listed in Table 10. Using these values, we investigated whether there was any relationship between CC_{50} and each of these descriptors (Fig. 20).

In HSC-2 and HSC-3 cells, there was some relationship between CC_{50} and eleven descriptors (Table 11 and Fig. 20, A1 to A11, C1 to C4). In $HepG_2$ cells, there was good close relationship between CC_{50} and descriptors except for electron affinity and E_{LUMO} (Fig. 20, B-1 to B-11). In HL-60 cells, there was good relationship only between CC_{50} and heat of formation, M.W. and molecular length (Fig. 20, E-1 to E-3).

It should be noted that a straight line was delineated when CC_{50} was plotted vs. log*P* in $HepG_2$ cells (Fig. 20, B-11) ($r^2 = 0.696$), in contrast to parabolic curves reported for derivatives of gallic acid [27] and coumarin [28]. The relationship between the CC_{50} and absolute hardness (η, χ, ω) was next investigated. There was also a good relationship between CC_{50} and the absolute

Table 10 CC_{50} and chemical descriptors for vitamin K_2 derivatives

Compd.	HL-60 CC_{50} (μM)	HSC-2 CC_{50} (μM)	HSC-3 CC_{50} (μM)	HSC-4 CC_{50} (μM)	$HepG_2$ CC_{50} (μM)	Heat of formation (Kcal/mol)	Electron affinity (eV)	Ionization potential (eV)	log P	E_{HOMO} (eV)	E_{LUMO} (eV)	η (eV)	χ (eV)	ω (eV)	M.W.	Length (Å)
1	46.0	130.0	104.0	129.0	55.0	−39.100	1.232	9.641	3.474	−9.641	−1.232	4.204	5.437	3.515	240.0	5.646
2	30.0	114.0	76.0	79.0	104.0	−45.093	1.239	9.447	4.952	−9.447	−1.239	4.104	5.343	3.478	308.0	6.557
3	13.0	159.0	137.0	93.0	193.0	−47.658	1.234	9.207	6.430	−9.207	−1.234	3.986	5.220	3.418	377.0	6.678
4	200.0	200.0	200.0	200.0	200.0	−53.287	1.242	9.346	7.909	−9.346	−1.242	4.052	5.294	3.458	445.0	9.739
5	200.0	200.0	200.0	200.0	200.0	−60.435	1.173	9.306	9.387	−9.306	−1.173	4.067	5.240	3.375	513.0	10.168
6	200.0	200.0	200.0	200.0	200.0	−65.590	1.157	9.294	10.866	−9.294	−1.157	4.068	5.226	3.356	581.0	10.389
7	200.0	200.0	200.0	200.0	200.0	−75.215	1.200	9.277	12.344	−9.277	−1.200	4.039	5.239	3.398	649.0	9.957

hardness (η) ($r^2 = 0.755$), electron negativity (χ) ($r^2 = 0.883$) and reactivity index (ω) ($r^2 = 0.694$). Since ω reflects the extent of reactivity, this results suggests a relationship between the reactivity and cytotoxicity of vitamin K_2 derivatives.

The relationship between electron structure and cytotoxicity of vitamin K_2 derivatives was next investigated, using absolute hardness (η) vs. absolute electron negativity (χ) (Fig. 21) [16–18]. We found that only $HepG_2$ cells showed a clear-cut relationship in the η–χ activity diagram. The cytotoxic

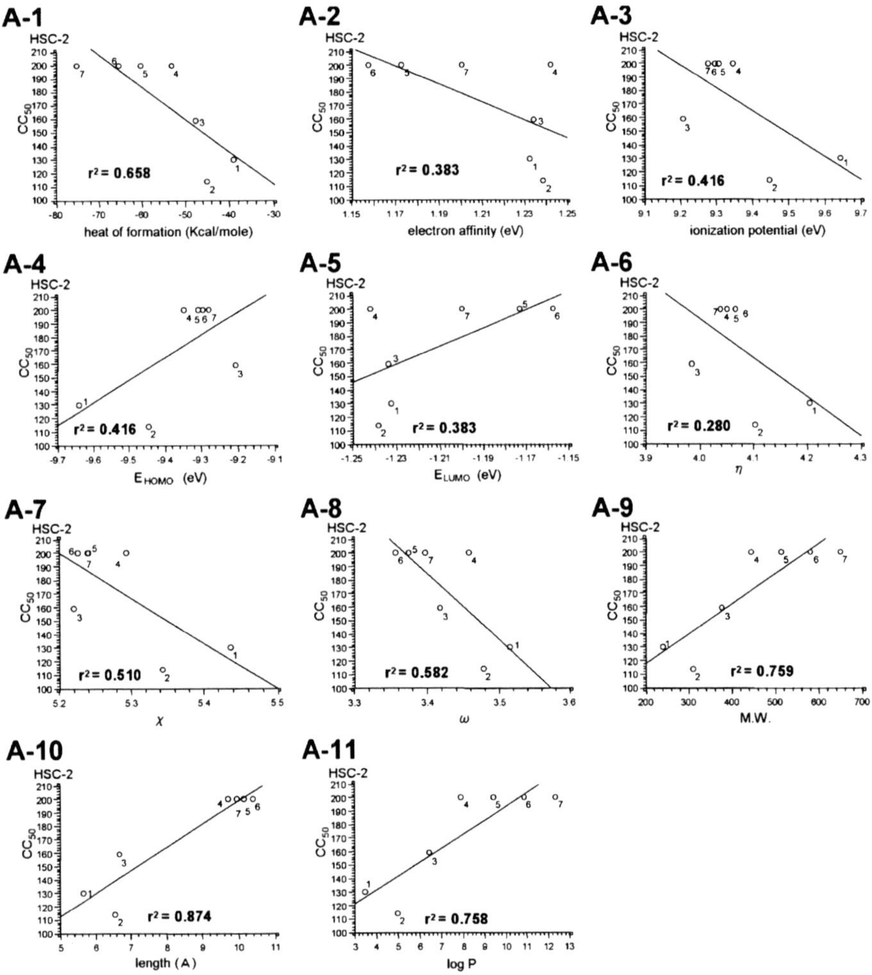

Fig. 20 Relationship between CC_{50} and each descriptor of vitamin K_2 derivatives against HSC-2 (**A-1** to **A-11**), $HepG_2$ (**B-1** to **B-11**), HSC-3 (**C-1** to **C-4**), HSC-4 (**D-1** to **D-3**), and HL-60 cells (**E-1** to **E-3**)

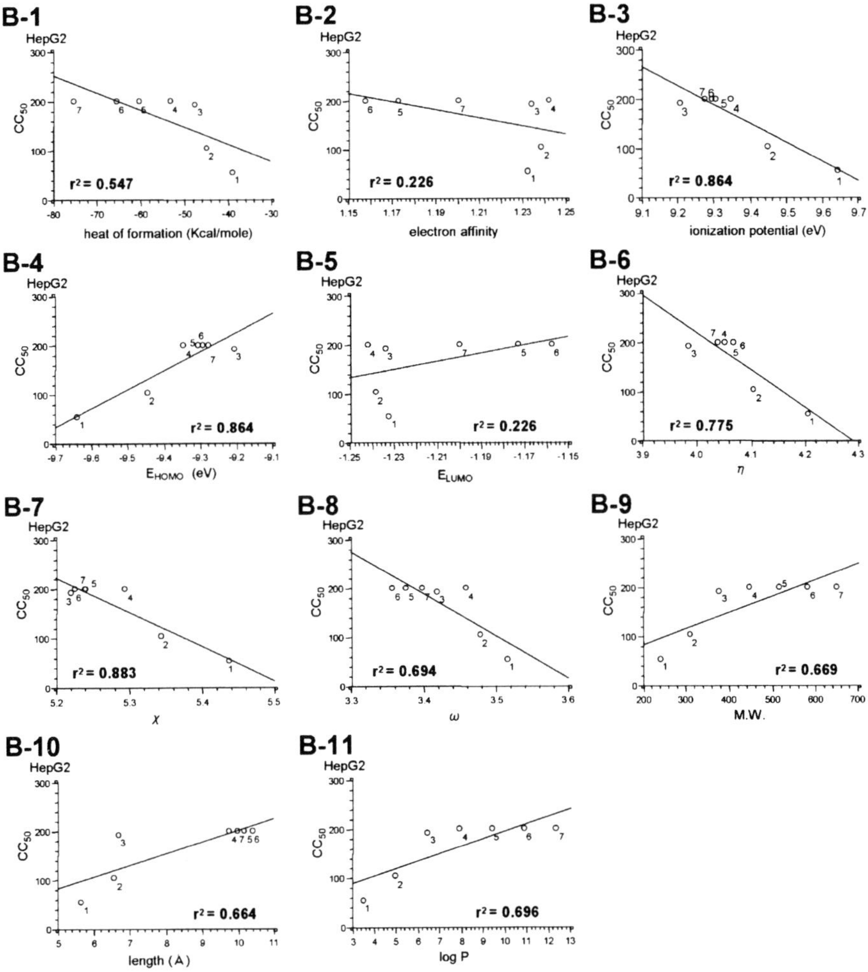

Fig. 20 (continued)

activity of vitamin K_2 derivatives depended on both η and χ. When the compounds were within the box ($\eta < 4.068$ and $\chi < 5.294$), no cytotoxicity was detected (Fig. 21).

The η–χ activity diagram determined by this calculation method may be useful for estimation of the cytotoxic activity of vitamin K_2 derivatives against $HepG_2$ cells. The relationship was confirmed between the cytotoxic activity against any of the cells used in this study and the molecular shape or molecular weight of vitamin K_2 derivatives.

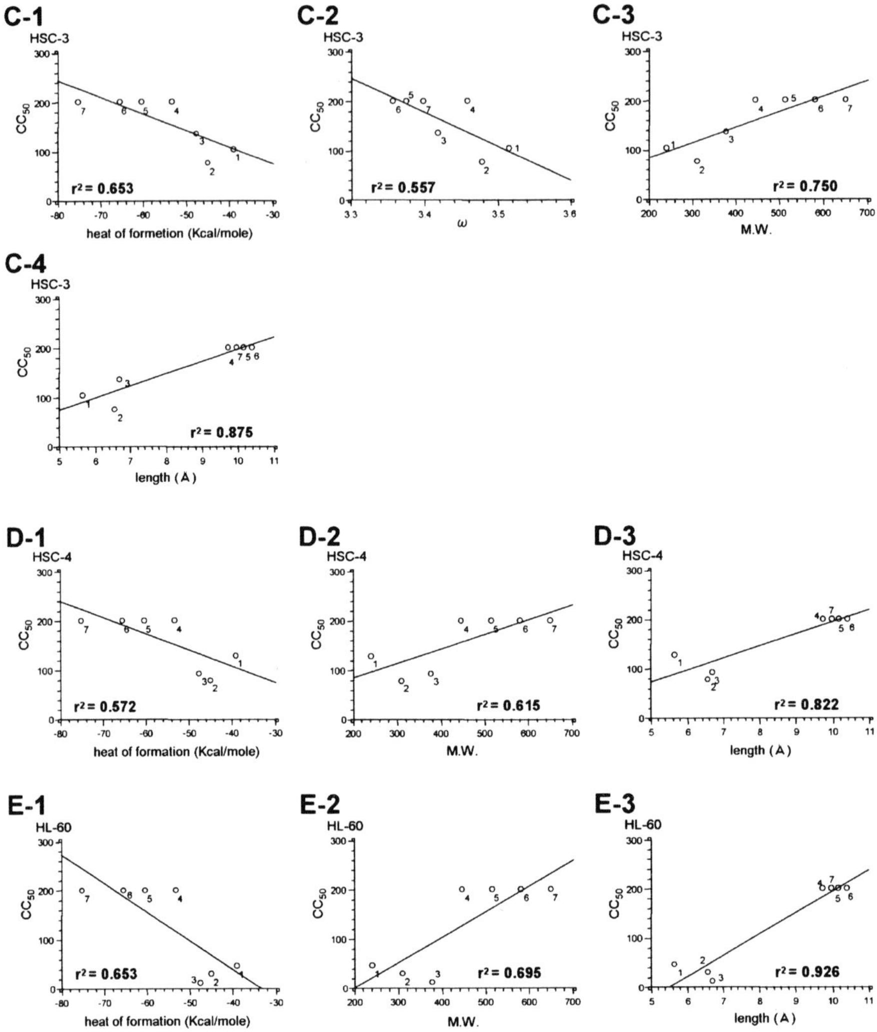

Fig. 20 (continued)

5
Conclusion

Evaluation of the toxicity for each chemical is an important research project. This review clarified the usefulness of calculation chemistry using the molecular orbital method, based on our own experimental data. QSAR analysis suggests the importance of selecting the best descriptor for each compound and each type of cell. A combination of CONFLEX/PM3 methods seems to be more advantageous than the PM3 method alone in determining

Table 11 Relationship coefficients between CC_{50} and each chemical descriptor of vitamin K_2 derivatives in five different cell lines

Chemical descriptor	Regression analysis (r^2) HSC-2	HSC-3	HSC-4	$HepG_2$	HL-60
Heat of formation	0.658	0.653	0.572	0.547	0.653
Electron affinity	0.383	0.390	0.407	0.226	0.418
Ionization potential	0.416	0.372	0.087	0.864	0.139
E_{HOMO}	0.416	0.372	0.087	0.864	0.139
E_{LUMO}	0.383	0.390	0.407	0.226	0.418
η	0.280	0.240	0.023	0.755	0.054
χ	0.510	0.468	0.163	0.883	0.226
ω	0.582	0.557	0.332	0.694	0.392
η–χ	No correlation	No correlation	No correlation	Good correlation	No correlation
M.W.	0.759	0.750	0.615	0.699	0.695
Length	0.874	0.875	0.822	0.664	0.926
log *P*	0.758	0.749	0.614	0.696	0.695

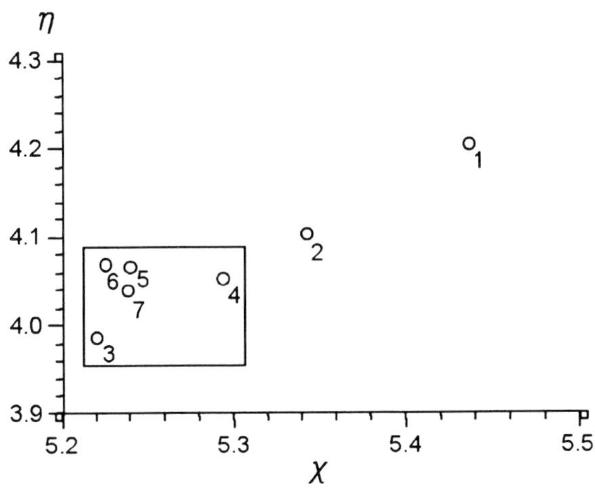

Fig. 21 η–χ activity diagram of vitamin K_2 derivatives in $HpeG_2$ cells. The *box* surrounds compounds for which no cytotoxicity was detected

the biomimetic structure with QSAR. Descriptors determined by the CONFLEX/PM3 method were found to be useful tools for the analysis of biological activity of test compounds. There was a good correlation between absolute chemical hardness and biological activity (CC_{50} in this article) of heterocyclic compounds. The concept of hardness is applicable to the estimation of the structure of more active compounds. This concept can be regarded as a new QSAR approach, and is expected to be applied in drug design.

References

1. Yoneda F, Nitta Y (1964) Chem Pharm Bull Jap 12:1264
2. Yamamoto K, Sawanishi H, Miyamoto K (1989) Biol Pharm Bull 210:356
3. Fujii A, Bush JH, Shores KE, Johnson RG, Garascia RJ, Cook ES (1977) J Pharm Sci 66:844
4. Hansch C, Fujita T (1964) J Am Chem Soc 86:1616
5. Hansch C, Maloney PP, Fujita T, Muir RM (1962) Nature 194:178
6. Bingham RC, Dewar MJ, Donald HL (1975) J Am Chem Soc 97:1294
7. Dewar MJ, Thiel W (1977) J Am Chem Soc 99:4899
8. Dewar MJ, Zoebisch EV, Healy EF, Stewart JJ (1985) J Am Chem Soc 107:3902
9. Stewart JJ (1989) J Comp Chem 10:209
10. Hohenberg P, Kohn W (1964) Phys Rev 136:B864
11. Kohn W, Sham LJ (1965) Phys Rev 140:A1135
12. Koppel LA, Bulrk P, Koppel I, Leito I, Sonoda T, Mishima M (2000) J Am Chem Soc 122:5114
13. Fujisawa S, Kadoma Y, Ishihara M, Atsumi T, Yokoe I (2004) J Liposome Res 14:39
14. Parr RG, Paarson RG (1983) J Am Chem Soc 105:7512
15. Sakar U, Parthasarathi R, Subramanian V, Chattaraji PK (2004) Internet Electronic J Mol Design IECMD 2004:1–24
16. Kobayashi S, Tanaka A, Sameshima K (1999) Chem Chem Indus 52:41
17. Kobayashi S, Sameshima K, Ishii Y, Tanaka A (1995) Chem Pharm Bull 43:1780
18. Kobayashi S, Hamashima H, Kurihara M, Miyata N, Tanaka A (1998) Pharm Bull 46:1108
19. Liu WK, Ho CK, Cheung KF, Lie PL, Ye WC, Che CT (2004) Eur J Pharmacol 498:71
20. Bennett JE (1996) Antimicrobial agents: antifungal agents. In: Goodman LS, Limbird LE, Milinoff PB, Ruddon RW, Gilman AG (eds) Goodman & Gilman's: The pharmacological basis of therapeutics, 9th edn. MacGraw-Hill, New York, pp 1175–1190
21. Bellina F, Cauteruccio S, Rossi R (2007) Tetrahedron 63:4571
22. Anil KC, Jayarama SB, Salimath BP, Rangappa KS (2007) Invest New Drugs 25:343
23. Takekawa F, Nagumo T, Shintani S, Hashimoto K, Kikuchi H, Katayama T, Ishihara M, Amano O, Kawase M, Sakagami H (2007) Anticancer Res (in press)
24. Kawase M, Saito S, Kurihara T (2001) Chem Pharm Bull 49:461
25. Kawase M, Saito S (2000) Chem Pharm Bull 48:410
26. Ishihara M, Takayama F, Toguchi M, Nakano K, Yasumoto E, Nakayachi T, Satoh K, Sakagami H (2000) Anticancer Res 20:4307
27. Ishihara M, Sakagami H (2003) Anticancer Res 23:2549
28. Ishihara M, Yokote Y, Sakagami H (2006) Anticancer Res 26:2883
29. Motohashi N, Mitscher LA, Meyer R (1991) Med Res Rev 11:239
30. Hara K, Okamoto M, Aki T, Yagita H, Tanaka H, Mizukami Y, Nakamura H, Tomoda A, Hamasaki N, Kang D (2005) Mol Cancer Ther 4:1121
31. Motohashi N (1982) Yakugaku Zasshi 102:646
32. Iwata A, Yamaguchi T, Sato K, Izumi R, Tomoda A (2003) Tohoku J Exp Med 200:161
33. Blank B, Baxter LL (1968) J Med Chem 11:807
34. Wadkins RM, Houghton PJ (1993) FEBS Lett 322:1
35. Klabunde T, Petrassi HM, Oza VB, Raman P, Kelly JW, Sacchettini JC (2000) Nat Struct Biol 7:312
36. Moosmann B, Skutella T, Beyer K, Ehl C (2001) Biol Chem 382:1601
37. Suzuki F, Hashimoto K, Kikuchi H, Kawase M, Motohashi N, Sakagami H (2007) Anticancer Res (in press)

38. Motohashi N (1993) Yakugaku Zasshi 103:364
39. Pippel DJ, Mapes CM, Mani NS (2007) J Org Chem 72:5828
40. Shimizu M, Hiyama T (2006) Angew Chem Int Ed Engl 45:5432
41. Kirk KL (2006) J Fluor Chem 127:1013
42. Singh SK, Vobbalareddy S, Kalleda SR, Casturi SR, Mullangi R, Ramanujam R, Yeleswarapu KR, Iqbal J (2006) Bioorg Med Chem Lett 16:3921
43. Kawase M, Miyamae H, Kurihara T (1998) Chem Pharm Bull 46:749
44. Kawase M, Saito S, Kurihara T (2000) Chem Pharm Bull 48:1338
45. Das U, Kawase M, Sakagami H, Ideo A, Shimada J, Molnar J, Dimmick JR (2007) Bioorg Med Chem 15:3373
46. Nawrot-Modranka J, Nawrot E, Graczyk J (2006) Eur J Med Chem 41:1301
47. Horton DA, Bourne GT, Smythe ML (2003) Chem Rev 103:893
48. Sabitha G (1996) Aldrichim Acta 29:15
49. Wang B, Yang Z-Y, Li T (2006) Bioorg Med Chem 14:6012
50. Havsteen BH (2002) Pharmacol Ther 96:67
51. Kawase M, Tanaka T, Kan H, Tani S, Nakashima H, Sakagami H (2007) In Vivo 21 (in press)
52. Murray RDH, Mendez J, Brown SA (1982) The natural coumarins: occurrence, chemistry and biochemistry. Wiley, New York, pp 282–289
53. Paya M, Goodwin PA, de las Heras, Hoult RS (1994) Biochem Pharmacol 48:445
54. Maucher A, Angerer E (1994) J Cancer Res Clin Oncol 120:502
55. Chu C-Y, Tsai Y-Y, Wang C-J, Lin W-L, Tseng T-H (2001) Eur J Pharmacol 416:25
56. Kawase M, Sakagami H, Hashimoto K, Tani S, Hauer H, Chatterjee SS (2003) Anticancer Res 23:3243
57. Roy D, Palangat M, Chen CW, Thomas RD, Colerangel J, Atkinson A, Yan ZJ (1997) J Toxicol Environ Health 50:1
58. Sakai I, Hashimoto S, Yoda M, Hida T, Ohsawa S, Nakajo S, Nakaya K (1994) Biochem Biophys Res Commun 205:1305
59. Kameda T, Miyazawa K, Mori Y, Yuasa T, Shiokawa M, Nakamura Y, Mano H, Hakeda Y, Kameda A, Kumegawa M (1996) Biochem Biophys Res Commun 220:515
60. Shibayama-Imazu T, Sonoda I, Sakairi S, Aiuchi T, An W-W, Nakajo S, Itabe H, Nakaya K (2006) Apoptosis 11:1535
61. Sakagami H, Hashimoto K, Suzuki F, Ishihara M, Kikuchi H, Katayama T, Satoh K (2007) Anticancer Res (in press)

Top Heterocycl Chem (2009) 16: 135–152
DOI 10.1007/7081_2007_096

Published online: 22 December 2007

The Chemistry of Bioactive Mesoionic Heterocycles

Masami Kawase[1] (✉) · Hiroshi Sakagami[2] · Noboru Motohashi[3]

[1]Faculty of Pharmaceutical Sciences, Matsuyama University, 4-2 Bunkyo-cho, Matsuyama-shi, Ehime 790-8578, Japan
kawase@cc.matsuyama-u.ac.jp

[2]Meikai University School of Dentistry, 1-1 Keyakidai, Sakado, Saitama 350-0238, Japan

[3]Meiji Pharmaceutical University, 2-522-1 Noshio, Kiyose, Tokyo 204-8588, Japan

Abstract This review focuses on the latest developments in biologically interesting mesoionic compounds, such as sydnones, sydnonimines, 1,2,3,4-oxatriazolium-5-aminates, and 1,3,4-thiadiazolium-2-aminides, mainly since 2000.

Keywords Mesoionic · Sydnones · Sydnonimines · 1,2,3,4-Oxatriazolium-5-aminates · 1,3,4-Thiadiazolium-2-aminides

1 Introduction

Mesoionic compounds have been known for more than a century [1]. The term "mesoionic" (mesomeric + ionic) has been generally restricted to a class of five-membered heterocycles in which both the positive and negative charges are delocalized. Mesoionic compounds cannot be satisfactorily represented by any one polar structure but only as hybrids of polar structures. This implies a certain degree of aromatic character in the positively charged ring, which is counterbalanced by a negatively charged exocyclic atom. However, the term mesoionic had been used to imply a variety of heterocyclic systems, such as heteropentalenes and six-membered heteroaromatic betaines.

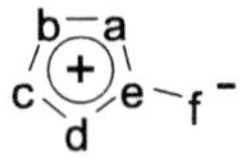

(1)

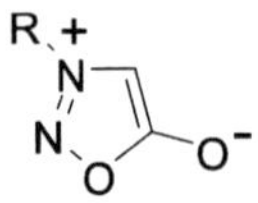

Sydnones (2)

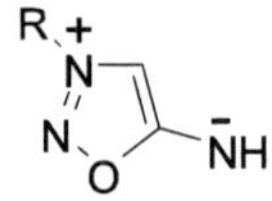

Sydnonimines (3)

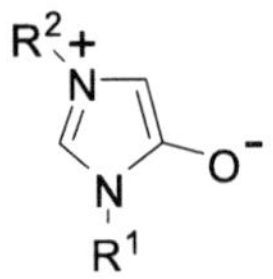

1,3-Diazolium-4-olates (4)

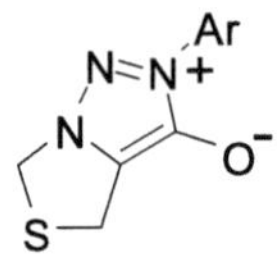

1,2,3-Triazolium-4-olates (5)

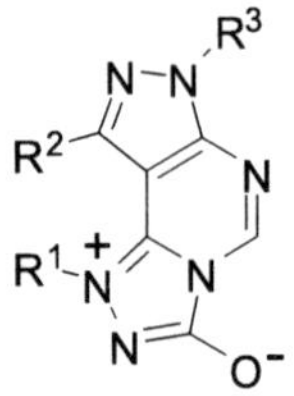

1,2,4-triazolium-3-olates (6)

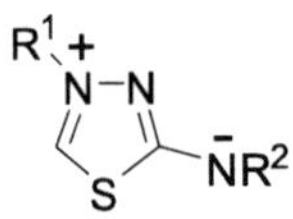

1,3,4-Thiadiazolium-2-aminides (7)

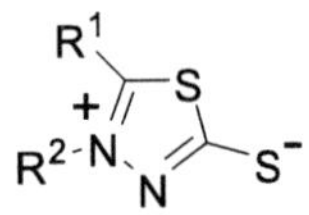

1,3,4-Thiadiazolium-2-thiolates (8)

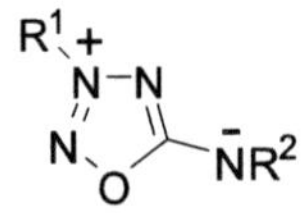

1,2,3,4-Oxatriazolium-5-aminides (9)

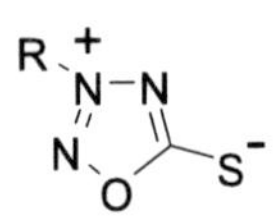

1,2,3,4-Oxatriazolium-5-thiolates (10)

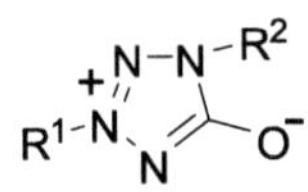

1,2,3,4-Oxatriazolium-5-olates (11)

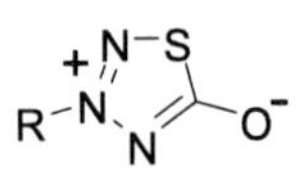

1,2,3,4-Thiatriazolium-5-olates (12)

A consensus is growing that the term mesoionic should be restricted to five-membered heterocycles. This restriction excludes betaines, five-membered heterocyclic *N*-oxides, and heteropentalenes, but it does not exclude five-membered rings fused to other heterocyclic systems. In accordance with the definition, mesoionic compounds are represented by formulae of general type (**1**), where it is noted that a–f are commonly C, N, O, S, or Se. Bieber has calculated that this combination of atoms could give possibly 228 five-membered mesoionic ring systems [2]. The number of mesoionic compounds actually synthesized to date may be less than 100.

Several mesoionic compounds have shown a wide spectrum of biological activity, and their biological activities up to 1980 have been summarized in several reviews [3–5]. Sydnones (**2**) have received the most attention and the following activities were of interest: antidepressant, antibacterial, analgesic, anti-inflammatory, antimalarial, and insecticidal [3]. The following claims for the biological activity of other mesoionic compounds have also been made: sydnonimines (**3**) as hypotensive agents; 1,3-diazolium-4-olates (**4**) as anthelmintics; 1,2,3-triazolium-4-olates (**5**) as herbicides; 1,2,4-triazolium-3-olates (**6**) as analgesics and anti-inflammatory agents; 1,3,4-thiadiazolium-2-aminides (**7**) as sedatives; 1,3,4-thiadiazolium-2-thiolates (**8**) as antibacterial agents; and 1,2,3,4-oxatriazolium-5-aminides (**9**), 1,2,3,4-oxatriazolium-5-thiolates (**10**), 1,2,3,4-oxatriazolium-5-olates (**11**), and 1,2,3,4-thiatriazolium-5-olates (**12**) as pesticides, herbicides and hypotensive agents [4, 5].

Their charged, yet net neutral electrical character and their chemical properties are valuable assets in their potential usefulness as medicinal agents. The structural characteristics of mesoionic compounds, which contain distinct regions of positive and negative charges associated with a polyheteroatomic system, enable them to cross the cellular membranes and interact strongly with biomolecules. However, their potential biological applications have been described for the limited mesoionic compounds [3], because most mesoionics are generally unstable and are not appropriate for biological assay in vitro. Therefore, their biological assessment is limited to several mesoionic compounds, such as sydnones (**2**), sydnonimines (**3**), 1,2,3,4-oxatriazolium-5-aminates (**9**), and 1,3,4-thiadiazolium-2-aminides (**7**).

This review will discuss mesoionic compounds with the main focus on the biologically active sydnones (**2**), sydnonimines (**3**), 1,2,3,4-oxatriazolium-5-aminates (**9**), and 1,3,4-thiadiazolium-2-aminates (**7**).

2 Sydnones

Sydnones were the first mesoionic molecules subjected to detailed study, and are named after the University of Sydney where they were first prepared by Earl and Mackney in 1935. Sydnones are generally prepared by dehydrative cyclization of *N*-nitrosamino acids, which are synthesized by N nitrosation of amino acids [6]. Sydnone derivatives have been extensively studied for their biological activities, such as antibacterial, antitumor, antimalarial, anti-inflammatory, and antihypertensive agents [3].

Mesocarb (sydnocarb) (**13**) and Feprosidnine (sydnofen) (**14**) are stimulants developed in Russia in the 1970s. Mesocarb is sold as a drug in Russia. However, it is almost unknown in Western countries and is not used in medicine. It has been shown to act as a dopamine reuptake inhibitor, antidepressant, and anticonvulsant [7, 8].

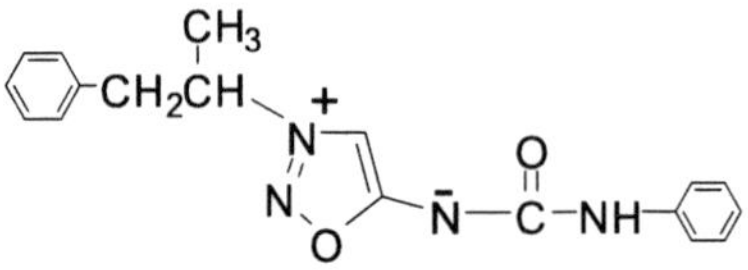

Mesocarb (**13**)
(sydnocarb)

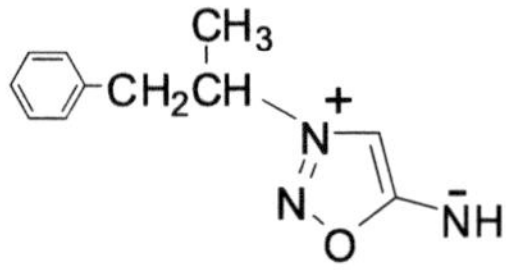

Feprosidnine (**14**)
(sydnofen)

In 1992, 1-(4′-substituted 3′-nitrophenyl)sydnones were synthesized and evaluated for their anticancer activity. The 4′-chloro (SYD-1) (**15**) and 4′-pyrrolidino (**16**) compounds significantly prolonged the survival times of sarcoma 180, Ehrlich carcinoma, and fibrous histiocytoma tumor-bearing mice [9]. 1-(4′-Substituted 3′-nitrophenyl)sydnones (**15–19**) were also tested for antitumor activity against a three-cell-line panel of MCF7 (breast), NCI-H460 (lung,) and SF-268 (CNS). 1-(4′-Fluoro-3′-nitrophenyl)sydnone (**19**) is shown to be more active than SYD-1 [10].

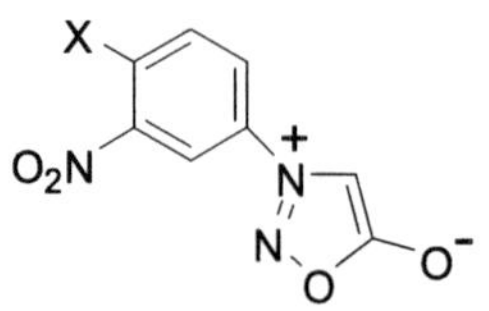

SYD-1 (**15**) X=Cl, **16**: X=, **17**: X=, **18**: X=, **19**: X=F

The antitumor effect of SYD-1 might be related to the enzyme-linked function of mitochondria, since it has been shown to depress the efficiency of electron transport and oxidative phosphorylation [11]. Thus, SYD-1 is now one of the most attractive compounds among sydnone derivatives which show antitumor activity.

The antimicrobial activity of 3-arylsydnones has been reported. 3-Arylsydones are known to be less toxic and more active than 3-alkylsydnones. Thus, several series of 3-phenylsydnone derivatives (**20**) (R^1 = H, EtO, MeCO, HO_2C, EtO_2C) and (**21**) (R^2 = Ph, 2-furyl, 4-ClC_6H_4, 4-$O_2NC_6H_4$, 2-Cl-5-$O_2NC_6H_3$), bearing a substituted phenyl ring at the 3-position, have been synthesized starting from the corresponding *N*-(carboxyphenyl)glycines. Nitrosation of the latter followed by cyclodehydration upon refluxing in acetic

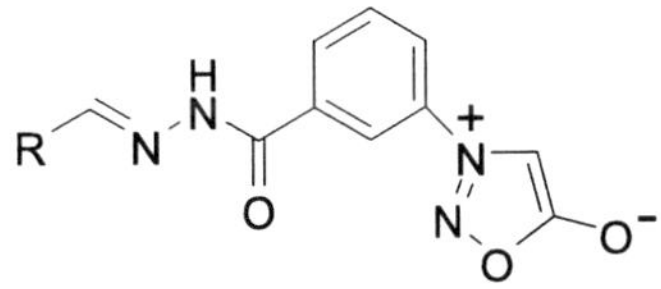

(20)

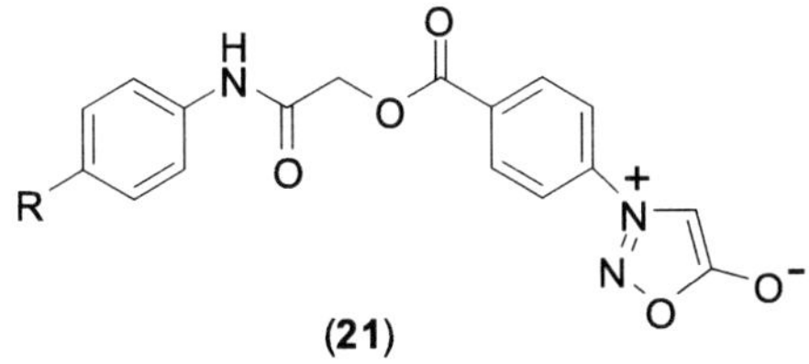

(21)

anhydride afforded 3-(carboxyphenyl)sydnones. The synthesized compounds were tested for their antibacterial activities against both Gram-positive and Gram-negative organisms. Some of the test compounds exhibited high activity; among them, **21** (R^2 = 2-Cl-5-$NO_2C_6H_3$) is considered to be a lead compound possessing a broad spectrum of antibacterial activity [12, 13].

The hybrid molecules, comprised of sydnones and other heterocyclic compounds, constitute one of the more promising approaches for the design of bioactive compounds. 3-Aryl-4-formylsydnones (**22**) are attractive starting compounds for several 4-heterocycle-substituted sydnones, such as thiazolidinone (**23**) and thiazoline (**24, 25**) [14, 15], imidazoles (**26**) [16], thiadiazolines (**27**), thiadiazoles (**28**) [17], and indoles (**29**) [18]. Some sydnonyl-substituted thiazolidinone (**23**) and thiazoline (**30**) derivatives exhibited a potent 1,1-diphenyl-2-picrylhydrazyl (DPPH) radical scavenging activity, comparable to that of vitamin E [14, 15].

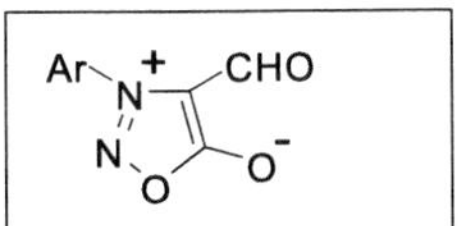

4-Formylsydnones (22)

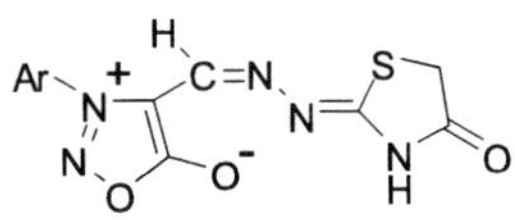

(23)

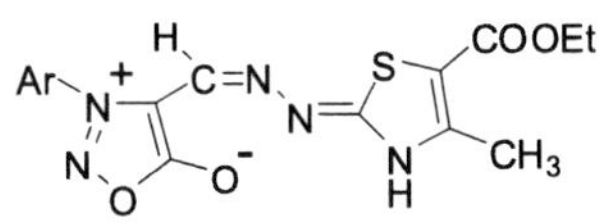

(24)

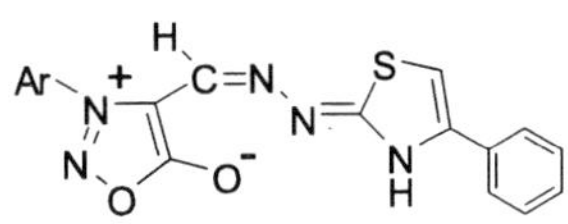

(25)

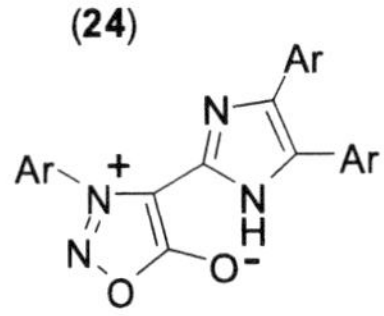

(26)

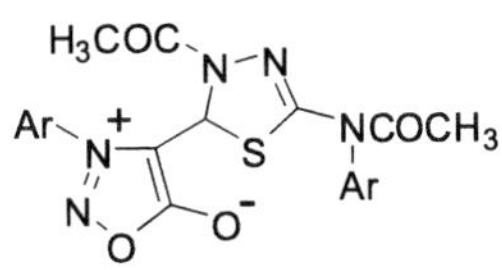

(27)

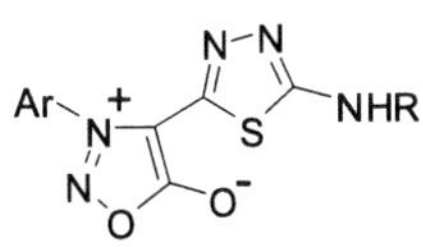

(28)

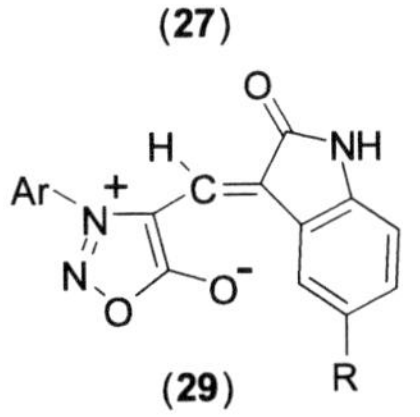

(29)

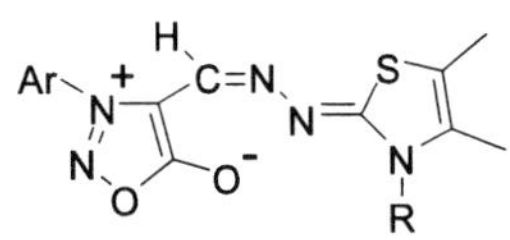

(30)

3
NO-Releasing Compounds

The biological role of nitric oxide (NO) has been extensively studied and several aspects of its roles in physiology and pathophysiology clarified. NO is a messenger molecule with several biological activities including inflammation, immunoregulation, and neurotransmission [19]. NO is also known to exert vasodilatory, bronchodilatory, and antiplatelet effects, and quantitative or functional NO deficiency has been implicated in various cardiovascular and airway diseases. Thus, replacement of endogenous NO with exogenous drug-derived NO has been a common medical practice for many years, and these pharmacological compounds present an attractive option in the treatment of cardiovascular diseases. NO donors, which release NO either spontaneously or tissue-dependently, represent a way of increasing NO [20]. Direct NO donors such as organic nitrates have been used for many years in cardiovascular therapeutics, but they exhibit limitations such as the development of nitrate tolerance and their inability to improve patient outcome among others. Cofactor-dependent and enzymically active prodrugs, such as molsidomine, mesoionic oxatriazoles, *S*-nitrosothiols, and diazeniumdiolates, and bifunctional NO donors such as NO-aspirin are potential alternatives to conventional nitrates [21].

It is noteworthy that NO has been implicated in both anti- and proapoptotic pathways depending on the cell type and conditions [22]. At high concentrations, NO induces cell death during ischemic injury or as a consequence of neurodegenerative diseases. On the other hand, NO seems to be cytoprotective in neuronal cell lines at lower concentrations. Recent findings showed that NO has a prosurvival effect on the interstitial cells of Cajal in the mouse stomach [23].

There are a number of reviews published in the field of NO donors [24–28]. This review will therefore focus on recent advances in NO-donor mesoionic compounds, sydnonimines (**3**), and 1,2,3,4-oxatriazolium-5-aminides (**9**), which have drawn the most interest today [26, 29].

4
Sydnonimines

Sydnonimines (**3**) are an important class of NO donors and this type of compound was first synthesized by Ohta [30] and Brookes [31] in 1957. The sydnonimine class of NO donors is largely represented by molsidomine (**31**), which has been used since the 1970s in European countries for the treatment of stable angina pectoris. Angina is a chest pain, pressure, and discomfort caused by insufficient blood flow to the heart muscle. However, molsidomine (**31**) itself is only poorly vasoactive in vitro. The active metabolite of molsidomine (**31**) is 3-morpholinosydnonimine (SIN-1) (**32**), which spontaneously releases NO and induces vasodilation via the guanylyl cyclase-3′,5′-cyclic guanosine monophosphate pathway.

Sydnonimines are unstable if not protonated to a salt or acylated to stable *N*-acylimino derivatives. At neutral conditions sydnonimines which are unsubstituted at the exocyclic nitrogen undergo a ring-opening reaction (Scheme 1). This instability may explain why sydnonimines and sydnones differ in their ability to release NO.

The proposed mechanism of NO release is shown in Scheme 1. Molsidomine (**31**) is converted to SIN-1 (**32**) by the action of liver esterases. SIN-1 (**32**) undergoes oxidation in the presence of oxygen or, in vivo, possibly by redox enzyme such as cytochrome *c* or by ferric myoglobin to release NO through a radical cation.

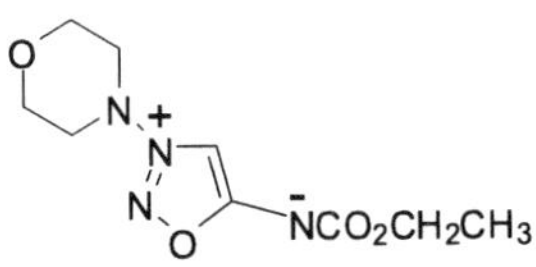

Molsidomine (**31**)

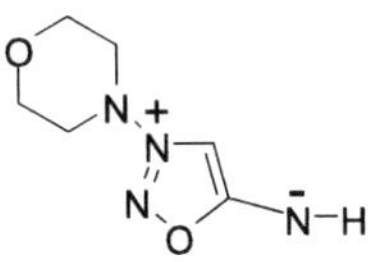

Linsidomine (SIN-1) (**32**)

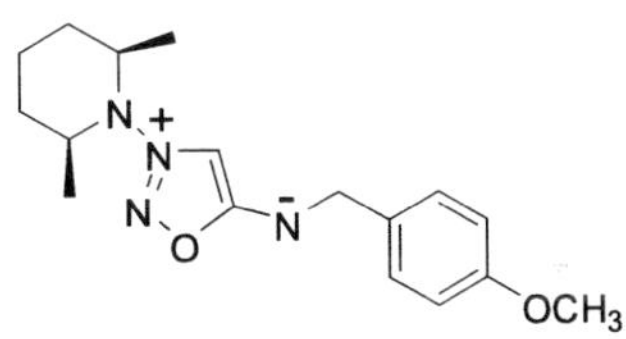

Pirsidomine (CAS 936) (**33**)

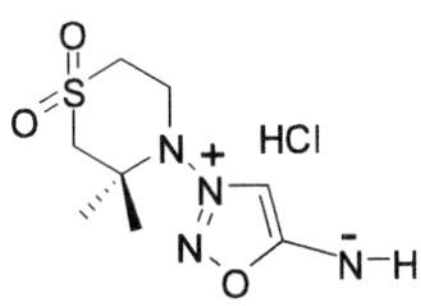

C89-4095 (**34**)

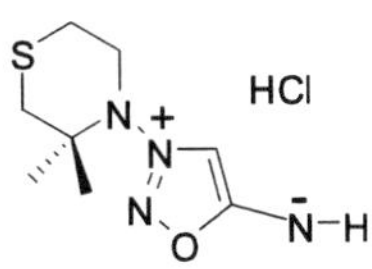

C4144 (**35**)

At alkaline or physiological pH, SIN-1 (**32**) undergoes rapid hydrolysis to yield the ring-opened form from which the NO radical is released. The mechanism of the inhibitory action of SIN-1 (**32**) on human neutrophil degranulation has been studied. SIN-1 (**32**) releases both NO and superoxide, which together form peroxynitrite. The NO-deficient metabolite of SIN-1, SIN-1C (**33**), inhibited human neutrophil degranulation in a concentration-dependent manner similar to that of SIN-1 (**32**) and reduced the FMLP-induced increase in the intracellular free calcium. C88-3934 (**36-1**), another NO-deficient sydnonimine metabolite, also inhibited human neutrophil degranulation. Thus, SIN-1 (**32**) inhibits human neutrophil degranulation in a cGMP-, NO-, and peroxynitrite-independent manner, probably due to the formation of more stable active metabolites such as SIN-1C (**33**) [32]. SIN-1 (**32**) activates soluble guanylyl cyclase (sGC), which is a cytosolic enzyme that catalyzes the formation of the second messenger cGMP from Mg^{2+} GTP. The nitrovasodilator action of SIN-1 (**32**) is related to the activation of sGC [33].

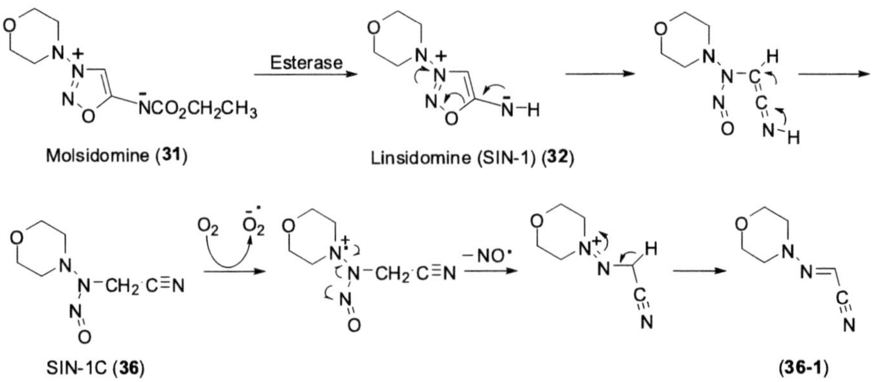

Scheme 1 Activation mechanism of molsidomine (31)

To achieve the site-specific delivery of SIN-1 (**32**), it is converted into stable prodrugs which can be specifically recognized by certain targets. Glucose conjugates (**37**), covalently coupled to SIN-1 (**32**) via a carbamate linkage at the anomeric position, are designed and are hydrolyzed by glycosidases to release NO [34]. Two vitamin E conjugates (**38**) and (**39**) were also prepared in order to achieve enzymatic activation prior to the release of NO [35]. These NO prodrugs may be used as enzyme-activated NO donors in biomedical research.

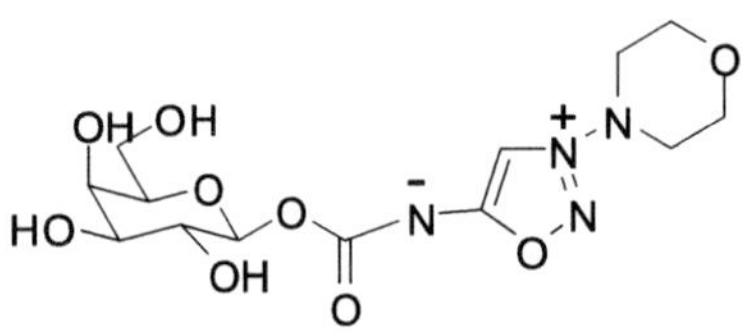

(37)

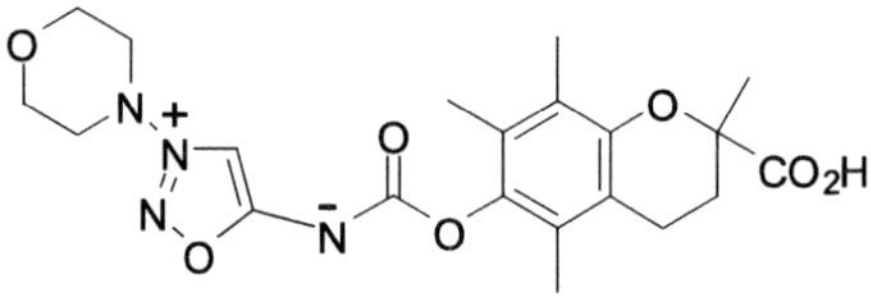

(38)

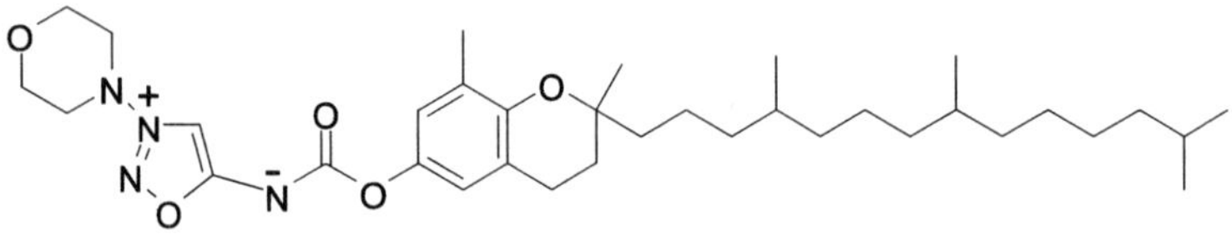

(39)

Recently, molsidomine (**31**) and SIN-1 (**32**) have been extensively utilized as a probe to elucidate the biomedical mechanisms of NO and/or peroxynitrite. Molsidomine has a high bioavailability, and the long-lasting duration of its action possibly stimulates the permeability through the blood–brain barrier. Molsidomine does not display any side effects at doses displaying an antiamnestic action during recognition memory tasks in the rat [36]. Therefore, molsidomine was used as a probe to investigate the role of memantine, an *N*-methyl-D-aspartate (NMDA) receptor antagonist, in recognition memory [37]. This study shows the efficacy of the coadministration of memantine and molsidomine, suggesting the involvement of the nitrergic system in facilitating the memantine effect on recognition memory.

Cardiovascular disease resulting from atherosclerosis and thrombosis remains a major cause of death and disability among adults. Therefore, there is a growing interest in therapeutic strategies that may lead to the selective and safe removal of macrophages within the atherosclerotic plaque. Molsidomine (**31**) eliminated macrophages in atherosclerotic plaques by the induction of endoplasmic reticulum stress, probably via inhibition of protein synthesis [38]. NO donors such as molsidomine are widely used by patients with coronary artery disease to relieve the symptoms of ischemia evoked by atherosclerosis and can be administered safely for many years [39]. Thus, NO donors are expected to offer new opportunities for long-term macrophage depletion in atherosclerotic plaques.

The loss of SIN-1-induced viability in lung epithelial cells was prevented by licorice compounds, glycyrrhizin and 18β-glycyrrhetinic acid, through suppressing the mitochondrial permeability transition, leading to the release of cytochrome *c* and activation of caspase-3. The licorice compounds seem to exhibit a protective effect against lung cell injury, which is mediated by the increased formation of nitrogen species [40].

To elucidate the mechanisms of NO-induced cellular stress, the effects of SIN-1 (**32**) on neuroblastoma cells were examined. SIN-1 induced a transient decline in ATP levels and the delayed loss of cell viability, with no significant increase in caspase-3 activity or DNA laddering. NO was suggested to be a potent toxin independent of peroxynitrite formation [41].

SIN-1 inhibited the MG132 (carbobenzoxyl-leucinyl-leucinyl-leucinal, a protease inhibitor)-induced change in the mitochondrial membrane permeability, radical oxygen species (ROS) formation, and the decrease in GSH content in PC12 cells [42]. SIN-1 induced an improvement in the sublingual microcirculatory blood flow through the improvement of intestinal perfusion (ileal–arterial PCO_2 gap) and oxygenation (serosal μPO_2). Sepsis is a disease of the microcirculation, in which NO donors may be used as a drug to improve the microvascular blood flow by increasing the driving pressure at the entrance of the microcirculation [43].

NO acts as a sympathoinhibitory substance within the central nervous system [44]. Using SIN-1 (**32**), central NO administration has been shown to enhance acute hypoxic pulmonary vasoconstriction (HPV) [45], and accentuation of the HPV by NO is exacerbated in spontaneously hypertensive rats (SHR) compared with Wister–Kyoto (WKY) rats [46].

NO donors such as SIN-1 were demonstrated to inhibit the 5-hydroxytryptamine (5-HT) transporter (SERT) [47]. As inhibitors of SERT are successfully used in pulmonary hypertension [48], NO donors could have a therapeutic future as vasodilators in pulmonary hypertension.

SIN-1 (**32**) potently increased the extracellular concentration of kynurenic acid, an endogenous glutamate receptor antagonist, in cortical slices [49]. In contrast, it did not alter the activity of kynurenine aminotransferases (KAT I and II) and kynurenic acid biosynthetic enzymes. Thus, SIN-1 may play a neuroprotective role through the enhanced formation of kynurenic acid, which is a neuroinhibitory compound.

The effect of SIN-1 on neonatal mouse cardiomyocytes was examined, and gene expression was evaluated using RNA labeled and hybridized to cDNA microarrays [50, 51]. SIN-1 induced significant changes in five caspases: 1, 6, 8, 11, and 14. In contrast, there were no changes in three genes involved in autophagy, namely beclin 1, Atg5, and Atg12. It is shown that NO is not involved in increased beclin 1 expression in ischemia/reperfusion injury in the heart, and therefore autophagy would not be involved in the hearts of patients with heart failure.

SIN-1 induced higher antiproliferation activity in BCR-ABL$^+$ leukemia K562 cells when compared to normal hemopoietic cells [52]. NO-mediated apoptosis is shown to be inhibited in myeloid leukemia cells by the elevation of intracellular iron.

Using SIN-1, the relationship between peroxynitrite and apoptotic signaling was studied [53]. Peroxynitrite, generated by the reaction of NO and mitochondrially derived superoxide (O_2^-), is a potent oxidant produced after brain trauma. Peroxynitrite can cause lipid peroxidation, DNA fragmentation, and protein nitration [54]. SIN-1, the peroxynitrite donor, blocked staurosporine-induced caspase-3 activation and its downstream effects including PARP-1 [poly(ADP-ribose)polymerase-1] cleavage and phosphatidylserine inversion. It is suggested that peroxynitrite could inhibit caspase-3-mediated apoptosis. However, cell death was not inhibited. It is concluded that cellular demise could be effectively shifted to a caspase-independent pathway through the exposure to peroxynitrite.

5
1,2,3,4-Oxatriazolium-5-aminides

1,2,3,4-Oxatriazolium-5-aminides (**9**) are structurally related to sydnonimines (**3**). As with sydnonimines, the 5-imine derivatives are stable only as salts or in their N-acylated forms. 3-Cyclohexyl-1,2,3,4-oxatriazole-5-imine hydrochloride was first synthesized by Finnegan and Henry in 1965 [55]. The biological activity was reported by Masuda et al. in 1971 [56]. Oxatriazole-5-imines (**9**) have been reported to be an important class of NO donors and potent antiplatelet, fibrinolytic, thrombolytic, and bronchiectatic agents [57]. However, an extensive investigation of the NO-releasing and other biologi-

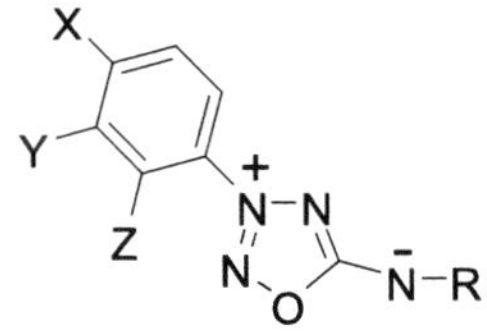

	X	Y	Z	R
GEA3162	Cl	Cl	H	H
GEA3175	H	Cl	CH_3	SO_2-C_6H_4-CH_3
GEA5024	H	Cl	CH_3	H
GEA5624	H	Cl	CH_3	$CONHCH_2C{\equiv}CH$

cal properties of the mesoionic oxatriazole derivatives has been started by researchers at the GEA Farmaceutisk Fabrik, Copenhagen, who synthesized a large number of this class of compounds.

The GEA compounds (GEAC) constitute a new class of NO donors, some of which stimulate selectively guanylate cyclase binding in either platelets, leukocytes, or lung tissues. In consequence, some GEAC are potent antiplatelet, fibrinolytic, thrombolytic, or broncholytic agents, both in vitro and in vivo. GEAC synergize with prostacyclin in their thrombolytic actions. GEAC also suppress the release of histamine and leukotriene B4, and prevent the degranulation of granulocytes. Some of the newly synthesized GEAC are likely to be potential candidates for treating either thrombotic or asthmatic disorders. Among the GEAC, GEA3162 and GEA3175 are the most studied compounds as NO donors [58].

The proposed mechanism of NO release is shown in Scheme 2, which is related to that postulated for sydnonimines (Scheme 1). Hydrolysis of GEAC with water yields acyclic nitroso semicarbazides (**40**) without enzymatic cleavage. The intermediates (**40**) release NO by an oxidative or thiol-mediated mechanism. However, the exact mechanism is not fully understood [59].

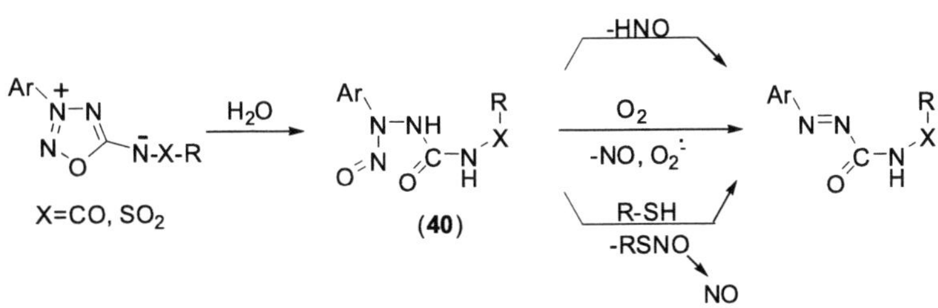

Scheme 2 Proposed mechanism of NO release from mesoionic 1,2,3,4-oxatriazolium-5-aminides

GEA3162 and GEA3175 were compared with SIN-1 (**32**) and *S*-nitroso-*N*-acetylpenicillamine (SNAP). GEA3162, GEA3175, SIN-1, and SNAP inhibited adenosine 5′-diphosphate-induced platelet aggregation (IC_{50} values 0.18,

0.39, 3.73, and 2.12 μM, respectively). All of these four compounds induced a dose-dependent and more than fourfold increase in cyclic GMP in platelets. These compounds converted oxyhemoglobin to methemoglobin and formed a paramagnetic NO–hemoglobin complex. All except GEA3175 formed nitrite and nitrate in phosphate buffer. The release of NO and NO_2 by GEA3175 was increased 140-fold in the presence of human plasma, as analyzed by ozone chemiluminescence. These results suggest that the mesoionic 3-aryl-substituted oxatriazole-5-imine derivatives GEA3162 and GEA3175, as well as SIN-1 and SNAP, release NO [60].

GEA3162 and GEA3175 inhibited mononuclear cell proliferation dose-dependently and more potently than SNAP [61]. GEA3162 and GEA3175 also inhibited proliferative responses in human lymphocytes in a cGMP-independent manner and may have an immunosuppressant action [62]. Thus, NO-releasing compounds have immunosuppressive actions which offer therapeutic possibilities.

The effects of the two NO donors GEA3175 and SNAP were studied on prostacyclin production in lipopolysaccharide (LPS)-stimulated human umbilical vein endothelial cells (HUVECs). GEA3175 (1–30 μM) inhibited LPS-induced production of prostacyclin in HUVECs dose-dependently and more potently than SNAP. NO affects the COX-2 pathway rather than having an overall effect on cyclooxygenases. NO-releasing compounds did not alter the level of COX-2 protein expression in LPS-treated HUVECs, as measured by Western blot analysis. NO donors inhibit the activity of COX-2 in human endothelial cells. A link between NO and the regulation of eicosanoid synthesis could represent an important mechanism in controlling vascular and inflammatory responses in pathophysiological states and during treatment with nitrovasodilators [63].

Current understanding of GEA3175 has been recently reviewed [64]. GEA3175 is a potent, stable, slow-release NO donor with important actions in various organ systems. It has been demonstrated to exert bronchodilatory, pulmonary vasodilatory, antiplatelet, and anti-inflammatory actions. GEA3175 offers advantages over existing drugs, such as SNAP and RS-NOs, because it is very stable in vivo and does not induce tolerance. The exact mechanism by which GEA3175 releases NO is, however, still unknown. In addition, most of the studies so far have been performed in isolated tissue preparations. Clearly, further in vivo studies involving animal models are required to clarify the safety issues and whether GEA3175 can be used for the treatment of pulmonary hypertension and/or airway diseases [64].

The NO-donating characteristics and pharmacology of 23 mesoionic oxatriazoles (MOTA) were compared with those of SIN-1, SNAP, sodium nitroprusside (NaNP), and glyceryl trinitrate (GTN). It is concluded that in vitro, the NO-donating profile of MOTA can hardly be used as a prediction measure for their pharmacological activities either in vitro or in vivo. For instance, MOTA 5-oxides were more potent thrombolytics than SIN-1,

SNAP, or NaNP. By manipulating the chemical structures of MOTA, hypotensive, antithrombotic, thrombolytic, or antiasthmatic drugs might be designed [65].

The cytopathological effects of GEA3162 and GEA3175, and a reference NO donor SIN-1, were investigated in proliferating human hematopoietic cells. GEA3162 and GEA3175 suppressed thymidine and uridine incorporation in a dose-dependent manner, reflecting the inhibition of DNA and RNA synthesis, respectively. In addition, the GEA compounds inhibited the growth of human bone marrow stem cells, CFU-GM colonies being more susceptible to the cytostatic action than BFU-E. The mesoionic oxatriazole derivatives have cytostatic action against human malignant and nonmalignant hematopoietic cells, supporting their therapeutic potential of NO-releasing and NO-inducing compounds as anticancer agents [66].

GEA3162 and GEA5624 are effective in inhibiting the growth of tumor cells and reducing blood pressure. GEA3162 is able to inhibit DNA synthesis and proliferation of aortic vascular smooth muscle cells. GEA5624 was able to inhibit mitogenesis but not proliferation. GEA3162 was found to be a more potent inhibitor of mitogenesis and cell proliferation than SIN-1 and SNAP [67].

Apoptotic cell death is regulated by NO and $ONOO^-$ in several cell types including myeloid-derived leukocytes such as neutrophils. The biological effects of NO and $ONOO^-$ have been recently reviewed [68]. GEA3162 is able to promote apoptotic cell death in human neutrophils in a caspase-dependent manner [69]. The tumor suppressor p53 is suggested to play a crucial role in apoptosis induced by oxidants. However, functional p53 is not a requirement for $ONOO^-$-mediated cell death, as GEA3162 induces mitochondrial permeability in a p53-deficient murine bone marrow cell line, Jaws II. GEA3162 activates caspases 3 and 2 which are important for apoptosis to proceed, with roles for caspases 8 and 9, and p38 MAP kinase [70].

GEA3162 can activate non-store-operated Ca^{2+} entry, but can inhibit store-operated Ca^{2+} entry pathways in rat neutrophils [71]. The dual effects of GEA3162 on the regulation of the external Ca^{2+} entry are mainly through the thiol modification of target protein(s) residing on the outside of the plasma membrane [72].

5-Hydroxytryptamine (serotonin, 5-HT) is an important messenger substance, which regulates colonic motility, secretion, or sensation by acting via the multiple receptor subtypes. Regulation of 5-HT release from the colonic mucosa may lead to new ways to control defecation or constipation in diseases such as irritable bowel syndrome (IBS). GEA3162 can suppress the 5-HT release via an action on colonic mucosa through a mechanism independent of the ODQ-sensitive cyclic GMP system or peroxynitrite generation [73].

GEA3162 was used as a NO donor in order to characterize the action of NO on rat myenteric neurons. It is suggested that NO stimulates the voltage-dependent Ca^{2+} channels in rat myenteric neurons [74].

6
1,3,4-Thiadiazolium-2-aminates

Among 1,3,4-thiadiazolium mesoionics, 4-phenyl-5-(4-nitrocinnamoyl)-1,3,4-thiadiazolium-2-phenylamine chloride (MI-D)(**42**) has been one of the most studied compounds.

In 1997, MI-D (**42**) was first reported to enhance the survival of Ehrlich and sarcoma 180 tumor-bearing mice, preventing the tumor growth [75]. MI-D (**42**) has also been demonstrated to be extremely cytotoxic to B16-F10 murine melanoma cells, when compared to fotemustine and dacarbazine reference compounds in melanoma treatment protocols describing the inhibition of tumors grown in vitro and in vivo [76, 77]. MI-D (**42**) decreased the viability and proliferation of MEL-85, SK-MEL, A2058, and MEWO cell lines in vitro, showing a considerable cytotoxic activity against these human cells. MI-D (**42**) decreased MEL-85 adhesion to laminin, fibronectin, and matrigel. The morphology and actin cytoskeleton organization of MEL-85 cells were also modified on MI-D (**42**) treatment. Therefore, MI-D (**42**) is an encouraging drug against melanoma, a tumor that is extremely resistant to chemotherapy.

The effect of a series of 4-phenyl-5-(2′-Y, 4′-X, or 4′-X-cinnamoyl)-1,3,4-thiadiazolium-2-phenylamine chlorides was evaluated on B16-F10 murine melanoma cells in vitro and the tumors resulting from implanted B16-F10 cells in C57BL/6 mice. These compounds differ from each other only at the cinnamoyl ring substituent (**42**, X=NO_2; **43**, X=OH; **44**, X=F; **45**, X=Y=F). Upon exposure of B16-F10 cells to MI-D (**42**), **44**, and **45**, all of them at the same micromolar concentration (50 μM) decreased the cell viability to 8, 50, and 22%, respectively, while compound **43** did not show any significant effect under the same conditions. However, doses as low as 10 μM MI-D were sufficient to impair the cell growth over 72 h, but for **44** and **45** the effect on

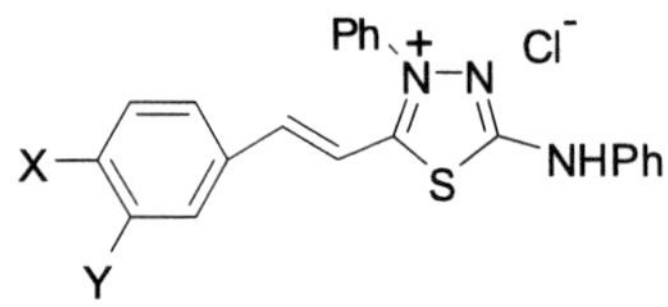

	X	Y
41	H	H
42 (MI-D)	NO_2	H
43	OH	H
44	F	H
45	F	F
46	OCH_3	H
47	Cl	H
48	Br	H
49	H	Cl
50	H	Br
51	H	OCH_3

B16-F10 proliferation was only observed at a concentration of 25 μM. Furthermore, **44** had a slightly better effect than **45** in vitro; its effect on tumor growth in vivo was not significant. MI-D (**42**) inhibited tumor growth by 77%. The greater effectiveness of MI-D (**42**) compared with **44**, **45**, and **43** against B16-F10 melanoma cells is probably due to its stronger electron-withdrawing group (NO_2), which increases the positive charge on the mesoionic ring and allows the extensive conjugation of the side chain with the exocyclic moiety. This property seems to be an important factor to determine the antitumor activity of these compounds [78].

MI-D (**42**) inhibited the oxidative phosphorylation and the respiratory chain in the liver mitochondria. ATPase activity was dramatically increased by MI-D in intact mitochondria, but inhibited in carbonylcyanide *p*-trifluoromethoxyphenylhydrazone (FCCP)-uncoupled mitochondria. These results suggest that MI-D (**42**) acts as an uncoupler, a group of compounds closely related to its structural characteristics [79].

MI-D (**42**) reduced the rate of swelling induced by valinomycin-K^+, as well as induced swelling in the presence of nigericin-K^+. Shrinkage was also affected, suggesting interference with the inner mitochondrial membrane, which would affect both the fluidity and elasticity. These alterations in the membrane fluidity are thus related to the effect of MI-D on the energy-linked functions of mitochondria [80].

MI-D (**42**) showed inhibition of superoxide anion production by macrophages and antinociceptive, antipyretic, and antiedematogenic properties, similar to nonsteroidal anti-inflammatory drugs (NSAIDs). The decrease of PGE2 level indicates that COX-2 is involved in the mechanism. Its effects on the hepatic metabolite, on O_2^- and NO production, and the possibility of in vivo models for potential antinociceptive, antipyretic, and anti-inflammatory activities were evaluated. In phorbol 12-myristate 13-acetate-stimulated macrophages, O_2^- generation was reduced (95%) by MI-D (**42**) (15 μM), whereas the production of NO was unaffected. MI-D (**42**) has a spectrum of activities similar to those of other NSAIDs, qualifying it as a potential anti-inflammatory drug [81].

Several 1,3,4-thiadiazolium-2-aminides were also tested against promastigote and amastigote forms of *Leishmania amazonensis*. Parasites were assayed with or without the drugs in axenic media, using pentamidine isethionate as a reference drug. The most active compounds were the

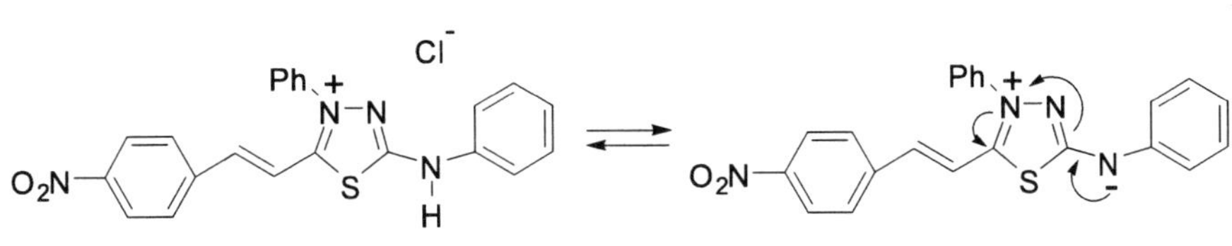

42
pH<7.0

pH>7.0

4′- and 3′-methoxy derivatives against promastigote forms, while 4′-fluor and 3′-bromo derivatives showed the highest activity against the amastigote forms. The target compounds included 3-phenyl-5-(phenylamino)-2-(2-phenylethenyl)-1,3,4-thiadiazolium chloride derivatives (**41–51**) [82].

A series of 1,3,4-thiadiazolium-2-phenylamine derivatives (**41–51**) was studied in *L. amazonensis*. The cytotoxic effects of these compounds on the host cells were investigated and the antileishmanial in vitro activity was compared with other species of *Leishmania* (*L. chagasi* and *L. braziliensis*). The compounds presented lower toxicity in murine macrophages than the reference drug pentamidine. The halogen derivatives (4-F, 4-Cl, 4-Br, and 3-Cl) (**44, 47–49**) were the most active compounds among all the species tested [83].

7 Conclusions

This review has attempted to highlight the bioactive mesoionic compounds and stimulate further development in this field. Mesoionic compounds are still considered as exotic structures even within the field of heterocyclic chemistry. Much attention has been directed toward the sydnones (**2**), sydnonimines (**3**), 1,2,3,4-oxatriazolium-5-aminates (**9**), and 1,3,4-thiadiazolium-2-aminates (**7**). In particular, their properties as NO-releasing compounds have been of interest.

References

1. Fischer E, Besthorn E (1882) Liebigs Ann Chem 212:316
2. Bieber TI (1955) Chem Ind (London) 910
3. Kier LB, Roche EB (1967) J Pharm Sci 56:149
4. Ollis WD, Ramsden CA (1976) Adv Heterocycl Chem 19:1
5. Newton CG, Ramsden CA (1982) Tetrahedron 38:2965
6. Azarifar D, Ghasemnejad-Bosra H (2006) Synthesis 1123
7. Anderzhanova EA, Afanasev II, Kudrin VS, Rayevsky KS (2000) Ann NY Acad Sci 914:137
8. Shpak AV, Appolonova SA, Semenov VA (2005) J Chromatogr Sci 43:11
9. Grynberg N, Gomes R, Shinzato T, Echevarria A, Miller J (1992) Anticancer Res 12:1025
10. Dunkley CS, Thoman CJ (2003) Bioorg Med Chem 13:2899
11. Halila GC, de Oliveira MB, Eshevarria A, Belem AC, Rocha ME, Carnieri EG, Martinez GR, Noleto GR, Cadena SM (2007) Chem Biol Interact 169:160
12. Moustafa MA, Nasr MN, Gineinah MM, Bayoumi WAH (2004) Arch Pharm (Weinheim) 337:164
13. Moustafa MA, Gineinah MM, Nasr MN, Bayoumi WAH (2004) Arch Pharm (Weinheim) 337:427
14. Shih MH, Ke FY (2004) Bioorg Med Chem 12:4633
15. Shih MH, Su YS, Wu CL (2007) Chem Pharm Bull 55:1126

16. Shih MH, Tsai CH, Wang YC, Shieh MY, Lin GL, Wei CY (2007) Tetrahedron 63:2990
17. Shih MH, Wu CL (2005) Tetrahedron 61:10917
18. Shih MH, Yeh MY (2003) Tetrahedron 59:4103
19. Moncada S, Higgs EA (1996) FASEB J 9:1319
20. Miller MR, Megson IL (2007) Br J Pharmacol 151:305
21. McGrowder D, Ragoobirsingh D, Brown P (2006) Int J Pharmacol 2:366
22. Wiley JW (2007) Neurogastroenterol Motil 19:541
23. Choi KM, Gibbons SJ, Roeder JL et al (2007) Neurogastroenterol Motil 19:585
24. Wang PG, Xian M, Tang X, Wu X, Wen Z, Cai T, Janczuk AJ (2002) Chem Rev 102:1091
25. Gryglewski RJ, Marcinkiewicz E, Robak J, Michalska Z, Madej J (2002) Curr Pharm Des 8:167
26. Gasco A, Schoenafinger K (2005) The NO-releasing heterocycles. In: Wang PG, Cai TB, Taniguchi N (eds) Nitric oxide donors. Wiley-VCH, Weinheim, pp 131–175
27. Herman AG, Moncada S (2005) Eur Heart J 26:1945
28. Miller MR, Megson IL (2007) Br J Pharmacol 151:305
29. Wang PG, Xian M, Tang X, Wu X, Wen Z, Cai T, Janczuk AJ (2002) Chem Rev 102:1091
30. Kato H, Hashimoto M, Ohta M (1957) Nippon Kagaku Zasshi 78:707
31. Brookes P, Walker J (1957) J Chem Soc 4409
32. Kankaanranta H, Knowles RG, Vuorinen P, Kosonen O, Holm P, Moilanen E (1997) Mol Pharmacol 51:882
33. Schrammel A, Pfeiffer S, Schmidt K, Koesling D, Mayer B (1998) Mol Pharmacol 54:207
34. Cai TB, Lu D, Tang X, Zhang Y, Landerholm M, Wang PG (2005) J Org Chem 70:3518
35. Vinatier V, Soulere L, Hoffmann P (2006) Nitric Oxide 15:363
36. Pitsikas N, Rigamonti AE, Cella SG, Sakellaridis N, Muller EE (2005) Neurobiol Aging 26:259
37. Pitsikas N, Sakellaridis N (2007) Eur J Pharmacol 571:174
38. Martinet W, Croons V, Timmermans JP, Herman AG, De Meyer GRY (2007) Br J Pharmacol 152:493
39. Herman AG, Moncada S (2005) Eur Heart J 26:1945
40. Lee CS, Kim YJ, Han ES (2007) Life Sci 80:1759
41. Meij JTA, Haselton CL, Hillman KL, Muralikrishnan D, Ebadi M, Yu L (2004) Mol Pharmacol 66:1043
42. Lee CS, Han ES, Park ES, Bang H (2005) Brain Res 1036:18
43. Siegemund M, Bommel JV, Sinaasappel M, Schwarte LA, Studer W, Girard T (2007) Acta Anaesthesiol Scand 51:693
44. Patel KP, Li YF, Hirooka Y (2001) Exp Biol Med 226:814
45. Schwenke DO, Pearson JT, Tsuchimochi H, Mori H, Shirai M (2005) Clin Exp Pharmacol Physiol 32:952
46. Schwenke DO, Pearson JT, Tsuchimochi H, Kangawa K, Shirai M (2007) Clin Exp Pharmacol Physiol 34:88
47. Bryan-Lluka LJ, Papacostas MH, Paczkowski FA, Wanstall JC (2004) Br J Pharmacol 143:63
48. Marcos E, Adnot S, Pham MH, Nosjean A, Raffestin B, Hamon M, Eddahibi S (2003) Am J Respir Crit Care Med 168:487
49. Luchowski P, Urbanska EM (2007) Eur J Pharmacol 563:130
50. Rabkin SW, Klassen SS (2007) Nitric Oxide 16:339
51. Rabkin SW (2007) Autophagy 3:347
52. Ferry-Dumazet H, Mamani-Matsuda M, Dupouy M, Belloc F, Thiolat D, Marit G, Arock M, Reiffers J, Mossalayi MD (2002) Leukemia 16:708
53. Lau A, Arundine M, Sun HS, Jones M, Tymianski M (2006) J Neurosci 26:11540

54. Hall ED, Detloff MR, Johnson K, Kupina NC (2004) J Neurotrauma 21:9
55. Finnegan WG, Henry RA (1965) J Org Chem 30:567
56. Masuda K, Kamiya T, Kashiwa K (1971) Chem Pharm Bull 19:559
57. Gryglewski RJ, Marcinkiewicz E, Robak J, Michalska Z, Madej J (2002) Curr Pharm Des 8:167
58. Corell T, Pedersen SB, Lissau B, Moilanen E, Metsa-Ketela T, Kankaanranta H, Vuorinen P, Vapaatalo H, Rydell E, Andersson R et al. (1994) Pol J Pharmacol 46:553
59. Holm P, Kankaanranta H, Metsa-Ketela T, Moilanen E (1998) J Pharmacol 346:97
60. Kankaanranta H, Rydell E, Petersson AS, Holm P, Moilanen E, Corell T, Karup G, Vuorinen P, Pedersen SB, Wennmalm A, Metsa-Ketela T (1996) Br J Pharmacol 117:401
61. Kosonen O, Kankaanranta H, Vuorinen P, Moilanen E (1997) Eur J Pharmacol 337:55
62. Kosonen O, Kankaanranta H, Lahde M, Vuorinen P, Ylitalo P, Moilanen E (1998) J Pharm Exp Ther 286:215
63. Kosonen O, Kankaanranta H, Malo-Ranta U, Ristimaki A, Moilanen E (1998) Br J Pharmacol 125:247
64. Laursen BE, Stankevicius E, Pilegaard H, Mulvany M, Simonsen U (2006) Cardiovasc Drug Rev 24:247
65. Gryglewski RJ, Marcinkiewicz E, Robak J, Michalska Z, Madej J (2002) Curr Pharm Des 8:167
66. Vilpo JA, Vilpo LM, Vuorinen P, Moilanen E, Metsa-Ketela T (1997) Anticancer Drug Des 12:75
67. Lahteenmaki T, Sievi E, Vapaatalo H (1998) Br J Pharmacol 125:402
68. Pacher P, Beckman JS, Liaudet L (2007) Physiol Rev 87:315
69. Taylor EL, Rossi AG, Shaw CA, Dal Rio FP, Haslett C, Megson IL (2004) Br J Pharmacol 143:179
70. Taylor EL, Li JT, Tupper JC, Rossi AG, Winn RK, Harlan JM (2007) Biochem Pharmacol 74:1039
71. Wang JP (2003) Eur J Pharmacol 458:243
72. Hsu MF, Chen YS, Huang LJ, Tsao LT, Kuo SC, Wang JP (2006) Eur J Pharmacol 535:43
73. Kojima S, Uchida K, Sasaki K, Sunagawa M, Ohno Y, Kamikawa Y (2006) Eur J Pharmacol 550:162
74. Sitmo M, Rehn M, Diener M (2007) Am J Physiol Gastrointest Liver Physiol 293:G886
75. Grynberg N, Santos AC, Echevarria A (1997) Anticancer Drugs 8:88
76. Senff-Ribeiro A, Echevarria A, Silva EF, Sanches Veiga SS, Oliveira MBM (2003) Melanoma Res 13:465
77. Senff-Ribeiro A, Echevarria A, Silva EF, Franco CRC, Veiga SS, Oliveira MBM (2004) Br J Cancer 91:297
78. Senff-Ribeiro A, Echevarria A, Silva EF, Veiga SS, Oliveira MB (2004) Anticancer Drugs 15:269
79. Cadena SMSC, Carnieri EGS, Echevarria A, De Oliveira MBM (1998) FEBS Lett 440:46
80. Cadena SMSC, Carnieri EGS, Echevarria A, De Oliveira MBM (2002) Cell Biochem Function 20:31
81. Cardoso JC, Cadena SMSC, Zampronio A, Arruda AMS, Carnieri EGS, Echevarria A, Constantin J, Bracht A, Oliveira MBM (2004) Drug Dev Res 61:207
82. da Silva EF, Canto-Cavalheiro MM, Braz VR, Cysne-Finkelstein LL, Leonor L, Echevarria A (2002) Eur J Med Chem 37:979
83. Rodrigues RF, da Silva EF, Echevarria A, Fajardo-Bonin R, Amaral VF, Leon LL, Canto-Cavalheiro MM (2007) Eur J Med Chem 42:1039

Top Heterocycl Chem (2009) 16: 153–231
DOI 10.1007/7081_2008_125

Published online: 18 June 2008

Bioactive Phenothiazines and Benzo[*a*]phenothiazines: Spectroscopic Studies, and Biological and Biomedical Properties and Applications

J. J. Aaron[1,2] (✉) · M. D. Gaye Seye[3] · S. Trajkovska[4] · N. Motohashi[5]

[1]ITODYS, Université Paris 7 – Denis-Diderot, Associé au CNRS UMR 70-86, 1, rue Guy de la Brosse, 75005 Paris, France
aaron@univ-paris-diderot.fr

[2]Laboratoire des Géomatériaux et Géologie de l'Ingénieur (G2I), Université Paris-Est Marne la Vallée, 5, boulevard Descartes, Champs-sur-Marne, 77454 Marne-la-Vallée Cedex 2, France

[3]Département de Chimie, Faculté des Sciences et Techniques, Université Cheikh Anta Diop, Dakar, Senegal

[4]Department of Medical and Experimental Biochemistry, Medical Faculty, University Sv Kiril i Metodij, 50, Divija 6, 1000 Skopje, Macedonia

[5]Department of Medicinal Chemistry, Meiji Pharmaceutical University, 2-522-1 Noshio, Kiyose-shi, 204-8588 Tokyo, Japan

Abstract Recent progress in spectroscopic, photophysical, photochemical and analytical studies, as well as in the biological and biomedical properties of bioactive phenothiazines and benzophenothiazines, is reviewed. Their electronic absorption and luminescence properties, and their complexation and interactions in organized media are discussed. Various applications, including analytical studies, relative to phenothiazines and benzophenothiazines are described. Among the important biological and biomedical properties of these compounds, their neurological effects, their antibacterial, antifungal, antiviral, antiparasitic and antitumour activities, and their cytotoxicity are particularly reviewed.

Keywords Benzophenothiazines · Biological and biomedical properties · Phenothiazines · Photophysical, photochemical and analytical studies · Spectroscopic properties

Abbreviations

ACV	Acyclovir
APS	4-Phenothiazin-10-yl-anisole
AZA	Azure A
BPD	Butaperazine dimaleate
BPHT	Benzophenothiazine
CD	Cyclodextrin
CJD	Creutzfeldt–Jakob disease
CL	Chemiluminescence
CNS	Central nervous system
CPH	2-Chlorophenothiazine
CPZ	2-Chlorpromazine hydrochloride
CSP	Chiral stationary phase
DH	Diethazine hydrochloride
DMMB	1,9-Dimethylmethylene blue
DNA	Deoxyribonucleic acid
DPPG	Dipalmitoyl phosphatidyl glycerol
EPR	Electron proton resonance (spectra)
EPZ	Ethopropazine
FAB-MS	Fast atom bombardment–mass spectrometry
FIA	Flow-injection analysis
FP	Fluphenazine hydrochloride
GC	Gas chromatography
GC-NPD	Gas chromatography–nitrogen–phosphorus detection

GSH	Glutathione
HBD	Hydrogen-bond donor
HMO	Hückel molecular orbital (method)
HOMO	Highest occupied molecular orbital
HP-β-CD	2-Hydroxypropyl-β-cyclodextrin
HPLC	High-performance liquid chromatography
HRP	Horseradish peroxidase
HSV-2	Herpes simplex virus type 2
ISC	Intersystem crossing
IPH	Isothipendyl hydrochloride
LOD	Limit of detection
LOQ	Limit of quantification
LPZ	Levomepromazine
LTP	Low-temperature phosphorescence
LUMO	Lowest unoccupied molecular orbital
MB	Methylene blue
MDR	Multidrug resistance
MFE	Magnetic field effect
MNDO	Modified-neglect-of-diatomic overlay (method)
MS	Mass spectrometry
MTPM	Methopromazine maleate
NACE	Non-aqueous capillary electrophoresis
NMB	New methylene blue
PAG	Photosensitive acid generator
PAS	1-(4-Phenothiazin-10-yl-phenyl)-ethanone
PBS	(4-Phenothiazin-10-yl-phenyl)-phenylmethanone
PCP	Prochlorperazine
PCS	4-Phenothiazin-10-yl-benzonitrile
PDT	Photodynamic therapy
PH	Phenothiazine
PIF	Photochemically induced fluorescence
PIM	Pimozide
PM3	Parametric method 3
PP	Perphenazine hydrochloride
PPP	Pariser–Parr–Pople (method)
PPS	*N*-Phenylphenothiazine
PZ	Promazine hydrochloride
REPE	Resonance energy per π electron
RSD	Relative standard deviation
SDS	Sodium dodecyl sulphate
TB	Toluidine blue
TFMPH	2-Trifluoromethylphenothiazine
TFMPZ	2-Trifluoromethylpromazine hydrochloride
TFZ	2-Trifluoromethylperazine dihydrochloride (or trifluoperazine)
TG	Triglyceride
TH	Thionine
TLC	Thin-layer chromatography
TMPZ	2-Thiomethylpromazine
TPZ	Thioproperazine
TR	Thioridazine hydrochloride

1 Introduction

Phenothiazine (PH), a three-ring fused heterocyclic system with an – NH – group and a sulphur atom (– S –) in the central six-membered ring, is the parent molecule of several series of bioactive, heterocyclic derivatives which present wide and various properties of biological and pharmaceutical interest (Fig. 1A). It is worthwhile noting that, unlike anthracene, acridine and phenazine, which constitute 14 π-electron systems, phenothiazine contains 16 π electrons and thus is not π-isoelectronic with the above compounds.

For many years, simple mono- and disubstituted phenothiazine derivatives have been mainly used as dyes, indicators and insecticides, and they have also been investigated as drugs with various psychotherapeutic properties [1–6]. In these phenothiazine derivatives, the type and intensity of neuroleptic activity was found to depend on the position and electron-withdrawing or electron-donating effects of the substituents. For instance, halogenated alkylaminophenothiazine derivatives, such as chlorpromazine and trifluopromazine, are known as strong psychotropic and sedative drugs [2, 3]. Promazines, piperazines and perazines have antidepressant properties (Fig. 1B and C). In the case of the latter derivatives, substitution at the 2-position of the phenothiazine group and the presence of an alkylamino chain at the 10-position could enhance the neuroleptic activity of these drugs [4–6]. Other phenothiazines, including thionine, azure A, methylene blue, toluidine blue, new methylene blue and 1,9-dimethylmethylene blue (Fig. 2), are dyes that have also been studied for their photochemotherapeutic activity against carcinomas [7]. Moreover, several research groups have investigated the spectroscopic, photophysical and photochemical properties of various substituted phenothiazines, such as 10-alkylamino-2-substituted phenothiazines (Fig. 1B and C) and *N*-phenylphenothiazines (Fig. 3), and they have demonstrated the existence of marked substituent and solvent effects on these properties [4, 8, 9]. Due to the presence of chemically active S and N atoms and of substituents located at the 2- and 10(N)-positions, the phenothiazine derivatives also exhibit several interesting analytical properties which have been the object of several studies [10, 11].

More recently, in the 1980s and 1990s new series of fused phenothiazine derivatives, the benzo[*a*, *b* or *c*]phenothiazines (BPHTs), were synthesized [3, 21 and references therein] and have received a great deal of attention, mainly because of their potential applications and their important biomedical properties [12–24]. Indeed, some BPHTs are coloured compounds and have been applied as polycyclic dyes or pigments for synthetic polymers, and also in optical recording media ([21] and references therein). Moreover, certain benzo[*a* or *c*]phenothiazine derivatives are potential antihelmintics, possess an antiviral activity, for example inhibiting the multiplication of encephalomyocarditis viruses in tissue cultures ([21, 22], and refer-

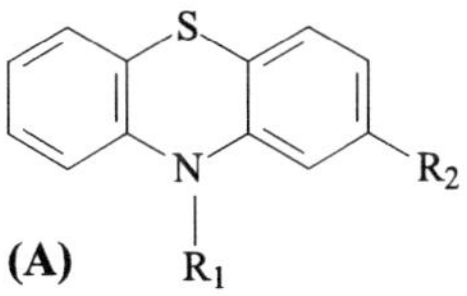

(A)

Abbreviations	Name	R_1	R_2
PH	Phenothiazine	H	H
PZ	Promazine hydrochloride	$(CH_2)_3$-$N(CH_3)_2$.HCl	H
CPZ	2-Chlorpromazine hydrochloride	$(CH_2)_3$-$N(CH_3)_2$.HCl	Cl
TFMPZ	2-Trifluoromethylpromazine Hydrochloride	$(CH_2)_3$-$N(CH_3)_2$.HCl	CF_3
TMPZ	2-Thiomethylpromazine	$(CH_2)_3$-$N(CH_3)_2$.HCl	SCH_3
PCP	Prochlorperazine	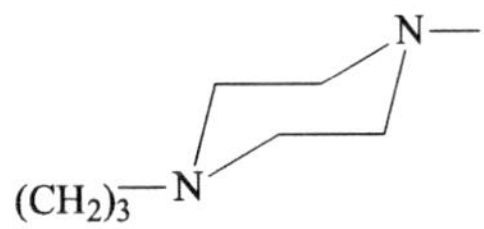	Cl
TFZ	2-Trifluoromethylpera zine dihydrochloride (or Trifluroperazine)	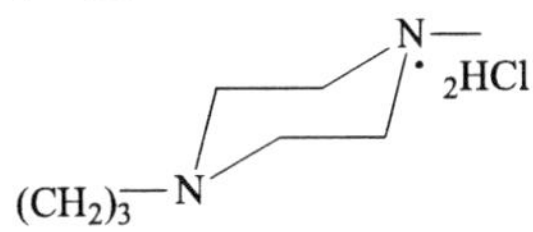	CF_3
TR	Thioridazine hydrochloride	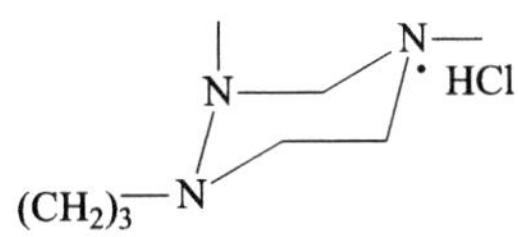	SCH_3
FP	Fluphenazine hydrochloride	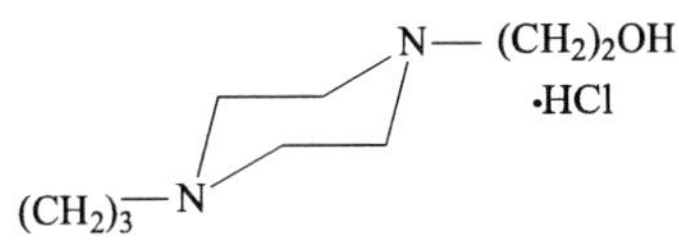	CF_3
PP	Perphenazine hydrochloride	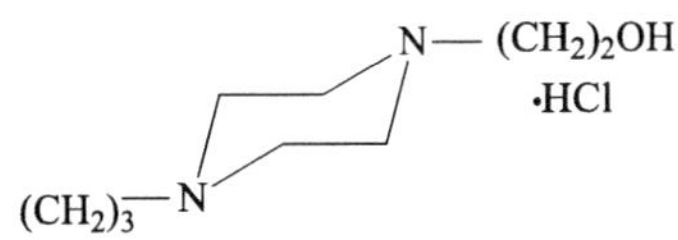	Cl

(B)

Fig. 1 Structure of the phenothiazine ring (**A**) and of 10-alkylated phenothiazine derivatives (**B** and **C**)

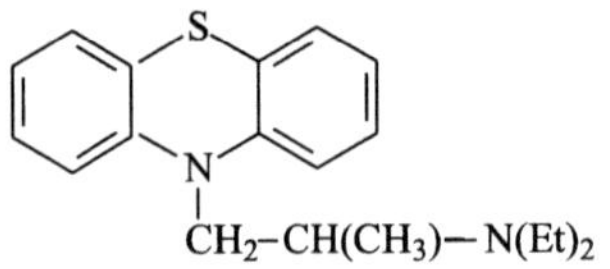

Ethopropazine (**EPZ**)

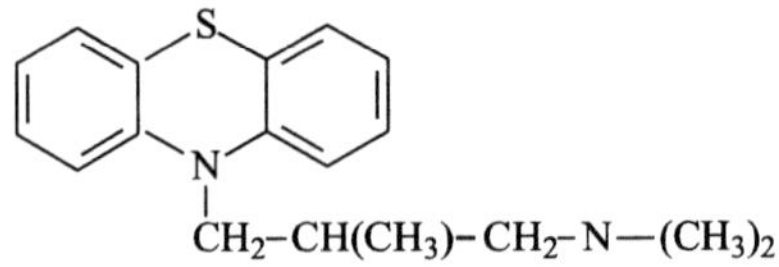

Levomepromazine (**LPZ**)

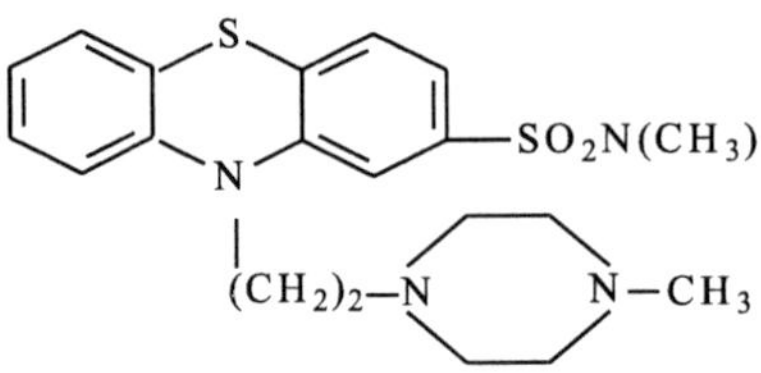

Thioproperazine (**TPZ**)

(C)

Fig. 1 (continued)

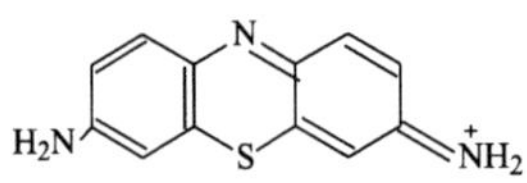

Thionine (**TH**)

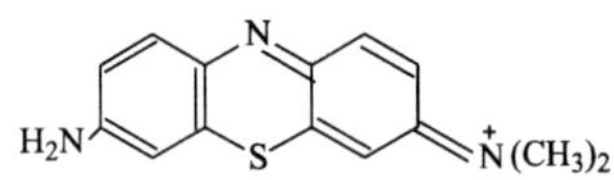

Azure A (**AZA**)

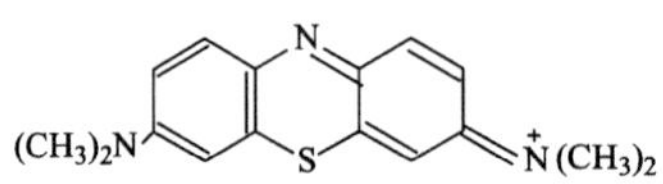

Methylene blue (**MB**)

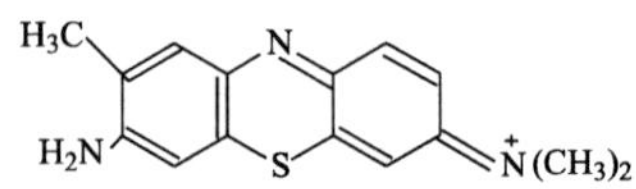

Toluidine blue (**TB**)

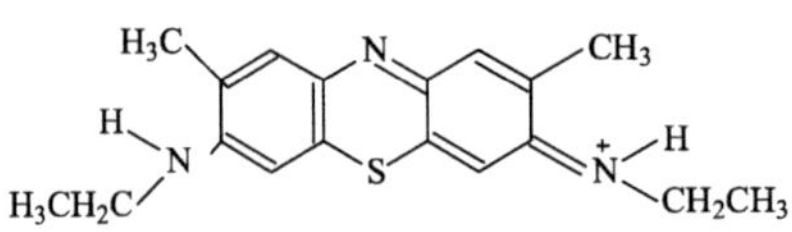

New methylene blue (**NMB**)

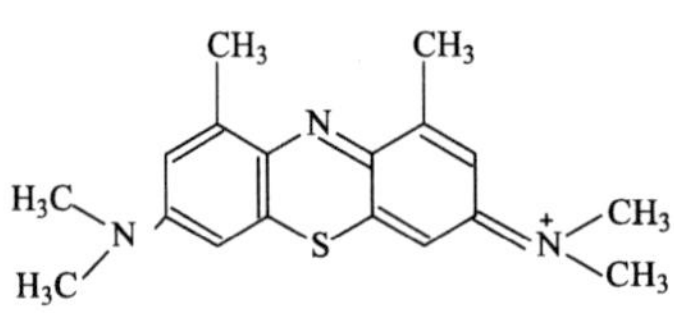

1,9-Dimethylmethylene blue (**DMMB**)

Fig. 2 Structure of phenothiazine dyes

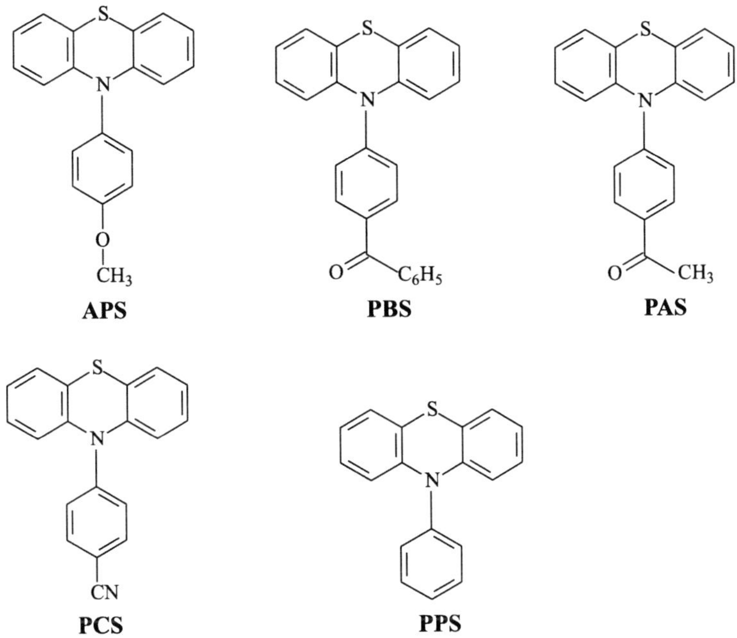

Fig. 3 Structure of *N*-phenylphenothiazine derivatives

ences therein), and have been proposed as antitumour drugs [13–15]. Also, benzo[*a*]phenothiazines and their hydro derivatives were shown to be strong leukotriene inhibitors, useful in the treatment of allergic inflammatory and cardiovascular conditions ([21] and references therein). Therefore, as already mentioned, BPHTs, and particularly benzo[*a*]phenothiazines (Fig. 4), have been the object of a number of physicochemical and biomedical studies, and they present interesting spectroscopic and photophysical properties as well as analytical applications [12–24]. In addition, due to their heteroaromatic, fused and planar molecular structure, BPHTs generally present a strong, native fluorescence in various organic solvents [16, 17] and in cyclodextrin (CD) aqueous media [18–23].

However, in spite of the increasing photophysical and analytical interest as well as the more and more numerous works concerning the biomedical and therapeutic role of these various compounds, to the best of our knowledge so far only two recent review articles have been devoted to the analytical applications of phenothiazines [11] and to the luminescence properties of BPHTs [25].

In this chapter, we will give a review of the most significant and recent literature data on the physicochemical, biomedical and biological properties of simple phenothiazines and fused benzo[*a*]phenothiazines (structures in Figs. 1–4). The first section will be concerned with the spectroscopic, pho-

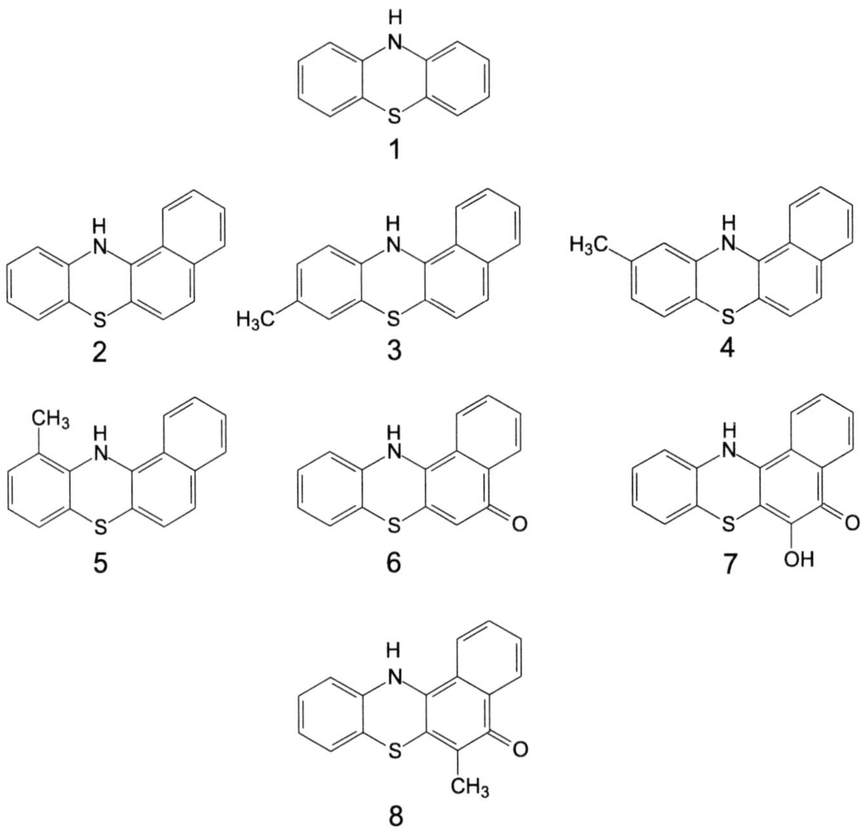

Fig. 4 Structure of phenothiazine (**1**) and of benzo[*a*]phenothiazine derivatives: 12*H*-benzo[*a*]phenothiazine (**2**); 9-methyl-12*H*-benzo[*a*]phenothiazine (**3**); 10-methyl-12*H*-benzo[*a*]phenothiazine (**4**); 11-methyl-12*H*-benzo[*a*]phenothiazine (**5**); 5-oxo-5*H*-benzo[*a*]phenothiazine (**6**); 6-hydroxy-5-oxo-5*H*-benzo[*a*]phenothiazine (**7**); 6-methyl-5-oxo-5*H*-benzo[*a*]phenothiazine (**8**)

tophysical, photochemical and analytical characteristics and applications of phenothiazines and BPHTs, while the second section will be devoted to the biomedical and biological properties of the same compounds.

2 Spectroscopic, Photophysical, Photochemical and Analytical Properties and Applications of Phenothiazines and Benzophenothiazines

Recently, several groups of researchers have been interested in the spectroscopic, photophysical and photochemical properties of phenothiazines and BPHTs [4, 9, 16–25]. For instance, Garcia et al. [4] found in 2005 that the notable modifications in the structure of phenothiazine drugs led to significant changes of interaction modes of each drug, as well as of their photophysical properties.

In this section, we will present the most significant and recent literature data concerning the UV–visible absorption and luminescence spectroscopies of a variety of phenothiazine derivatives and BPHTs (Figs. 1–5). We will also describe the photophysical and photochemical properties as well as the characteristics of organized media or molecular complexes formed between a number of phenothiazine derivatives or BPHTs and either micelles or CDs. Finally, we will examine several analytical methods which have been developed to determine phenothiazines in biological samples and pharmaceutical formulations, due to biomedical interest, and other recent applications of phenothiazines and BPHTs in various fields.

2.1 Phenothiazine Derivatives

2.1.1 UV–Visible Absorption Spectral Properties

The electronic spectral absorption properties of several 2-substituted, 10-alkylated phenothiazine derivatives, including promazine hydrochloride (PZ), 2-chlorpromazine hydrochloride (CPZ), 2-trifluoromethylpromazine hydrochloride (TFMPZ), 2-thiomethylpromazine (TMPZ), prochlorperazine (PCP), 2-trifluoromethylperazine dihydrochloride (trifluoperazine, TFZ), thioridazine hydrochloride (TR), fluphenazine hydrochloride (FP) and perphenazine hydrochloride (PP) (Fig. 1B), were investigated in different media (acetonitrile, methanol and pH 7.4 phosphate buffer solution) by Garcia et al. [4].

In all these derivatives, substitution by a Cl, CF_3, or SCH_3 group occurred at the phenothiazine 2-position, while an alkylamino chain or ring was located at the 10-position. The amino group in the 10-position was either fixed at the extremity of an alkyl chain, like in PZ, CPZ, TFMPZ and TMPZ, or included into an alkylcycloamino ring system, as in PCP, TFZ, TR, FP or PP (Fig. 2). The effects of the 2-substituent and 10-alkylamino chain or ring on the UV–visible absorption spectra of these compounds was investigated in acetonitrile, methanol and a pH 7.4 phosphate buffer solution [4]. For all compounds in this study, their spectra presented two main absorption bands in the 250–265 and 300–320 nm wavelength regions (Table 1). The short-wavelength band was attributed to a π–π^* transition, whereas the long-wavelength band was mainly due to an n–π^* transition. The interpretation of the latter band was supported by the presence of the sulphur lone-electron pairs in the phenothiazine ring – as pointed out by Aaron et al. [17]. Moreover, the blue shift of the 300–320 nm band observed in polar solvents and

Table 1 Electronic absorption spectral properties of phenothiazine and its 10-alkylated derivatives (adapted from [4], with permission)

Compound	λ (nm); $\varepsilon_{max} \times 10^{-4}$ (M^{-1} cm^{-1}) Acetonitrile	Methanol	PBS [a]
PH	253; 4.98	254; 5.1	
	318; 0.48	318; 0.43	[b]
PZ	258; 3.48	255; 3.3	252; 2.04
	305; 0.42	307; 0.42	302; 0.27
CPH	256; 7.5	256; 7.4	[b]
	323; 0.70	323; 0.74	
CPZ	258; 3.46	258; 3.66	255; 2.7
	312; 0;44	311; 0.46	307; 0.35
TFMPZ	258; 3.27	259; 3.31	256; 2.45
	310; 0;40	310; 0.37	306; 0.32
TFMPH	258; 4.02	259; 4.09	[b]
	322; 0.41	324; 0.43	
TMPZ	264; 3.83	264; 3;32	262; 3.44
	315; 0.49	314; 0.39	312; 0.42
PCP	259; 3.31	258; 3.48	255; 2.99
	314; 0.46	313; 0.45	309; 0.38
TFZ	259; 3.31	260; 3.8	257; 2.97
	310; 0.41	310; 0.46	308; 0.37
TR	264; 3.36	263; 3.99	263; 3.11
	316; 0.44	314; 0.54	312; 0.43

[a] PBS = phosphate buffer solution, pH 7.4
[b] Compound insoluble

the relatively small ε_{max} values of this band, when compared to those of the first band, could confirm its n–π^* nature. Quantum theoretical calculations – based on optimization of the geometry by a combination of molecular mechanics, molecular dynamics and quantum dynamics calculations – also showed a dependence of the n–π^* transition on the nature of the substituent at the phenothiazine 2-position, to a much greater extent than on nitrogen substitution by the 10-dimethylaminopropyl group [4].

Borowicz et al. [9] have studied a series of 4′-substituted-*N*-phenylphenothiazine derivatives, including 4-phenothiazin-10-yl-anisole (APS), *N*-phenylphenothiazine (PPS), (4-phenothiazin-10-yl-phenyl)-phenylmethanone (PBS), 1-(4-phenothiazin-10-yl-phenyl)-ethanone (PAS) and 4-phenothiazin-10-yl-benzonitrile (PCS) (Fig. 3). The three latter compounds are characterized by the occurrence of an acceptor–donor (A–D) system in which the benzophenone, acetophenone or benzonitrile group plays an electron-acceptor role, whereas the phenothiazine moiety is acting as an electron donor through a single bond at the N-position. In the two former *N*-phenylphenothiazine derivatives (APS and PPS), no A–D system was present. As expected, these distinct electronic features led to two different spectral behaviours for both types of compounds.

The room-temperature absorption spectra recorded in methylcyclohexane solution exhibited two absorption bands, respectively centred around 250–256 and 316–336 nm, for all compounds. Both bands are related to electronic transitions mainly localized in the phenothiazine subunit. Similar molar absorption coefficients (ε_{max}) for the first, short-wavelength band ranging from 1.3×10^4 to 1.7×10^4 M^{-1} cm^{-1} were found in the case of PCS, PBS and PAS. These values were higher than those found for phenothiazine (4.0×10^3 M^{-1} cm^{-1}) [26] and PPS or APS, whereas an opposite trend was observed for the second, long-wavelength band. This finding was attributed to the changes in the conformation of phenothiazine induced by a strong electron-acceptor substituent, located at the N_{10} nitrogen atom, for PCS, PBS and PAS. In addition, the absorption spectra of A–D compounds, including PAS, PBS and PCS, presented a long-wavelength shoulder of relatively low intensity attributed to a charge transfer (CT) transition [27].

2.1.2 Luminescence Spectral Properties

The pharmacological activity as well as the side effects of phenothiazine derivatives have been shown to often depend on the nature and position of substituents [4]. Since there is also a correlation between the molecular structure and the photophysical properties of these compounds, it is worthwhile to thoroughly examine their first excited singlet- and triplet-state characteristics.

2.1.2.1 Fluorescence Spectral Properties

The room-temperature fluorescence properties of a series of 2-substituted, 10-alkylated phenothiazine derivatives, including emission spectra, Stokes shifts, fluorescence quantum yields (Φ_F) and lifetimes (τ_F), were investigated by Garcia et al. [4, 28] in the same solvents as for the absorption spectroscopy study. Comparative values were also reported by Elisei et al. [29] for two similar promazines, namely fluphenazine hydrochloride and perphenazine hydrochloride. All these data are gathered in Table 2.

The emission spectra of all 2-substituted, 10-alkylated phenothiazine derivatives were characterized by the presence of a single broad band with a maximum between 440 and 470 nm in acetonitrile, methanol and phosphate buffer solution (Table 2). For these derivatives, the Stokes shift was larger than 10^4 cm^{-1} which is a considerable magnitude. The 2-substituents, including strong electron-withdrawing (CF_3) and electron-donating (SCH_3) groups, provoked significant red shifts ($\Delta\lambda = 10$–25 nm) of the emission

Table 2 Fluorescence emission spectral properties of 10-alkylated phenothiazine derivatives (adapted from [4], with permission)

Compound	λ_{max}^{em} (nm); Stokes shift (cm^{-1}); $\phi_F \times 10^3$; τ_F (ns) Acetonitrile	Methanol	PBS [a]
PZ	444; 8924; 20.0; 1.85	444; 10050; 4.5; 1.75 [b]	452; 10449; 14.3; 2.04 [b]
CPZ	451; 9372; 0.89; 0.92	449; 9474; 0.95; 0.89 [b]	453 [c]; 9772; 3.6; 0.35 [b]
TFMPZ	470; 10982; 1.1; 2.66	463; 10659; 1.0; 2.47	475; 11627; 1.4; 3.17
TMPZ	460; 10007; 3.0; 1.16	455; 9869; 3.0; 1.15	475; 11102; 3.0; 1.17
PCP	449; 8688; 8.4; 0.91	449; 9175; 46.6; 0.89	457 [c]; 9174; 3.3; –
TFZ	470; 10981; 1.0; 2.68	467; 10844; 1.0; 2.52	482; 11720; 1.4; 3.01
TR	455; 9668; 5.0; 1.03	453; 9772; 4.1; 0.96	466; 10592; 6.0; 1.47

[a] PBS = phosphate buffer solution, pH 7.4
[b] Values from [28]
[c] Measured in PBS, pH 6.9

maximum relative to the unsubstituted phenothiazine derivative. Moreover, the emission maxima were found to be more dependent on solvent than the corresponding absorption maxima. Contrarily to the effects observed for the absorption spectra, the emission was shifted towards the red with increasing solvent polarity, which is characteristic of the π^* nature of the first excited singlet state (S_1) of phenothiazine derivatives. This red-shift behaviour was explained by Garcia et al. [4] in terms of the involved molecular orbitals, which are located on the phenothiazine moiety. It can be concluded that there is no contribution of the *N*-alkyl chain to the emission properties of the phenothiazine derivatives, which is in agreement with the fact that the emission maxima of the parent phenothiazines are exactly the same as those of the corresponding 10-alkylated derivatives [30].

The fluorescence quantum yield values were found to be very low for these phenothiazine derivatives, ranging between about 2×10^{-2} and 10^{-3}, according to the compound [4, 29] (Table 2). The 2-CF_3 phenothiazines exhibited practically similar Φ_F values in all three solvents under study. The 2-SCH_3 analogues presented Φ_F values three to six times larger than those of the corresponding CF_3-substituted compounds, but significantly lower than those of unsubstituted promazine. In terms of 2-substitution, the Φ_F values for the methylthio group (SCH_3) larger than those found for the CF_3 and Cl groups can be explained by a higher planarity of the methylthio derivatives, and the related better orbital interaction of the SCH_3 substituent with the phenothiazine tricyclic ring. Nevertheless, the small Φ_F values for all these 10-alkylated phenothiazine derivatives indicate that fluorescence is not the major energy-dissipation pathway for these compounds, and that other deactivation mechanisms for the S_1 state are in competition with fluorescence, which represents only a 0.1 to 0.6% energy dissipation. The flexibility of the

10-alkylamino chain, giving several stable conformations to all molecules, and the interconversion between these conformations should contribute to diminish the Φ_F values. Therefore, Garcia et al. [4] attributed the low fluorescence yield values to the high efficiency in the various thermal deactivation processes (internal conversion and/or intersystem crossing (ISC)).

The fluorescence lifetime values (τ_F) of the 10-alkylated phenothiazine derivatives presented the same solvent and substituent dependency noted for quantum yields [4], and matched very closely those provided by Elisei et al. [29] (Table 2). It was found that the nature of the 10-alkyl chain had practically no significant effect on the τ_F values, as shown by the fact that these lifetimes had the same values, within experimental error, as those reported for the corresponding non-alkylated phenothiazines: PH ($\tau_F \leq 1$ ns), CPH ($\tau_F \leq 1$ ns) and TFZF ($\tau_F = 2.5$ ns) [30]. Garcia et al. [4] noted that the 2-CF_3-substituted derivatives gave similar τ_F values and presented the same solvent dependency, a τ_F value of 2.4–2.7 ns being measured in acetonitrile and methanol, whereas in a more polar medium, such as the phosphate buffer solution, a 3.1-ns value was obtained. This behaviour can be attributed to the solvatochromic effect of the polar S_1 state of these phenothiazine derivatives. The smaller τ_F values (<1.0 ns) obtained for the chlorinated derivatives, such as CPZ, PCP and TR, resulted from the presence of the chlorine atom, which is well known to enhance the spin–orbit coupling through a heavy-atom effect in the ISC process [31]. The same explanation – based on the presence of a sulphur atom at the 2-position – can be provided for the lifetimes of the 2-SCH_3-substituted phenothiazines, which are about 50% lower than those obtained for the 2-CF_3-substituted analogues.

Garcia et al. [4] concluded from their important photophysical study that the fluorescence emission properties of these 10-alkylated phenothiazine derivatives are very sensitive to the solvent and the 2-substituent, but practically not to the nature of their 10-alkylamino chain.

The fluorescence properties of another series of phenothiazines, namely the 4′-substituted *N*-phenylphenothiazine derivatives (Fig. 3), were investigated by Borowicz et al. [9, 27]. For all the compounds under study, only a single emission was observed at room temperature. For PPS and APS, there was no change of the spectral position of the emission maximum with the solvent polarity. In contrast, a considerable red shift of the emission maximum, and a parallel increase of the Stokes shift and of the emission bandwidth with the increasing solvent polarity were observed in the case of the A–D compounds, including PAS, PBS and PCS. Moreover, the energies corresponding to the fluorescence maximum position of the latter compounds were correlated by the difference in the standard redox potentials $[E^{\circ}_{ox}(D) - E^{\circ}_{red}(A)]$ in a given solvent. These findings led the authors to conclude the formation of a distinct charge transfer complex (CT complex) in the fluorescent states.

2.1.2.2
Phosphorescence and EPR Spectral Properties

In the same work, Borowicz et al. [9] recorded the low-temperature phosphorescence (LTP) spectra (77 K) and the electron proton resonance (EPR) spectra of the 4′-substituted *N*-phenylphenothiazine derivatives under study in an *n*-propanol (PrOH) glass. They found that the PPS and APS phosphorescence spectra presented a position and shape very similar to that obtained for phenothiazine, which seems to indicate that the triplet excitation is localized in the phenothiazine subunit. In contrast, the LTP spectra of PCS, PAS and PBS, compounds containing an electron-acceptor group (– CN, – $COCH_3$ or – COC_6H_5 at the 4′-position), were characterized by energies of the lowest (very probably 0–0) bands significantly higher (about 2000–3000 cm^{-1}) than those of PPS and APS. This suggests that the triplet excitation is mainly localized in the acceptor subunits of the A–D compounds (PCS, PAS and PBS). The EPR spectral results also indicated that, in PCS, the spin density in the lowest T_1 state is essentially distributed over the π system of the acceptor subunit. Similarly to the above-reported UV–visible absorption spectral results, this behaviour could be related to changes in the conformation of the phenothiazine moiety, induced by the presence of the strong electron acceptor subunit connected to the N-10 nitrogen atom [9].

The phosphorescence lifetime (τ_P) values measured in PrOH glass at 77 K confirmed the postulated nature of the T_1 states in rigid media [9]. Indeed, the PPS and APS τ_P values (∼90 ms) were found to be very close to the value reported for phenothiazine ($\tau_P \approx 80$ ms) [32]. On the other hand, PAS and PBS exhibited τ_P values of 0.44 and 0.12 s in PrOH at 77 K, respectively, which were significantly longer than those measured for the parent carbonyl compounds (acetophenone or benzophenone), but were rather similar to those obtained for 4′-aminoacetophenone ($\tau_P = 0.72$ s) [33] and 4′-aminobenzophenone ($\tau_P = 0.20$ s) [34]. Borowicz et al. [9] explained these findings by a change of the phosphorescent triplet state from $^3(n,\pi^*)$ to $^3(\pi,\pi^*)$, resulting from the substitution of an electron-donor group at the 4′-position.

2.1.2.3
Triplet-State Lifetimes and Quantum Yields

Both Garcia et al. [4, 28] and Elisei et al. [29] also investigated the triplet-state properties, including the triplet-state extinction coefficients (ε_T), lifetimes (τ_T) and quantum yields (Φ_T), of several 2-substituted, 10-alkylated phenothiazine derivatives. The ε_T values were determined from the respective maximum triplet absorbances by using the energy-transfer method to energy acceptors, either merocyanine-540 (MC 540) in deaerated methanol and acetonitrile, or to β-carotene in deaerated acetonitrile or benzene. These ε_T

values were in the range (1.5 – 6) × 10^4 M^{-1} cm^{-1} according to the compound and the solvent.

Generally, the triplet lifetime is considered as a very sensitive parameter of the substituent and solvent effects [4]. However, the 10-substitution yielded a very small effect, as shown by the absence of significant dependence of τ_T on the nature of the 10-alkylated chain in a given solvent. For instance, the τ_P values of the chlorinated derivatives, CPH, CPZ, PCP and PP, in acetonitrile were 0.80, 0.91, 0.80 and 0.66 μs, respectively. In contrast, the 2-substitutions produced a larger variation in the τ_T values, as noted for the promazine series in methanol (PZ: 61 μs; CPZ: 2.2 μs; TFMPZ: 91 μs; TMPZ: 59 μs). The solvent effect on the τ_T values was prominent. Indeed, except for PZ, the triplet lifetime values were found to decrease drastically when going from acetonitrile or methanol to water (for example, for TFZ from 111 μs in methanol to 2.13 μs in phosphate buffer solution), the triplets even being undetectable in water for the 2-chlorinated derivatives [4]. Therefore, these results confirm that the triplet state of halogenated promazines is efficiently quenched by a very fast proton-transfer mechanism [28]. An interesting feature is that the triplet quenching rate could correlate very satisfactorily with the phototoxicity of promazine drugs, which indicates that the triplet-state species might be directly related to the phototoxic side effect of neuroleptic drugs [4].

The triplet quantum yields (Φ_T) were measured by means of nanosecond laser flash spectroscopy [4]. Φ_T values ranged between 0.2 and 0.9, according to the compound, and also depended to some extent on the solvent. The derivatives having a 2-methylthio substituent generally formed their triplets in very good yield, with Φ_T values larger than 0.7. 2-CF_3 phenothiazine derivatives, such as TFMPZ and TFZ, had smaller triplet quantum yield values (0.39 and 0.22 in phosphate buffer solution, respectively), whereas the 2-SCH_3 analogues presented greater Φ_T values (TMPZ: $\Phi_T = 0.89$; TR: $\Phi_T = 0.74$, in methanol). Chlorinated promazines had markedly different Φ_T values (0.90 for CPZ and 0.35 for PCP, in methanol). This large difference was attributed to the bulky perazine group in PCP. For the limited number of data available, the perazines were found to have smaller Φ_T values than the corresponding promazines. It could be concluded that the N-side-chain substitution exercises an important effect on the triplet quantum yield, as supported by the higher Φ_T value determined for phenothiazine ($\Phi_T = 0.97$) [12]. This confirms that the 10-alkylated chain could probably provoke the deactivation of the excited singlet state of phenothiazine derivatives, thus reducing the ISC efficiency.

After these photophysical studies, Elisei et al. [29] investigated, in the case of PP, FP and TR, the photoionization process in aqueous solutions, which led to the formation of phenothiazine radical cations and solvated electrons. Then, the phototoxicity of these three drugs was tested on various biological substrates through a series of in vitro assays under UVA irradiation. Photohaemolysis of mouse erythrocytes and phototoxicity on cultured murine fibroblast cells were examined for all three phenothiazine derivatives. The photosensitization capability of these phenothiazines towards lipids was also studied by investigating the photo-oxidation of linoleic acid, chosen as an unsaturated lipid model, and isolated red blood cell membranes. Moreover, the drug-induced photodamage was estimated on proteins by measuring the photosensitizing crosslinking in erythrocyte ghosts. The combined approach was found to be useful in understanding the mechanism by which these phenothiazines might induce skin photosensitization. It is worthwhile stressing that the photophysical properties of the compounds in the study and the results of the phototoxicity tests are in good agreement with a mechanism that involves the radical cation of the drugs as a main intermediate.

2.1.3 Complexation and Interactions of Phenothiazine Derivatives in Organized Media

Because of the potentially useful pharmacological and biological applications of phenothiazine derivatives, it is very important to investigate the physicochemical aspects of their interactions and aggregations within organized media, such as CDs and micelles of surfactants. As a consequence, several literature studies have been reported in this field [35–44].

A number of years ago, it was shown that the phenothiazine dyes often underwent dimerization in aqueous media [45–49]:

$$2M \rightleftarrows \mathrm{D} \qquad K_\mathrm{D} = [D]/[M]^2 ,$$

with K_D = the dimerization constant for the monomer (M)/dimer (D) equilibrium.

In the literature, there are important differences between the K_D values obtained for phenothiazine dyes, probably because of the ambiguity in the UV–visible spectroscopic data generally used for the K_D measurements and the possible adsorption of the phenothiazines on glassware [41, 45–49]. The existence of this dimerization phenomenon generally might complicate the biomedical and pharmaceutical applications involving the phenothiazine dyes.

Moreover, this dimerization equilibrium also plays a role in the formation of the inclusion complexes of phenothiaziness with CDs [41]. The CDs are cyclic oligosaccharides obtained from the degradation of starch by the glucosyltransferase enzyme [50]. CDs are composed of six, seven or eight glucose units, designated as α-, β- or γ-CD, respectively. The number of glucose units might determine the dimension and size of the cavity. The cavity interior surface is lined by hydrogen atoms and glycosidic oxygen bridges, which make it relatively hydrophobic. In contrast, the external face is hydrophilic since all hydroxyl groups are located outside the cavity. The main interest in CDs is their ability to include organic hydrophobic molecules in their empty cavities. Several groups have studied the effects of the position of the monomer/dimer equilibria and inclusion of phenothiazine dyes in CDs on the electronic absorption and fluorescence spectra [35–41], as well as the photochemical [35–42] and electrochemical [43] properties of these compounds.

In a notable paper, Lee et al. [41] examined the dimerization and inclusion complexation equilibria of six phenothiazine dyes, including thionine (TH), azure A (AZA), methylene blue (MB), toluidine blue (TB), new methylene blue (NMB) and 1,9-dimethylmethylene blue (DMMB) (Fig. 2), with CDs (α-, β-, and γ-CDs) in aqueous medium. These phenothiazine dyes presented their structural differences in the position and number of methyl substituents located on the same phenothiazine skeleton. By means of the UV–visible absorption and fluorescence spectra, the authors studied the factors which were responsible for the dimerization of phenothiazine dyes with CDs, in terms of a multiple equilibrium system (Scheme 1). In addition, the authors also determined the K_D values and the corresponding association constants of the monomers (M) and dimers (D) with the CDs [41].

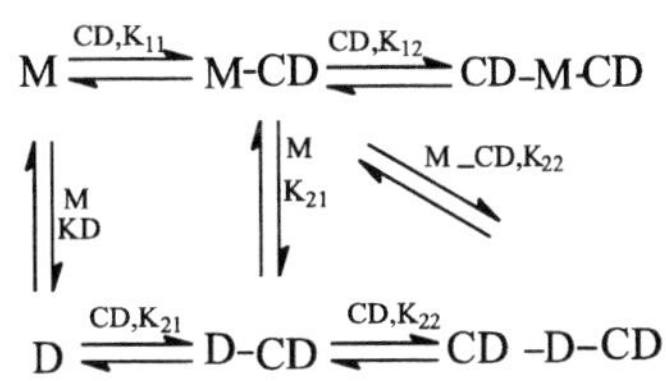

Scheme 1 Multiple equilibrium system for the association of phenothiazine dye monomers (M) and dimers (D) with the cyclodextrins (CD) (with permission from [41])

According to the authors, the dyes having two methyl substituents on the phenothiazine ring, such as NMB and DMMB, considerably increased the dimerization constant K_D relative to those possessing unsubstituted rings, whereas the presence of methyl substituents on the amino groups affected the K_D values much less [41]. Also, the position of the monomer/dimer equilibria did not change significantly with the presence of α-CD, while the dimerization of dyes was enhanced by the addition of γ-CD but was inhibited by the presence of β-CD. In another interesting feature underlining the selective role of the CD cavity, the dye monomers could fit well to β-CD, and the dimers were snugly included into γ-CD, except DMMB which possesses methyl groups at the 1- and 9-positions of the fused phenothiazine ring. Also, fluorescence spectral studies showed that the phenothiazine dye dimers were not fluorescent, and that the inclusion of the monomer in β-CD produced a 3–5 times enhancement of fluorescence intensity. Finally, the equilibrium constants of the multiple equilibrium system of the dyes in CD media were determined, and the equilibrium constants could be used, in conjunction with the dye fluorescence properties, to control the aggregation of the phenothiazine dyes as well as their photophysical and photochemical properties [41].

The interactions of two phenothiazines, chlorpromazine (CPZ) and trifluoperazine (TFZ), with anionic sodium dodecyl sulphate (SDS) monomers and/or micelles were investigated by Caetano et al. [44] by means of electronic absorption and fluorescence spectroscopy. The authors evaluated the pK_a values of these drugs in the micelles and their binding constants (K_b) to micelles, by using the red shifts of the maximum absorption band and absorbance changes upon alkalization or increase of SDS concentration. The binding of TFZ and CPZ to SDS was characteristic of a biphasic interaction, depending on the detergent concentration and the respective protonation state of both drugs. In the low surfactant concentration, the binding constant values higher than 2×10^4 M^{-1} were estimated for TFZ at pH 5.0 and 2.0, whereas for CPZ, for which no protonation change took place in this pH range, the binding constants were a factor of three lower than for TFZ. At pH 7.0, the binding constants were 2×10^4 and 3.6×10^3 M^{-1} for TFZ and CPZ, respectively. In the higher SDS concentration range, the binding constants were also pH dependent, although the binding equilibrium was more complex [44]. Static fluorescence probe analysis based on pyrene was used to study the nature of the phenothiazine–surfactant pre-micellar and self-aggregates, and the steady-state anisotropy of the fluorescent probe dipyridamole was applied to monitor the changes of the self-aggregates and micellar packing, which was found to become more rigid in the presence of both CPZ and TFZ.

2.1.4 Analytical Methods and Analytical Applications

Because of the biological and pharmaceutical interest in phenothiazine derivatives, these compounds have been used in a variety of analytical studies and applications. For a number of years, several analytical methods have been proposed by various groups for the determination of phenothiazine and its derivatives in pharmaceutical formulations and biological fluids, including thin-layer chromatography (TLC) [51–54], high-performance liquid chromatography (HPLC) [55–58], spectrofluorimetry, either direct or with derivatization [59–64], flow-injection analysis (FIA) [10,65–73] and other techniques [11,74–76].

Prabhakar et al. [56] developed and validated six HPLC assay methods which permitted resolution of tailing problems for the separation of phenothiazine bioactive components and preservatives in liquid pharmaceutical formulations by using cyano-, C-8, C-18, and cation-exchange phases. The mass spectrometry (MS) of phenothiazines in biological samples is important in forensic analysis. However, since most phenothiazines having long piperazinyl side chains are thermolabile, the phenothiazines are not suitable for conventional gas chromatography (GC)/MS analysis. Therefore, a system of capillary HPLC/fast atom bombardment–mass spectrometry (FAB-MS) was applied by Sato et al. [57] to analyse 17 phenothiazines. Quasi-molecular peaks along with an adequate number of fragment peaks were detected for all compounds, with detection limits of less than 1 ng on column. Four representative phenothiazines and their metabolites could be also identified in human sera by the capillary HPLC/FAB-MS method, after oral administration. This system would seem to be useful for searching a minor, but new metabolic pathway of a drug in pharmacokinetics, in addition to its use in forensic toxicology.

Tesarova and Bosakova [58] proposed an HPLC method for the enantioselective separation of some phenothiazine and benzodiazepine derivatives on six different chiral stationary phases (CSPs). These selected CSPs, with respect to the structure of the separated compounds, were either based on β-CD chiral selectors (underivatizized β-CD and hydroxypropyl ether β-CD) or on macrocyclic antibiotics (vancomycin, teicoplanin, teicoplanin aglycon and ristocetin A). Measurements were carried out in a reversed-phase separation mode. The influence of mobile phase composition on retention and enantioselective separation was studied. Enantioselective separation of phenothiazine derivatives, including levopromazine (LPZ), promethazine and thioridazine, was relatively difficult to achieve, but it was at least partly successful with both types of CSPs used in this work (CD-based and glycopeptide-based CSP), except for levomepromazine for which only the β-CD-based CSP was suitable.

Most simple, substituted phenothiazines generally possess too weak a native fluorescence for analytical purposes. Therefore, in most cases the application of fluorimetric methods to phenothiazines requires the use of a preliminary step to convert them into strongly fluorescent derivatives. Earlier, chemical oxidation [60,61] and complexation reactions [62] were proposed to fulfil this goal, but they are rather time consuming and tedious processes. A more simple, rapid, precise and efficient approach consists of photochemically transforming the phenothiazines into strongly fluorescent phenothiazine sulphoxides by means of UV irradiation, which significantly improves their fluorimetric detection [63,64]. The latter photochemical–fluorimetric method was found to be analytically useful for all phenothiazine derivatives under study, including several aminophenothiazines and phenothiazine dyes. Indeed, the optimum irradiation times were relatively short, ranging between 2 and 1020 s, and the photochemically induced fluorescence enhancement factors were rather high, between 7 and 21 according to the compound. Moreover, this approach allowed one to quantify these phenothiazine derivatives at the ng/mL level in biological liquids (urine) and pharmaceutical samples, with satisfactory recovery values of 88–119% according to the compound.

FIA, combined with UV absorption, fluorimetric or chemiluminometric detection, has also been applied to the determination of some phenothiazine derivatives by using either chemical oxidation [65–67,73] or photo-oxidation [10,68–72]. Martinez Calatayud et al. [65,66] and other groups [67, 68] determined promethazine in pharmaceutical formulations by means of a FIA method, based on chemical oxidation by cerium (Ce(IV)) and UV absorption spectrophotometric detection. In these studies, linear calibration graphs were obtained over a promethazine concentration range of 5–400 μg/mL. Chen et al. [69] used a photochemical reaction to simultaneously determine chlorpromazine and promethazine by irradiating the flow-injection manifolds with UV light. The simultaneous quantification was based on the difference of pH of the solutions where the photochemical transformation of each phenothiazine derivative into a fluorescent photoproduct occurred. Under these conditions, chlorpromazine and promethazine were simultaneously quantified at the μg/mL level with relative standard deviation (RSD) values between 2 and 4% and sampling frequencies in the range 30–40/h. Also, a combined photochemical–spectrofluorimetric FIA system allowed Martinez Calatayud et al. [70] and Mellado Romero et al. [71] to determine chlorpromazine and promethazine in pharmaceutical formulations.

Laassis et al. [10, 72] developed a simple, rapid and sensitive method combining FIA with photochemically induced fluorescence (PIF) for the determination of unsubstituted phenothiazine, phenothiazine dyes, such as thionine, azure A and methylene blue, and aminophenothiazines, including ethopropazine (EPZ), levomepromazine (LPZ), thioproperazine (TPZ) and trifluoperazine (TFZ). In the case of unsubstituted phenothiazine and the phenothiazine dyes, low limits of detection (LODs) ranging between 13 and 35 ng/mL, according to the compound, were obtained, and the application of this FIA-PIF method to their determination in urine samples gave recovery values of 94 to 117% [72]. For aminophenothiazine derivatives, after optimization of the working analytical parameters, their flow rate, injected volume and photoreactor length could result, satisfactorily demonstrating the benefit of using this method for routine analysis of these compounds in pharmaceutical samples [10]. Indeed, the linear calibration curves over two to three orders of magnitude, LOD and limit of quantification (LOQ) values in the 60–90 and 200–400 ng/mL ranges, according to the compound, and RSD values of 1.2–4.5% demonstrated the good applicability of this FIA-PIF method to the determination of aminophenothiazines. In pharmaceutical formulations (majeptil, nozinan, terfluzine), recovery values in the ranges of 91–93% for TPZ in majeptil, 98–117% for LPZ in nozinan and 95–118% for TFZ in terfluzine were found (Table 3) [10]. In the case of biological media such as urine, the standard addition procedure was used to quantify these compounds. The standard addition results suggested that the possible presence of endogenous substances in urine should only interfere in a limited manner with the FIA-PIF measurements. Although satisfactory recovery values were obtained, in the ranges of 97–118% for EPZ, 88–109% for TFZ, 95–102% for LPZ and 87–102% for TPZ, it was recommended by the authors to perform a preliminary clean-up step prior to the application of FIA-PIF to the determination of aminophenothiazines in urine (Table 4) [10].

Table 3 Determination of aminophenothiazines in pharmaceutical formulations (from [10], with permission)

	Levomepromazine (LPZ)				Thioproperazine (TPZ)		
Nozinan added (mL)[a]	LPZ added (ppm)	LPZ found (ppm)	Recovery (%)	Majeptil added (μL)[b]	TPZ added (ppm)	TPZ found (ppm)	Recovery (%)
1		6.69	–	5	–	0.34	–
1	1.00	7.80	101	5	0.50	0.76	91
1	10.0	15.4	92	5	1.00	1.00	83
1	30.0	43.2	117	5	2.00	2.06	88
1	50.0	63.9	112	5	3.00	2.92	87
1	70.0	79.3	103	5	5.00	4.69	88
1	90.0	97.4	100	5	7.00	6.41	87
1	110	115	98	5	10.0	9.17	88

	Trifluoperazine (TFZ)		
Terfluzine added (mL)[c]	TFZ added (ppm)	TFZ found (ppm)	Recovery (%)
1	–	14.9	–
1	0.10	15.5	103
1	5.00	20.3	102
1	10.0	27.7	111
1	30.0	46.3	103
1	50.0	76.7	118
1	70.0	91.4	107
1	100	112.5	98
1	200	239.7	111
1	300	322.2	102
1	400	395	95

[a] Nozinan solution prepared by dissolving 44 mg of pharmaceutical preparation in 100 mL of distilled water. Number of mL of solution added to a 10-mL calibrated flask
[b] Majeptil solution prepared by dissolving 25 mg of pharmaceutical preparation in 250 mL of distilled water. Number of mL of solution added to a 10-mL calibrated flask
[c] Terfluzine solution prepared by dissolving 48 mg of pharmaceutical preparation in 100 mL of distilled water. Number of mL of solution added to a 10-mL calibrated flask

Aly et al. [73] proposed a rapid and sensitive flow-injection chemiluminometric method for the determination of three phenothiazine derivatives: fluphenazine hydrochloride, levomepromazine hydrochloride and trimeprazine tartrate. The method was based on the chemiluminescence (CL) induced by the oxidation of the drugs with cerium (Ce(IV)) in an acidic medium. The CL intensity was greatly enhanced when rhodamine B was used as a sensitizer in the case of levomepromazine hydrochloride and trimeprazine tartrate. The linear dynamic ranges were 0.5–90 μg/mL for

Table 4 Determination of aminophenothiazines in urine samples (from [72], with permission)

Levomepromazine (LPZ) Urine added (μL)[a]	LPZ added (ppm)	LPZ found (ppm)	Recovery (%)	Thioproperazine (TPZ) Urine added (μL)[a]	TPZ added (ppm)	TPZ found (ppm)	Recovery (%)
400	–	16.87	–				
400	1	17.51	97	50	–	3.28	–
400	2	19.20	101	50	0.10	3.34	98
400	4	20.37	97	50	4.00	7.46	102
400	5	20.37	95	50	6.00	8.34	98
400	7	22.41	102	50	8.00	10.75	95
400	10	27.12	100	50	10.0	11.63	87
Trifluoperazine (TFZ) Urine added (μL)[a]	**TFZ added (ppm)**	**TFZ found (ppm)**	**Recovery (%)**	**Ethopropazine (EPZ) Urine added (μL)[a]**	**EPZ added (ppm)**	**EPZ found (ppm)**	**Recovery (%)**
5	–	0.43	–	20	–	0.37	–
5	0.20	0.69	109	20	0.50	0.97	110
5	0.50	0.82	88	20	1.00	1.62	118
5	0.70	1.08	88	20	4.00	4.80	109
5	1.00	1.28	90	20	10.0	11.5	110
5	1.50	1.85	96	20	15.0	15.0	97

[a] Number of μL of urine added to a 10-mL calibrated flask

fluphenazine and 0.1–6.5 μg/mL for levomepromazine and trimeprazine. The limits of detection (3σ) were 0.01 μg/mL for fluphenazine hydrochloride and 0.1 μg/mL for the other two drugs. The method was applied successfully to the determination of the drugs in dosage forms as well as in biological fluids.

An improved solid-phase extraction method for the simultaneous identification and quantitative determination of various drugs, including phenothiazines, tricyclic antidepressants and benzodiazepines, in biological fluids (blood, urine, bile and gastric contents) was developed by Soriano et al. [74]. This procedure involved solid-phase extraction with Bond-Elut Certify columns followed by analysis by gas chromatography–nitrogen–phosphorus detection (GC-NPD) and confirmation by GC-MS, after derivatization, when necessary. It was found that systematic toxicological analysis could be reliably performed using the optimized extraction conditions. The recovery rates for the compounds under study were always higher than 61%.

A horseradish peroxidase (HRP) immobilized carbon composite electrode has been developed by Petit et al. [75] for the amperometric study of phenothiazine analogues. FIA and batch experiments were realized in acetate buffer in the presence of hydrogen peroxide. Cyclic voltammetry and amperometry using a thin-layer flow cell (dual configuration, serial mode) permitted proposal of the mechanisms governing the biosensor signal at – 0.1 V (vs Ag/Ag^{+}), in the presence of hydrogen peroxide, by addition of a phenothiazine derivative. It was inferred that the electron transfer mediation by the phenothiazine occurred at HRP/graphite adsorbed sites, in addition to peroxidation by a dispersed HRP with substrate recycling at the graphite array-like structure of the biosensor. By means of these processes, high sensitivities were achieved especially in the batch configuration with the amperometric detection capabilities down to 10^{-8} M in acetate buffer of pH 4.7. Application of the biosensor to the determination of phenothiazines in drug formulations was effectively realized.

Madej et al. [76] used a non-aqueous capillary electrophoresis (NACE) method for the screening and quantification of seven phenothiazine derivatives in blood. The optimal medium for dissolving the examined blood extracts was tested. The linear dynamic ranges were between 0.25 and 4.00 μg/mL (correlation coefficients higher than 0.996), the RSD values going from 1.21 to 9.15%, according to the compound. The detection limits were 0.08 μg/mL for promazine and 0.15 μg/mL for the rest of the drugs under study. Finally, the proposed method was applied to two forensic blood samples, and concentrations of the examined phenothiazines determined by the HPLC and NACE methods were found to be comparable.

A recent review has reported the application of a number of phenothiazine derivatives, including CPZ, PMH, TFZ, TR, PCP, diethazine hydrochloride (DH), PZ, TFMPZ, butaperazine dimaleate (BPD), EPZ, propionyl promazine phosphate, FP, methopromazine maleate (MTPM) and isothipendyl, as analytical reagents for the UV–visible absorption spectrophotometric quantification of several metals, such as Cr(VI), V(V), Fe(III), Cd(II), Hg(II), Ag(I), Au(III), Pd(II), Ru(III), Pt(IV), Ir(IV) and others in various samples [11]. This method was based on the reversible one-electron oxidation of phenothiazines into coloured radical cations by metal ions, which further undergo irreversible oxidation to yield sulphoxides. The probable mechanism of oxidation of phenothiazine derivatives is presented in Scheme 2. This reaction was carried out in acidic or buffer media at room temperature, and the formation of the coloured species was monitored spectrophotometrically. The experimental conditions (phenothiazine derivatives, media etc.) were optimized in order to obtain the coloured species with maximum stability and maximum absorbance. The concentration linear dynamic ranges were about 0.1–30 ppm and depended on the reagents and metals. The effect of foreign ions was also investigated. The developed method was applied to the quantitative analysis of metal ions in several samples, such as steel, alloys, minerals,

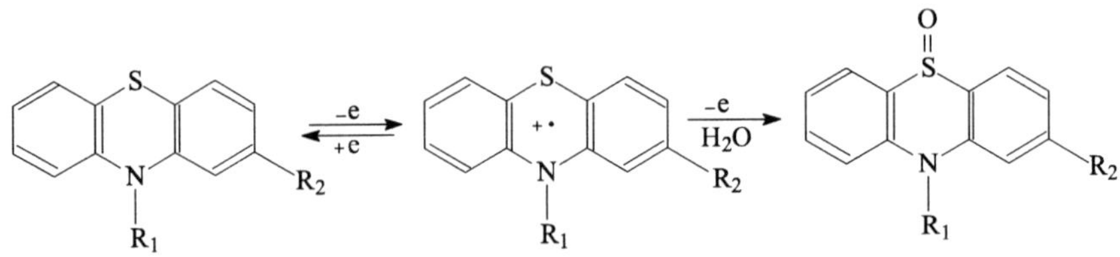

Scheme 2 Probable mechanism of the oxidation of phenothiazine derivatives (from [11], with permission)

soils, water and spiked urine samples [11]. However, it should be pointed out that this spectrophotometric method presents the disadvantage of being rather slow, poorly selective and not very sensitive for the determination of metals, especially when compared to the modern atomic spectrometric analytical methods.

2.1.5 Other Applications

Several applications have been proposed recently for the use of phenothiazines in various fields, such as materials chemistry [77, 78], magnetic field effects on fullerene–phenothiazine linked compounds [79, 80] and fluorescent labelling [81].

Gessner et al. [77, 78] provided two patents describing the use of phenothiazine derivatives of phenothiazine-*S*-dioxide and phenothiazine-*S*,*S*-dioxide as host matrix materials for organic light-emitting devices, especially in the light-emitting layer of organic light-emitting diodes, or as emitter substances or hole and exciton blockers in organic light-emitting diodes. Organic light-emitting diodes based on phenothiazine derivatives, devices with displays employing the diodes, and light-emitting layers and hole- and exciton-blocking layers containing these derivatives were also described.

Magnetic field effects (MFEs) and photoinduced intramolecular electron-transfer studies were performed in fullerene–phenothiazine and C_{60}–phenothiazine linked compounds [79, 80]. Transient absorption spectra of fullerene–phenothiazine linked compounds indicated that the photoinduced intramolecular electron transfer took place in benzonitrile, but not in benzene. In benzonitrile, the lifetime of a photogenerated biradical was very long, in spite of being around the top region in the Marcus theory. The lifetime of the biradical increased in the presence of the magnetic fields. The MFEs were verified when the intramolecular electron transfer from phenothiazine to the triplet excited state of fullerene occurred. The long lifetime could be ascribed to the spin multiplicities of the biradical [79]. In a further investigation, photoinduced electron-transfer reactions and MFEs on the decay of a photogenerated biradical in a phenothiazine–C_{60} linked compound with six methylene groups were examined in benzonitrile at various temperatures (283–343 K) [80]. The decay rate constants of the biradicals were dramatically decreased in the lower magnetic fields (0–0.2 T) and increased in the higher magnetic fields (0.2–1.0 T). These reverse phenomena on the MFEs were more clearly observed with the increase of the temperature, and were explained by the contribution of a spin–lattice relaxation mechanism due to anisotropic Zeeman interaction. The temperature dependence of the MFEs supported the above mechanism.

A patent describing the use of fluorescent phenothiazine dyes for the labelling of biomolecules was proposed by Lukhtanov et al. [81]. In this patent, phenothiazine, benzophenothiazine, xanthene, benzoxanthene and coumarin lactone dye reagents useful in labelling biological materials were provided along with the methods for their use. The phenothiazine dyes were found to be suitable for their multiple detection processes and might be readily converted to reactive, although stable, fluorescent labelling reagents for biomolecules.

2.2 Benzophenothiazine Derivatives

Mainly based on our experimental findings, this part of the review consists of three sub-sections concerning essentially the electronic absorption and luminescence spectral and photophysical properties of benzo[*a*]phenothiazines [25]. First, we present the electronic spectral and photophysical studies on BPHTs. Solvent and substituent effects on the electronic and fluorescence spectra, as well as the dipole moments in both ground and first excited states, are discussed. Second, we examine the hydrophobic interactions occurring between BPHTs and CDs such as β-CD and 2-hydropropyl-β-CD (HP-β-CD) in aqueous media, using electronic absorption and fluorescence spectroscopy. Finally, in the third sub-section we also discuss some literature results on the physicochemical properties and analytical applications of BPHTs. The compounds in our study include 12*H*-benzo[*a*]phenothiazine (**2**), 9-methyl-12*H*-benzo[*a*]phenothiazine (**3**), 10-methyl-12*H*-benzo[*a*]phenothiazine (**4**), 11-methyl-12*H*-benzo[*a*]phenothiazine (**5**), 5-oxo-5*H*-benzo[*a*]phenothiazine (**6**), 6-hydroxy-5-oxo-5*H*-benzo[*a*]phenothiazine (**7**) and 6-methyl-5-oxo-5*H*-benzo[*a*]phenothiazine (**8**) (Fig. 4).

2.2.1 Spectroscopic and Photophysical Studies

The electronic absorption spectra as well as the fluorescence spectral properties of BPHTs have been investigated by several research groups [16–24]. The spectroscopic results, obtained in several organic solvents of different polarities, in water and in CD media, provided important information on the electronic transitions occurring in these BPHT derivatives, their electronic structure, solvatochromic relationships and CD effects. Study of the solvent effects led to the determination of ground-state and excited singlet-state dipole moments, whereas CD effects demonstrated the formation of inclusion complexes.

2.2.1.1
Electronic Absorption Spectral Properties

The electronic absorption spectra of the unsubstituted BPHT and its methyl and oxo derivatives were recorded at room temperature (298 K) in organic solvents of different polarity, including cyclohexane, ethyl ether, ethyl acetate, tetrahydrofuran, ethanol, dimethylformamide (DMF), acetonitrile and dimethyl sulphoxide (DMSO), and in CD aqueous solutions by the groups of Aaron, Parkanyi et al. [16–23]. All electronic absorption spectra showed several strongly absorbing peaks in the 250–280 and 320–380 nm regions for BPHTs and methyl derivatives, and at 240–320 and 360–700 nm for the oxo derivatives, in organic solvents [17]. The detailed literature data have been presented in our recent review [25]. High molar absorption coefficients ($\log \varepsilon = 4.7$–3.9) were obtained for the shorter wavelength peaks and were attributed to π–π^* (^{1}B) transitions. In contrast, the longer wavelength bands with smaller molar absorption coefficients ($\log \varepsilon = 4.0$–3.3) were attributed to π–π^* (^1L_a and ^1L_b) transitions. All the 5-oxo-5*H*-BPHT derivatives presented an additional band in the visible region (λ_{max} higher than 460 nm), which was due to the particular quinone-like electronic structure of these heterocycles [16, 17].

Relatively small shifts of the absorption maxima, ranging between 2 and 22 nm according to the compound and the specific band, were observed with the change of solvent polarity. Moreover, it is worthwhile noting that the absorption spectra of unsubstituted and methyl-substituted benzophenothiazines (BPHT, 9-MeBPHT, 10-MeBPHT, 11-MeBPHT) are similar to those of unsubstituted phenothiazine and promethazine. This demonstrates similar electronic transitions in both series of heterocycles [24].

The effects of CDs (HP-β-CD and β-CD) on the electronic absorption spectra of 9-MeBPHT, 10-MeBPHT and 11-MeBPHT in aqueous media [18, 20, 22, 23] showed that all these compounds exhibited similar absorption spectra, with three main bands located in the 220–223, 274–282 and 321–324 nm regions, suggesting that the electronic transitions are identical. Their spectra were poorly resolved in water, whereas a fine vibrational structure was obtained in presence of CDs. The highest molar absorption coefficients (ε_{max}) for the shorter wavelength band in the presence of CDs were larger than $10^4\ M^{-1}\ cm^{-1}$, and similar to those found in the case of organic solvents. This suggests that the corresponding electronic transitions are also of the π–π^* type in the CD media. In the case of the longer wavelength band, ε_{max} values were close to 10^3 M^{-1} cm^{-1}, indicating a partial overlap of n–π* and π–π* transitions. No significant variation of the ε_{max} values occurred from ethanol to the CD media. In contrast, a blue shift (about 8–9 nm) of the π–π* absorption bands of the spectra in the presence of CDs was obtained relative to water, probably due to the decrease in polarity of the CD cavities when compared to water. An increase of the ε_{max} values of the 220, 270 and 280 nm bands in ethanol and CD media was also noted relative to water. These ε_{max} value increases were more important in HP-β-CD than in β-CD.

2.2.1.2
Excited Singlet State and Fluorescence Spectral Properties

Aaron et al. [16, 17] studied the fluorescence spectral properties of BPHTs in various organic solvents. All BPHTs in our study exhibited a strong native fluorescence in the investigated organic solvents, except for 5-oxo-5*H*-6-OH-benzo[*a*]phenothiazine and 5-oxo-5*H*-6-Me-benzo[*a*]phenothiazine. In most cases, the fluorescence excitation spectra included three main bands at about 240–270, 270–285 and 330–340 nm, and a shoulder located at about 380 nm, whereas the emission spectra showed only one principal band appearing approximately between 470 and 525 nm, corresponding to the π–π* transition. For the less polar solvents, a shoulder also appeared at higher wavelengths. The detailed fluorescence literature data have been presented in our recent review [25]. Generally, the fluorescence intensity increased with the solvent polarity, suggesting that the BPHT singlet excited state was stabilized by dipolar interactions within the solvent cage [16]. A 26–70-nm red shift was observed for the emission maxima of BPHTs, relative to unsubstituted phenothiazine, which is in agreement with the extent of the π electron system due to the presence of an additional phenyl ring in BPHTs [24]. When varying the solvent polarity from cyclohexane to DMSO, a 40-nm red shift of the emission maxima was observed, except for 11-MeBPHT for which there was only a 16-nm red shift. The authors attributed these red shifts to the stabilization of the BPHT π* molecular orbitals resulting from more efficient solvation by the more polar solvents. This would decrease the energy of the π–π* transition of the fluorophore [17].

The fluorescence spectral properties of BPHTs and three methyl derivatives were also studied in a predominantly aqueous medium (water/ethanol 99 : 1, v/v) in the presence and absence of CDs (HP-β-CD or β-CD), and in ethanol [18–22]. The excitation and emission spectra presented a similar shape in the various media. The excitation spectra appeared in the 320–340 nm region for all compounds under study, whereas for 10- and 11-MeBPHTs, a strong band located at 276–279 nm and a weaker one around 340–350 nm were obtained in all media. For unsubstituted BPHT, 9-, 10- and 11-MeBPHT, slight blue shifts (3–15 nm) occurred in a predominantly aqueous medium (water/ethanol 99 : 1, v/v) mixture relative to ethanol [18–21]

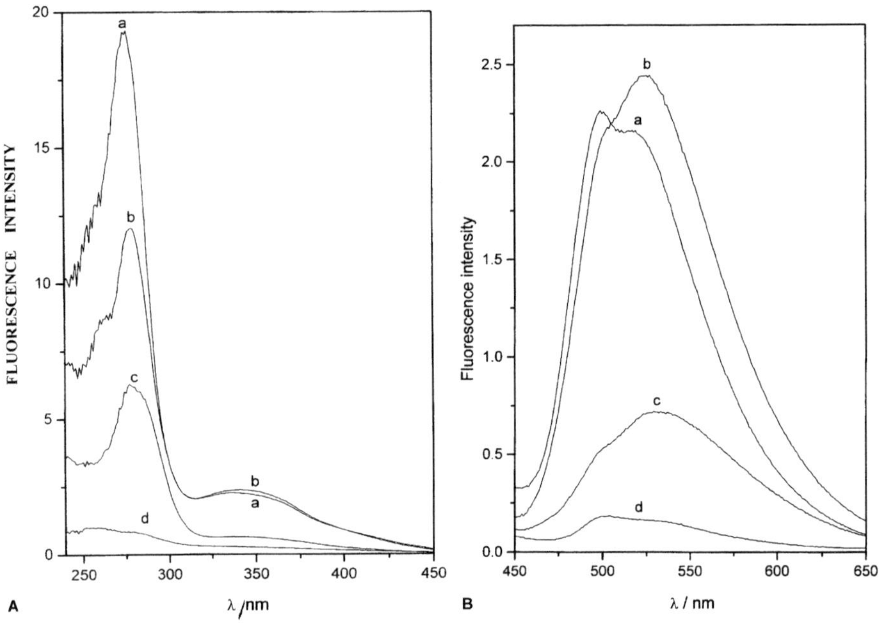

Fig. 5 Fluorescence excitation (**A**) and emission (**B**) spectra of 10^{-5} M 10-methyl-12*H*-benzo[*a*]phenothiazine in: (*a*) ethanol; (*b*) 10^{-1} M HP-β-CD water/ethanol (WE) mixture, 99 : 1, v/v; (*c*) 10^{-2} M β-CD WE mixture; (*d*) WE mixture (from [18], with permission)

(Fig. 5A). The emission spectra of benzo[*a*]phenothiazine and the three methyl derivatives exhibited generally a broad and structureless band located at 500–527 nm in water/ethanol mixture (99 : 1, v/v), 500–507 nm in ethanol and 512–519 nm in HP-β-CD (Fig. 5B). Enhancement of the Stokes shift in HP-β-CD solutions was interpreted by a greater stabilization of the $\pi-\pi^*$ solvated excited singlet state [21], due to the reduced rotational and conformational motions of the BPHT molecules within the HP-β-CD cavity [82]. Comparable weak excitation or emission spectral shifts upon adding CD to an aqueous medium have also been reported in the case of unsubstituted phenothiazine [21], azure A [83] and several types of aromatic molecules [84, 85].

The fluorescence intensity of benzo[*a*]phenothiazine and the methyl derivatives was found to be higher in ethanol than in the predominantly aqueous medium. For the authors, this is at least partially due to the fact that the hydrophobic BPHTs are poorly solvated in water and, consequently, the radiationless deactivation process of the excited singlet state is generally faster in water than in organic solvents [21]. The use of CDs was also found to increase significantly the fluorescence intensity of BPHTs relative to water [18, 19, 22, 23]. For example, it was found that the fluorescence quantum yields (ϕ_F) of 10-MeBPHT were about 22 times and 6 times larger in HP-β-CD and β-CD media, respectively, than in water [18]. These results predicted that the MeBPHT molecules should be included in the low-polarity CD cavities [18, 86]. The stoichiometry, formation constants and proposed molecular structures of MeBPHT:CD complexes will be examined in more detail in Sect. 2.2.2.

2.2.1.3 Solvatochromic Effects and Dipole Moments

a) Solvatochromic Correlations

The Aaron and Parkanyi groups [17] studied the solvatochromic correlations of BPHTs by plotting the Stokes values $(\bar{\nu}_A - \bar{\nu}_F)$ and $(\bar{\nu}_A + \bar{\nu}_F)/2$ against the solvent functions F_1 and F_2 for the different solvents under study, using the Bakhshiev and Kawski–Chamma–Viallet solvatochromic equations ([25] and references therein). These equations are based on the shifts of the maxima in electronic absorption and fluorescence spectra, measured in solvents of different polarities. The majority of solvents gave linear plots, with correlation coefficients larger than 0.91.

The solvatochromic shifts obtained for the BPHT electronic absorption and fluorescence emission maxima were also applied to evaluate the contributions of the polarity/polarizability of the solvent and its hydrogen-bond donor (HBD) ability to solute–solvent interactions in the ground and first excited singlet states of BPHTs. For this purpose, a simplified form of the classical Kamlet–Abboud–Taft solvatation energy relationship was used as follows [87, 88]:

$$XYZ = XYZ_0 + s\pi^* + a\alpha$$

where XYZ is a solvatochromic property (e.g. a shift), π^* is the solvent polarity/polarizability, α is the solvent HBD ability, and s and a are the respective coefficients. Satisfactory correlation coefficients (0.900 and higher) were obtained for this analysis based on the available absorption and fluorescence spectral data of BPHTs [17].

The fact that negative values were obtained for s coefficients indicates that the first excited singlet state of BPHTs should become more stabilized with increasing solvent polarity. This correlates well with the red-shifted spectra and the higher dipole moments found in the first excited singlet state of these compounds [17]. The same conclusion was also reached previously in the case of unsubstituted phenothiazine and its simple substituted derivatives [24].

The a coefficients presented negative values in the case of fluorescence wavenumbers, but most of them had positive values for the absorption wavenumbers. The authors [17] suggested that an increase of the solvent HBD ability should lead to a blue shift in absorption spectra, corresponding to a decrease in the formation of solute–solvent hydrogen bonding in the excited singlet state relative to the ground state. With the exception of 5-oxo-6-OH-BPHT, the a coefficients were significantly smaller than the s coefficients for electronic absorption and for fluorescence data, which suggests that the HBD ability is significantly weaker than the solute–solvent polar interactions taking place in the excited state of most BPHTs [17].

b) Ground-State and First Excited Singlet-State Dipole Moments

Permanent dipole moments constitute one of the important characteristics of polar compounds with an uneven distribution of electronic charge in their ground state, and are very useful parameters to study the molecular geometry (differentiation between isomers, conformational analysis), various electronic interactions (donor–acceptor, charge transfer), hydrogen bonding, polar character of the respective molecules, etc. [89]. Some of these physicochemical properties could provide important information for the study of several parameters of biologically active molecules, such as the solubility in different solvents, partition coefficients, permeability through membranes, and others.

Aaron, Parkanyi and co-workers [16, 17, 24, 90] have widely investigated the ground-state and the excited singlet-state dipole moments of unsubstituted phenothiazine, some susbstituted phenothiazine derivatives and BPHTs. The ground-state dipole moment of the phenothiazine parent molecule (**1**) measured in dioxane was 2.93 D (D = Debye), whereas the calculated value was 2.58 D [24, 25]. However, the experimental excited singlet-state dipole moment ranged between 3.10 and 3.90 D, and the calculated value was 2.40 D [24, 25]. The various substituted phenothiazines have experimental ground-state dipole moments ranging from 2.75 to 4.44 D, with the corresponding first excited singlet-state values ranging between 3.36 and 5.16 D [24, 25].

Also, Aaron, Parkanyi and co-workers ([25] and references therein) measured the ground-state dipole moments of BPHT derivatives (compounds **2–8**) in dioxane with a dipole meter at 20 °C (Table 5). While the dipole moments of compounds **5** and **8** were lower than that of phenothiazine (**1**), the remaining derivatives displayed higher ground-state dipole moment values, ranging between 3.4 and 5.53 D.

The authors compared the ground-state dipole moments of the BPHTs and concluded that their magnitude depended on both the nature of the substituent (electron-donating, electron-withdrawing) and, to a somewhat lesser extent, on the position of the substituent in the benzo[*a*]phenothiazine ring. Indeed, the dipole moment experimental values of unsubstituted benzo[*a*]phenothiazine (**2**) and its 9-methyl and 10-methyl derivatives (**3** and **4**, respectively) were higher than the dipole moment of the 11-methyl derivative (**5**). On the other hand, the two 5-oxo-substituted BPHTs (compounds **6** and **7**) possessed significantly larger dipole moments than those obtained for unsubstituted or methyl-substituted BPHTs. This indicates the existence of stronger electronic interactions (oxo group) in these compounds, with a less symmetrical distribution of electronic charge, and the presence of possible dipolar resonance structures [16, 17].

To determine the experimental first excited singlet-state dipole moments of the BPHTs, the authors established the Bakhshiev and Kawski–Chamma–Viallet solvatochromic equations, as described in previous works ([25] and references therein) (Table 5). For non-fluorescent compounds **7** and **8**, no excited singlet-state dipole moment value could be obtained.

While two distinct sets of experimental excited singlet-state dipole moments were actually determined, either by Bakhshiev or by Kawski–Chamma–Viallet equations, the values given in Table 5 corresponded to average values obtained from the above correlations. The average experimental values of the excited singlet-state dipole moments were found to vary between 5.2 and 7.5 D, and were significantly higher than the ground-state values for all compounds under study. The authors concluded that BPHTs are more polar in the first excited singlet state than in the ground state, because of a redistribution of electronic charge in the excited state [17]. This conclusion

Table 5 Ground-state and first excited singlet-state dipole moments of phenothiazine and benzo[*a*]phenothiazines [a] (from [25], with permission)

Compound no.	Ground-state μ (D)		First excited singlet-state μ (D)	
	Experimental [b]	Calculated [c]	Experimental [d]	Calculated [e]
1	2.93 [f]	2.58 [g] (2.24)	3.50	2.40
2	3.69 [h]	2.65 (2.42)	6.0	6.31 (7.86)
3	3.47	2.60 (2.35)	5.9	6.99 (10.30)
4	3.91	2.72 (2.54)	6.4	6.84 (11.09)
5	2.87	2.73 (2.67)	5.2	6.56 (9.38)
6	5.53	4.99 (3.68)	7.5	6.66 (6.51)
7	5.39	1.89 (3.52)	– [i]	6.12 (7.32)
8	2.58	4.47 (3.10)	– [i]	6.67 (8.43)

[a] Values in Debye units. The value for **1** is from [24], the values for **2–8** are from [17]
[b] In dioxane
[c] The dipole moment is a vector sum of the π moment calculated by the PPP method and the σ moment obtained from σ-bond moments. The value in parentheses is obtained by the PM3 method [91]; the reference for **1** is [92]
[d] Average value obtained from those determined by using the Bakhshiev equation and the Kawski–Chamma–Viallet equation
[e] π Moment obtained by the PPP method, σ moment assumed to be the same as in the ground state. The values in parentheses represent the first excited state comprising one triplet and three singlet states (MECI algorithm) [91]
[f] Other values: 2.50 D in dioxane and 2.14–2.16 D in benzene (see [25])
[g] Other values: 1.80–2.73 D (see [25])
[h] Other value: 1.99 D in benzene (see [25])
[i] Nonfluorescent

appears to be in good agreement with the results obtained for unsubstituted phenothiazine and simple substituted phenothiazines [25].

The ground-state dipole moments of BPHTs were calculated by evaluating the vector sum of the π moment (computed by the Pariser–Parr–Pople (PPP) method) and the σ moment (from σ-bond moments) [17], a separate set of dipole moment values being obtained by the parametric method 3 (PM3) ([25] and references therein) (Table 5). By comparison, the calculated ground-state dipole moment values were considerably lower than the experimental ones. In some cases, the agreement between the experimental and calculated ground-state dipole moments was rather poor [17].

In the calculation of the first excited singlet-state dipole moments, the σ moment was assumed to be the same as in the ground state and only changes of the π moment were considered. The authors also found a relatively good agreement between the calculated excited-state dipole moments and the average values of the experimental excited-state dipole moments (Table 5) [17]. For comparison, another set of calculated excited-state dipole moments is presented in Table 5 [25, 91, 92].

c) Dipole Moments and Biological Activity

Since dipole moments constitute one of the most important physicochemical characteristics of the biologically active heterocycles, it was of interest to establish relationships between the dipole moments of substituted benzo[*a*]phenothiazines and their biological activity.

Pusztai et al. [93] studied the exposition of human adenovirus, oncogene-type 12 infected Hep-2 cells to six benzo[*a*]phenothiazines, including compounds **2–4** and **6–8**. The authors investigated the relationship between the substituent effects and the tumour (T) antigen expression. The experimental ground-state dipole moments of compounds **6** ($\mu_g = 5.53$ D) and **7** ($\mu_g = 5.39$ D), which presented a moderate T antigen-positive expression, were found to be significantly higher than those of the four inactive benzo[*a*]phenothiazines **2**, **3**, **4** and **8** ($\mu_g = 3.69$, 3.47, 3.91 and 2.58 D, respectively), which indicated a relatively satisfactory positive correlation between the two quantities [93]. In the case of the first excited singlet-state dipole moments, among the four fluorescent benzo[*a*]phenothiazines with differentiation-inducing ability, compound **6** which possessed both the T antigen-positive expression and differentiation activity had the highest first excited singlet-state dipole moment ($\mu_e = 7.97$ D, value of the Bakshiev equation), followed by compounds **4**, **2** and **3** ($\mu_e = 7.42$, 7.01 and 6.88 D, respectively, values of the Bakshiev equation). Also, the authors noted that in the case of two non-fluorescent compounds, such as **7** and **8** without differentiation-inducing activity, compound **7** reduced the T antigen expression whereas compound **8** increased the antigen expression [93]. In other related studies, the same research group investigated the relationship between the experimental dipole moments (μ_g and μ_e) and antimutagenicity for seven benzo[*a*]phenothiazines (compounds **2–8**) [94]. In an analogous manner, among the four benzo[*a*]phenothiazines **2–5**, the μ_g experimental values were higher for compounds **2–4** with moderate antimutagenicity than for the weakly antimutagenic one, 11-MeBPHT (**5**). Similar conclusions were reached in the case of three oxo benzo[*a*]phenothiazine derivatives (**6–8**) [94].

Another attempt to find correlations with biological activity was made for a series of substituted benzo[*a*]phenothiazines including the same compounds **2–8**, 6-methyl-12*H*-benzo[*a*]phenothiazine and 5*H*-benzo[*a*-1,4]benzothiazino[3,2-*c*]phenothiazine, by using dipole moment values calculated with the PM3 method ([25] and references therein). 9-MeBPHT (**3**), with the smallest value of the ground-state dipole moment, was found to inhibit most the growth of Ehrlich mouse carcinoma solid cells, whereas 5-oxo-5*H*-BPHT (**6**), with the highest ground-state dipole moment, displayed the smallest antineoplastic activity (Table 5). Finally, Motohashi et al. [95] also investigated the possible relationships between the in vitro and in vivo antibacterial activity of some novel "half-mustard type" phenothiazines and several theoretical parameters (calculated by the PM3 method), such as ground-state and first excited singlet-state dipole moments, π-electron densities, highest occupied molecular orbital (HOMO) and lowest unoccupied molecular orbital (LUMO) energies and HOMO and LUMO coefficients, lone pair electron orbital (n) energies and lone pair electron densities on the urea site of these phenothiazines. Apparently, satisfactory correlations were found between antibacterial activity and dipole moments, n energies and lone pair electron densities of N-1, O and N-3 atoms on the urea site, which suggests that the urea sites should greatly contribute to the antibacterial activity of half-mustard type phenothiazines.

2.2.1.4
Triplet and Singlet Excited-State Luminescence Quenching

The photochemical and photophysical properties of BPHTs and other phenothiazine derivatives have been investigated by means of a combination of time-resolved laser techniques, including laser flash photolysis and steady-state fluorescence spectroscopy [30, 96]. The main goal of these studies was to develop novel photosensitizers for photoresist technology in the photoinduced crosslinking of polymers.

Barra et al. [30] studied the mechanisms of quenching of BPHTs and other phenothiazine derivatives by 1,3,5-tris(2,3-dibromopropyl)-1,3,5-triazine-2,4,6-(1*H*,3*H*,5*H*)-trione, the phenothiazines and BPHT acting as photosensitizers in these mechanisms. Laser excitation of BPHT in diglyme [bis(2-methoxyethyl) ether] solutions exhibited an intense signal, attributed to a triplet–triplet absorption centred at $\lambda_{max} = 560$ nm. A triplet lifetime of 1.5 μs was found. On the other hand, a steady-state fluorescence study of BPHTs in diglyme solution revealed a 500-nm emission band and a fluorescence lifetime of 22 ns. The quenching of BPHT fluorescence by 1,3,5-tris(2,3-dibromopropyl)-1,3,5-triazine-2,4,6-(1*H*,3*H*,5*H*)-trione obeyed a Stern–Volmer relationship, yielding a bimolecular rate constant (k_Q) value of 4.0×10^8 M^{-1} s^{-1}. The authors concluded that the ability of molecules such as BPHTs to act as photosensitizers might be related more to their singlet than their triplet state [30].

In a further study, Pohlers et al. [96] also measured the rate constants for the triplet-state quenching of BPHTs by three triazines, including 2-methyl-, 2-(2′-furylethylidene)- and 2-[(4′-methoxy)styryl]-4,6-bis(trichloromethyl)-1,3,5-triazine, in deaerated acetonitrile at room temperature. For these three compounds, the authors obtained k_Q values of 9.9×10^{10}, 12.2×10^{10} and 15.9×10^{10} M^{-1} s^{-1}, respectively, which were larger than the diffusion-limited rate constant in acetonitrile ($k_{DIFF} = 2 \times 10^{10}$ M^{-1} s^{-1}), and exceeded by about two orders of magnitude those determined in diglyme [30]. In contrast, the fluorescence properties of BPHTs, such as the emission maximum and lifetime in acetonitrile (509 nm and 26.4 ns, respectively) [96] were very close to those reported in diglyme. The k_Q values, obtained for BPHTs in acetonitrile by means of the Stern–Volmer plots, ranged between 1.1 and 1.7×10^{10} M^{-1} s^{-1}, according to the respective triazine derivative, which indicates that the BPHT fluorescence quenching process was diffusion-controlled [96]. However, these k_Q values were also about two orders of magnitude larger than those found in diglyme, which shows the effect of solvent on the mechanism. Clearly, all these literature data suggest that the sensitization mechanism of BPHTs involves an electron transfer from the BPHT excited states to the triazine derivatives.

2.2.2
Inclusion Complexes of BPHTs with Cyclodextrins

Because BPHTs possess a strongly hydrophobic structure, these compounds are sparingly soluble and rather unstable in water. This particular behaviour constitutes an important problem for biological, physicochemical and analytical studies. Indeed, for instance, it is very difficult to detect fluorescent BPHTs in aqueous media by spectrofluorimetry at low concentrations and with good precision, although the fluorimetric determination of the same compounds in organic solutions is generally very easy and precise. Therefore, it has been proposed to add CDs as complexing reagents in BPHT aqueous solutions, in order to increase the solubility and stability of the hydrophobic structures of the BPHTs, as well as to facilitate their quantitative analysis [21].

CDs are cyclic oligosaccharides obtained from the degradation of starch by glucosyltransferase enzyme ([25] and references therein). They have internal cavities made of six, seven or eight glucose units, designated by the symbols α, β and γ, respectively. The number of glucose units determines the dimension and size of the cavity. The cavity interior surface is lined by bridges between hydrogen atoms and glycosidic oxygens, which makes the CDs relatively hydrophobic. In contrast, the external face is hydrophilic because all hydroxy groups are located outside the cavity. The principal interest in CDs might rise from their ability to include organic hydrophobic molecules in their empty cavities and to form inclusion complexes. According to the size and type of the CD host and guest molecular structures, the CD complexes could present different host–guest stoichiometries. The formation of these inclusion complexes (CD complexes), obtained with physical forces and without covalent bonding, could be either totally or partially dependent on the geometrical factors rather than the chemical parameters of the guest molecules, such as phenothiazines or BPHTs.

Recently, Aaron et al. [18–23] investigated the hydrophobic interactions of BPHT and three methyl derivatives (9-Me, 10-Me and 11-MeBPHT) with two CDs, i.e. HP-β-CD and β-CD, in a predominantly aqueous medium by fluorescence spectroscopy. The stoichiometry and thermodynamic characteristics of the inclusion complexes were determined. Also, AM1 semi-empirical calculations of the geometry of BPHT and the methyl derivatives were performed to propose a molecular structure of these complexes.

2.2.2.1 Fluorescence Enhancement of Benzo[*a*]phenothiazines in the Presence of CDs

A significant and progressive increase of the fluorescence intensity of benzo[*a*]phenothiazine and its methyl derivatives was observed upon addition of increasing concentrations of HP-β-CD and β-CD, from 0 to 10^{-1} M and to 10^{-2} M, respectively. When plotting BPHT fluorescence intensity (I_F, measured at the emission maximum wavelength), versus [CD], levelling-off curves were obtained (see, for example, Fig. 6) [18–20, 22, 23]. A plateau region was reached for HP-β-CD concentration values of 0.02–0.1 M, depending on the BPHT derivative, whereas no plateau was observed for β-CD because of its lower solubility limit. These curves suggest that a great majority of BPHT molecules could form inclusion complexes with CDs in the ground state. This conclusion appears to be in good agreement with the molar fraction values of complexed molecules, calculated by the Flamigni equation [97], which were found to range between 0.90 for 9-MeBPHT and 0.93 for BPHT [19, 21]. Also, it is worthwhile noting that the effect of CD concentration did not produce any noticeable spectral changes, except for 10-MeBPHT because a progressive 12-nm red shift of the maximum emission wavelength of 10-MeBPHT was observed at CD concentrations larger than 5×10^{-3} M, indicating the probable formation of more than one type of host–guest complex (1 : 1) for the latter compound.

The fluorescence enhancement of BPHTs observed in the presence of CDs was probably due to the following processes. When BPHT molecules were included in the apolar HP-β-CD cavity, the collisions with solvent (water)

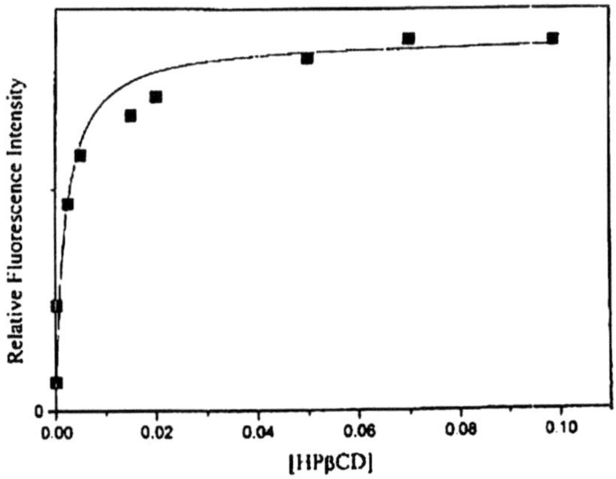

Fig. 6 Influence of HP-β-CD concentration on the fluorescence intensity (measured at $\lambda_{em} = 513$ nm) of a 10^{-5} M 9-methyl-12*H*-benzo[*a*]phenothiazine water/ethanol 99 : 1, v/v, mixture. The *solid line* was calculated through the use of Eq. 4, assuming a 1 : 1 stoichiometry and using F_{∞} and K_f values obtained from a non-linear regression analysis (from [19], with permission)

molecules would be practically suppressed, and the non-radiative deactivation of the guest molecules' singlet excited state would become less important, which would prevent the fluorescence quenching of the BPHTs by water molecules [21, 22, 98, 99]. Indeed, several authors [99, 100] have shown earlier that water molecules could quench the singlet excited states of solute molecules possessing an N – H group because of the formation of exciplexes between the fluorophore excited state and water molecules.

2.2.2.2
Stoichiometry and Association Constants of BPHT:CD Inclusion Complexes

The stoichiometry and association constants of the inclusion complexes formed between BPHTs and CDs were classically calculated by means of the Scatchard and Benesi–Hildebrand methods [101, 102]. Assuming a 1 : 1 stoichiometry ratio, the following equilibrium 1 is established:

$$[\mathrm{CD}] + [\mathrm{BPHT}] \rightleftarrows [\mathrm{CD} : \mathrm{BPHT}] \,. \tag{1}$$

The complex formation constant (K_f) can be expressed by Eq. 2:

$$K_f = \frac{[\mathrm{CD} : \mathrm{BPHT}]}{[\mathrm{CD}] \cdot [\mathrm{BPHT}]} \,, \tag{2}$$

where [CD], [BPHT] and [CD] : [BPHT] are the corresponding equilibrium concentrations of these species.

Taking into account a large excess of initial CD concentration, $[\mathrm{CD}]_0$, over the complex concentration and the mass balance, it is possible to establish the following form of the Benesi–Hildebrand equation between the fluorescence intensity and the CD concentration, for an inclusion complex possessing a 1 : 1 stoichiometry [102]:

$$\frac{1}{F - F_0} = \frac{1}{F - F_0} + \frac{1}{(F_\infty - F_0) \cdot K_f \cdot [\mathrm{CD}]_0} \tag{3}$$

where F_0 and F_∞ denote, respectively, the BPHT fluorescence intensity in the absence of CD and when all fluorophore molecules are essentially complexed, F is the measured fluorescence at each CD concentration tested and $[\mathrm{CD}]_0$ is the initial CD concentration.

For a complex with a 1 : 1 stoichiometry, a linear plot of $1/(F - F_0)$ vs $1/[\mathrm{CD}]_0$ should be obtained. Also, an alternative equation exists for a 2 : 1 stoichiometry, in which a plot of $1/(F - F_0)$ vs $1/[\mathrm{CD}]_0{}^2$ should provide a straight line. In both cases, the linear plot can be used to estimate a K_f value by simply dividing the intercept by the slope.

For all BPHT derivatives studied by Aaron and co-workers [18–23], a linear relationship was established when plotting $1/(F - F_0)$ vs $1/[\mathrm{CD}]_0$ according to Eq. 3, with correlation coefficient values larger than 0.99, which indicates that the inclusion complex stoichiometry is 1 : 1. In contrast, in all cases a downward concave curvature was observed when the data were fitted to the equation corresponding to a 2 : 1 complex, which means that in the conditions of these studies, no 2 : 1 host–guest complex was formed. However, the occurrence of the sequential, equilibrated formation of 1 : 1 plus 2 : 1 complexes cannot be excluded since a linear plot should also be obtained.

In addition, it is worthwhile noting that the use of the Benesi–Hildebrand equation provides only approximate K_f values, because more emphasis is placed on the lower concentration values than on the higher ones, and the data are not weighted properly [85, 103, 104]. Therefore, a better estimation of K_f can be made by using Eq. 4 based on a non-linear regression (NLR) analysis [83, 86]:

$$F = F_0 + \frac{(F_\infty - F_0) \cdot K_f \cdot [\mathrm{CD}]_0}{1 + K_f \cdot [\mathrm{CD}]_0} \, . \tag{4}$$

The K_f values of the BPHT : CD (1 : 1) inclusion complexes, determined by means of Eq. 4, are presented in Table 6. It can be seen that these K_f values are low, ranging between about 14 and 460 M^{-1} at 20 °C, which suggests that the complexation ability of CDs with BPHT molecules should be very weak.

The effect of temperature on the K_f values was also investigated in the case of the 10-MeBPHT : HP-β-CD inclusion complex, the measurement temperatures ranging between 5 and 45 °C [18]. As expected, the K_f values were found to increase with temperature from about 40 to 76 M^{-1} (Table 6). Using these results, it was possible to determine the formation enthalpy (ΔH^0), entropy

Table 6 Association constants (K_f) of BPHT : CD inclusion complexes, determined by fluorescence and UV–visible absorption spectroscopy at different temperatures (from [25], with permission)

Complex [a]	K_f (M^{-1}) [b]	T (°C)	Refs.
BPHT : HP-β-CD	100 ± 20	20	[21, 22]
9-MeBPHT [c] : HP-β-CD	460 ± 100	20	[19]
10-MeBPHT [d] : HP-β-CD	43 ± 17 [e]	20	[18]
10-MeBPHT : β-CD	14 ± 7	20	[18]
10-MeBPHT : HP-β-CD	40 ± 7	5	[18]
10-MeBPHT : :HP-β-CD	50 ± 11	20	[18]
10-MeBPHT : HP-β-CD	66 ± 9	35	[18]
10-MeBPHT : HP-β-CD	76 ± 18	45	[18]
11-MeBPHT : HP-β-CD	141 ± 11	20	[20]

[a] [BPHT] = 10^{-5} M, unless otherwise noted. In H_2O/EtOH (99 : 1, v/v), with [HP-β-CD] varying between 0 and 10^{-1} M, or [β-CD] between 0 and 1.5×10^{-2} M
[b] Determined by fluorescence spectroscopy, unless otherwise noted
[c] [9-MeBPHT] = 5×10^{-5} M
[d] [10-MeBPHT] = 5×10^{-5} M
[e] Determined by UV–visible absorption spectroscopy

(ΔS^0) and free energy (ΔG^0) from the Vant'Hoff equation, by plotting $\ln K_f$ vs $1/T$. A linear relationship was obtained, indicating that the Vant'Hoff equation was verified. A ΔH^0 value of 12.40 ± 0.08 kJ mol^{-1} was found, showing that the complexation process is endothermic, whereas a large positive value of ΔS^0 (75.0 ± 0.3 J mol^{-1} K^{-1}) was calculated, indicating an increase of the molecular freedom degree during the 10-MeBPHT : HP-β-CD inclusion complex formation. The latter result for entropy was attributed by the authors to the variation of the contribution of water molecules, associated in solvent cages around the 10-MeBPHT molecules, occurring during this process. The negative value of ΔG^0 (– 10.1 kJ mol^{-1}) demonstrated that complex formation was energetically favourable.

2.2.2.3
Proposed Molecular Structure of the BPHT:CD Inclusion Complexes

By considering the respective molecular size of both entities which are present in the inclusion complexes, it is possible to propose a reasonable model for the structure of these inclusion complexes [18, 21, 22]. This model was obtained by means of simple calculations of the BPHT molecular structure, based on the AM1 semi-empirical method.

When taking into consideration the calculated dimensions of BPHT and 10-MeBPHT as well as the HP-β-CD cavity dimensions [105, 106], it was

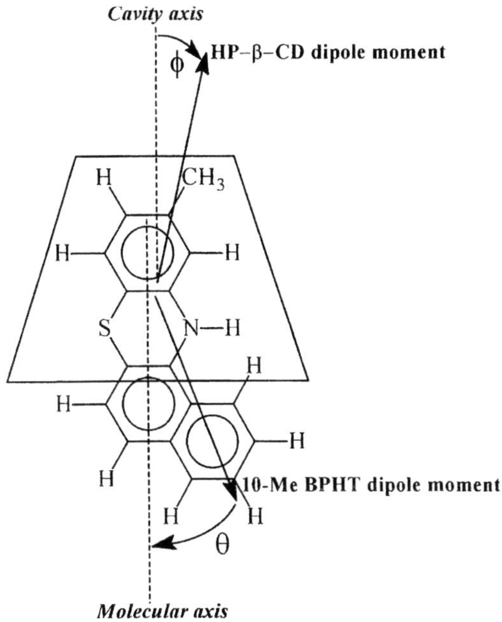

Fig. 7 Molecular geometry of the inclusion complex formed between 10-MeBPHT and HP-β-CD (from [18], with permission)

shown that an equatorial inclusion could not occur in the HP-β-CD cavity because of the BPHT longitudinal dimensions, which are larger than the CD cavity length. Therefore, the inclusion should be partially axial for these two molecules, two different complexation sites being theoretically possible in the BPHT molecules, located on the naphthalene moiety and on the aromatic ring linked to the N and S heteroatoms (Fig. 7).

In the case of the 10-MeBPHT:CD complex formation, the calculated interatomic distances and additional factors, including the effect of the 10-methyl group van der Waals radius and the non-planar shape of the molecule, clearly supported the hypothesis that the insertion in the HP-β-CD cavity should very likely take place on the heteroatom-linked phenyl side group. Moreover, when taking into account the 1 : 1 stoichiometry of the complex, only one site among the two possible ones should be active during the inclusion process, and the smaller K_f value found for 10-MeBPHT than for unsubstituted BPHT (Table 6) confirmed that the presence of the methyl group had the effect of hindering and decreasing the complexation process rate. The existence of important dipole–dipole interactions occurring between HP-β-CD and 10-MeBPHT, according to an antiparallel orientation of both dipoles, might also support the insertion in the HP-β-CD cavity of 10-MeBPHT by the heteroatom-linked phenyl side group, as could be seen in the proposed molecular structure of the inclusion complex (Fig. 7).

2.2.3
Theoretical Calculations and Various Applications

In this section, we present theoretical calculations on BPHTs related to their antitumour therapeutic properties and various applications concerning photoresist technology and analytical chemistry.

2.2.3.1
Theoretical Calculations and Antitumour Therapeutic Properties

In several papers, Motohashi et al. [107–109] have attempted to established correlations between the antitumour activity of BPHTs and their π-electron density, evaluated by quantum-chemical calculations. The prediction of the BPHT antitumour effect and carcinogenicity was performed by calculating the resonance energy per π electron (REPE) values for these compounds and comparing these REPE values with those of their analogues, constituted by the parent molecule skeleton in which a double bond is lacking, in the K region, L region or M region, respectively. Calculations were based on the Hückel molecular orbital (HMO) method and the modified-neglect-of-diatomic overlay (MNDO) orbital method. Aihara's graph theory of aromaticity was used to delineate the resonance energy (RE) and other analogous parameters [107–109]. The authors obtained REPE resonance integral (β) values higher for BPHTs than for their analogues, which suggests a greater stability of BPHTs relative to these analogues.

The respective charges located on the nitrogen atom (Q_N) and the L region (Q_L), the M region (Q_M) and the K region (Q_K) in the BPHTs were calculated (Fig. 8) [109]. The comparison of the Q_N, Q_K and Q_M values of unsubstituted BPHT with those of 9-Me-, 10-Me- and 11-MeBPHT showed no significant differences in the Q_K values for these compounds, whereas Q_N was found to be the highest in the case of 9-MeBPHT. Also, several 5-oxo-BPHT derivatives displayed Q_N, Q_K and Q_M values lower than those of 9-MeBPHT. All the obtained data indicated that the stability of the π electrons in one benzene

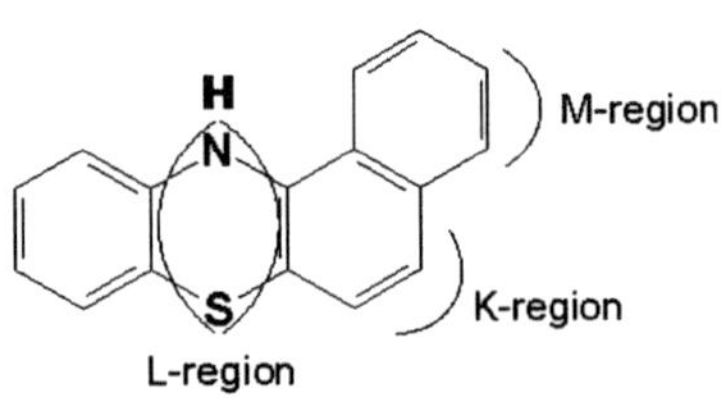

Fig. 8 Regions of charges of the BPHT molecules (from [25], with permission)

and one naphthalene ring in the BPHT skeleton might be responsible for the induction of anticancer activity.

In another work, Motohashi et al. [108] have reported that, among several BPHT derivatives, 9-MeBPHT presented the greatest antitumour activity against a solid type of Ehrlich carcinoma implanted in dd/Y mice, with a rate inhibition larger than 60%. As a consequence, the authors concluded that the K, L and M regions of 9-MeBPHT might be responsible for the potent anticancer activity of this compound. The effects of methyl substitution on the carcinogenic activity of BPHT have been also studied by means of the Aihara theory [107]. Several correlations have been established between the REPE β values for the cationic species, corresponding to the parent arenes which do not possess a heteroatom, and the carcinogenic activity.

2.2.3.2 Application as Photosensitizers for Photoresist Technology

BPHT and other phenothiazine derivatives have been widely applied as photosensitizers to the development of photoresist technology [30, 96, 110]. Photoresist systems have been established to achieve the differential solubility of thin polymer films [30]. A key element in these photoresist systems is represented by the so-called photosensitive acid generator (PAG). The PAG can be defined as a compound that, under UV irradiation, liberates an acid (HX) and yields other photolysis photoproducts, according to the following reaction [30]:

$$\text{PAG} \xrightarrow[\text{room temperature}]{h\nu(\text{UV})} \text{Photoproducts} + \text{HX}\,.$$

Several experiments have been performed on the use of BPHT and other phenothiazine derivatives as photosensitizers for an important class of PAGs, including triazines such as trichloromethyl-1,3,5-triazine and 2,3-dibromopropyl-1,3,5-triazine [30, 96, 110].

An electron-transfer mechanism from the photosensitizer excited state to triazines has been proposed, which has permitted a satisfactory interpretation of the observed fluorescence quenching. The steps of the photosensitization mechanism of PAGs with bromine-containing compounds by BPHTs are summarized below:

$$\text{BPHT} \xrightarrow{h\nu} (\text{BPHT})^*$$
$$(\text{BPHT})^*(S_1, T_1) + \text{RBr} \xrightarrow{\text{Electron transfer}} (\text{BPHT})^{+\cdot} + \text{RBr}^{-\cdot}$$
$$(\text{BPHT})^{+\cdot} \rightarrow (\text{BPT})^{\cdot} + \text{H}^+$$
$$\text{RBr}^{-\cdot} \rightarrow \text{R}^{\cdot} + \text{Br}^-\,.$$

The rate constants for the reactions of both BPHT lowest excited singlet (S_1) and triplet (T_1) states with bromo- and chloro-substituted compounds were found to be well correlated to the electron-transfer free energy.

2.2.3.3 Analytical Applications

Until now, in contrast to the case of simple phenothiazines (see Sect. 2.1.4), very few analytical methods have been applied to the determination of BPHTs. A simple CD-enhanced spectrofluorimetric method has been developed for the quantitative analysis of methyl-substituted BPHTs, including 9-Me-, 10-Me- and 11-MeBPHT, by Aaron et al. [19, 20, 23]. The analytical figures of merit of the three compounds were evaluated in ethanol and in CD aqueous media. The linear calibration curves were established with their correlation coefficients close to unity. The linear dynamic ranges (LDRs) were spread over two to four orders of magnitude, according to the compound. The RSDs ranged from 4 to 9%, which showed a good reproducibility of the spectrofluorimetric measurements. The LOD values were generally lower in CD media than in ethanol and water. For example, LOD values of 0.07 and 1.9 ng/mL for 9-MeBPHT, 9 and 33 ng/mL for 10-MeBPHT and 0.06 and 1 ng/mL for 11-MeBPHT were reported in the presence of β-CD and HP-β-CD, respectively.

By means of this CD-enhanced fluorescence method, the authors could determine 11-MeBPHT in spiked human urine samples [23]. The standard addition procedure, based on the addition of a constant volume of the spiked urine sample to solutions of increasing methyl-BPHT concentration, was applied using spectrofluorimetric measurements. The linear calibration curves and standard addition linear plots were parallel, which suggested the absence of matrix interferences in the urine samples and permitted the validation of the method. Good linearity and precision were also obtained for the standard addition plots, with correlation coefficients close to unity. Satisfactory recovery percentage values ranging from 97 to 108% were found in the urine samples for the determination of 11-MeBPHT concentrations at low μg/mL levels.

3 Biological and Biomedical Properties of Phenothiazines and Benzo[*a*]phenothiazines

The remarkable photochemical, photophysical and physicochemical properties of the phenothiazine and benzo[*a*]phenothiazine substituted derivatives should also result in a variety of biological activities and a number of applications in medicine and related fields.

The presence of two heteroatoms, i.e. nitrogen and sulphur atoms, in the phenothiazine fused rings plays an important role in the occurrence of various and extensive interactions of these molecules with biological systems. Moreover, the chemical synthesis of new molecules based on the phenothiazine and benzo[*a*]phenothiazine heteroaromatic cycles has widely opened several research fields to be investigated in connection with their biological and therapeutic properties. Technological developments and improvements have allowed one to elucidate, in several cases, the interaction mechanism of these novel compounds with tissues, cells, membranes, nucleus and subcellular structures. Because this type of investigation constitutes an important part of pre-clinical research, it is also necessary to determine the positive medical effects of phenothiazines and benzo[*a*]phenothiazines as well as their negative and/or unwanted ones. Indeed, the overdose and/or long-term use of phenothiazine-based medicaments might provoke unwanted consequences, such as difficulties in neuromuscular coordination, photosensitivity and haemolytic anaemia [111, 112]. Nowadays, these problems represent the goal of a number of research groups because the mechanisms of the side effects of these phenothiazine-based drugs are still not completely understood.

In this part of the chapter, we try to systematically present the most recent, biological effects and biomedical applications of the phenothiazine and benzo[*a*]phenothiazine derivatives.

3.1
Antibacterial, Antifungal, Antiviral and Antiparasitic Activities of Phenothiazines

Phenothiazine and many of its substituted derivatives possess important and various biological activities, and have been shown to present significant insecticidal, antifungal, antibacterial and anthelmintic properties. Unsubstituted phenothiazine has also been used for vermifugal application, which has earned this molecule a good place alongside penicillin and DDT for its great impact on mankind, as recently pointed out by Mitchell [111].

The mechanism of the antibacterial activities of phenothiazines was investigated on several biological systems, including *Escherichia coli* and mouse skin. For example, Komatsu et al. [113] tested 17 different phenothiazines and demonstrated that these compounds had no antibacterial effect in direct contact with *E. coli*, using the disc diffusion method, except for chlorpromazine which displayed a weak bacterial growth inhibitory action. In the next step, these authors performed a pre-treatment on mice using the same 17 phenothiazines. It was found that all phenothiazines could induce an activity against *E. coli* at different intensities, depending on the compound. On the basis of their experiments, the authors concluded that the protective effect might be due either to the immuno-potential of the macrophage inducing activity, or to the inactivation of bacteria-induced lymphokines. The damage of the *E. coli* membranes was probably provoked by the phototoxicity of some of these phenothiazines. Indeed, the latter ones were lipophilic and could induce surface pressure changes on lipid monolayers, which are mimetic of bacterial membranes, allowing the penetration of these molecules through the biological membranes. The bacteriolysis was found to be not significantly different on light exposure and in the dark [114]. Recently, the treatment of infectious diseases based on the use of chemo-sensitizing compounds, including thioxanthene and phenothiazine derivatives, and also in combination with an anti-infective agent was also the object of a patent [115]. In another study, Kristiansen et al. [116] examined the effect of phenothiazines on the resistance of methicillin-resistant *Staphylococcus aureus* (MRSA) strains to oxacillin. The mechanisms of antibiotic resistance of bacteria included efflux pumps which might extrude the antibiotic prior to reaching its target. Phenothiazines inhibited the activity of some efflux pumps, thereby altering the susceptibility of bacteria. Chlorpromazine and thioridazine were found to reduce the susceptibility of MRSA, although not that of methicillin-susceptible *S. aureus* (MSSA) strains to oxacillin. Reserpine, an inhibitor of antibiotic efflux pumps, could also reduce the resistance of MRSA strains to oxacillin.

Antibacterial photodynamic therapy is based on the combination of a non-toxic photoactivatable dye (or photosensitizer) and harmless visible light. In these conditions, singlet excited-state oxygen and free radicals are generated, and the microbial cells are killed. The cationic phenothiazinium dyes, such as methylene blue and toluidine blue O, have been studied for this purpose, and clinically applied to antibacterial photodynamic therapy. However, the question of using this therapy supplementary to and/or in combination with other multidrug resistance inhibitors is still not clear [117]. The experiment using light-activated phenothiazinium-based photosensitizers in order to induce the photodamage of *S. aureus* DNA was called the "comet assay". The major site of photosensitizer attack was found to be DNA nucleosides other than guanosine [118]. Because fungi are eukaryotic cells, they might share many pathways with human cells, thus increasing the probability of antifungal activity of "non-fungal drugs". The potential of these drugs for the treatment of fungal infections has been sporadically investigated, using the drugs either alone or in combination with "classic" antifungal agents. Therefore, the role of these antifungal agents in the epidemiology and clinical manifestations of fungal infections, as well as the potential of certain drugs for treating invasive fungal infections, might require further investigation.

The inhibition of human hepatic enzyme cytochrome P4502E1 by azole antifungal, CNS-active drugs and non-steroidal anti-inflammatory agents was also investigated by Tassaneeyakul et al. [119]. Several antifungal azoles, CNS-active drugs (anticonvulsants, antidepressants, antipsychotics and benzodiazepine hypnosedative anxiolytics) and non-steroidal, anti-inflammatory agents were shown to inhibit the human liver microsomal 4-nitrophenol (4-NP) hydroxylation, a reaction used as a marker of CYP2EI activity. The imidazoles bifonazole, clotrimazole, econazole and miconazole could be un- or non-competitive inhibitors of 4-NP hydroxylation, whereas fluonazole, itraconazole and ketoconazole caused minor or negligible inhibition. Among the CNS-active drugs screened, a significant inhibition occurred only with the tricyclic antidepressants, including phenothiazines and two benzodiazepines, flurazepam and medazepam. These data indicate that, although these CNS-active drugs are not substrates for CYP2EI, some clinically used drugs also had the capacity to inhibit this enzyme and hence they have the potential to modulate the toxicity of non-drug xenobiotics metabolized by CYP2E1.

In the treatment of paludism, a series of newly synthesized phenothiazine compounds were found to inhibit the growth of both chloroquine-sensitive and chloroquine-resistent strains of *Plasmodium falciparum*. The antimalarial activity of these phenothiazines increased with the number of alkaline groups located in the alkylamino side chain. Also, phenothiazine, chloropromazine and other novel phenothiazine compounds inhibited the formation of β-hematin, in a way similar to chloroquine [120].

In another type of medical study, some phenothiazine derivatives were prepared in order to obtain a drug permitting the prevention and/or treatment of hearing impairment [121, 122]. The efficiency of these phenothiazine compounds (Fig. 9) in the treatment of damage to the cochlea was investigated. The developed drug was protected by patents [121, 122].

Rashidi et al. [123] examined the interactions of some phenothiazines and purines with the metabolism of famciclovir catalysed by guinea pig liver aldehyde oxidase. The activity of famciclovir was analogous to that of acyclovir, a drug used against herpes and hepatitis B viruses. In their study, the authors showed that both famciclovir and 6-deoxypenciclovir served as substrates of hepatic aldehyde oxidase. This in vitro study based on spectrophotometric and HPLC methods indicated that aldehyde oxidase was partially purified from guinea pig liver, and that it induced interactions of chlorpromazine, promethazine, phenothiazine, azathioprine, 6-mercaptopurine, theophylline, caffeine and allopurinol with famciclovir or 6-deoxypenciclovir. It is worthwhile noting that only phenothiazine derivatives yielded a marked inhibition of oxidation of both substrates, the treatment with chlorpromazine (CPZ) giving the highest inhibition. In contrast, the oxidation of famciclovir and 6-deoxypenciclovir was not significantly affected by allopurinol (the xanthine

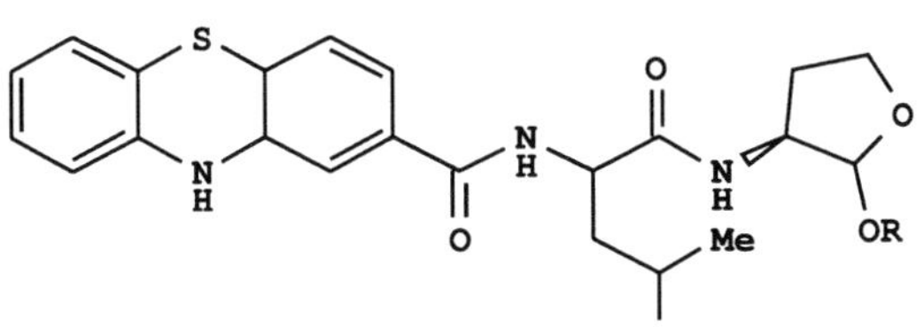

I

Fig. 9 Structure of phenothiazine compounds (R = H, alkyl, aralkyl, etc.) used in medicine for preventing and/or treating hearing loss

oxidase inhibitor). This means that aldehyde oxidase, not xanthine oxidase, is the major molybdenum hydroxylase in the oxidation of both substrates in the guinea pig liver fractions.

The antibacterial activity of two novel compounds, phenothiazinium new methylene blue N and phenoxazinium brilliant cresyl blue, was investigated [124]. It was found that the two compounds exhibited their bactericidal activity at lower concentrations when compared to the well-established phenothiaziniums, such as methylene blue and toluidine blue O, as well as a recently synthesized blood photovirucidal agent of 1,9-dimethyl-methylene blue. The photoactivity of these compounds was undiminished in the presence of red blood cells [124].

In another study, human peripheral blood monocyte derived macrophages (HPBMDMs) and human macrophage cell line THP-1 were used at extracellular concentrations to investigate the effect of clinical concentrations of several phenothiazines, including thioridazine, against phagocytosed *S. aureus* [125]. These macrophages have their nominal killing activity against bacteria, and therefore they would not influence any activity that the phenothiazine compounds might have against intracellular localized *S. aureus*. The results indicated that concentrations of phenothiazines ranging from 0.1 to 10 mg/L were non-toxic to the macrophages, although thioridazine presented an in vitro minimum inhibitory concentration of 18 mg/L against the strains of *Staphylococcus aureus* and did not induce cell death. In the former conditions, thioridazine did not cause the cellular expression of activation marker CD69 or the induction of CD3+ T cell production of IFN-γ, but blocked cellular proliferation and down-regulated the production of T-cell-derived cytokines such as IFN-γ and IL-5. This suggests that thioridazine might induce an intracellular bactericidal activity, independently of its capacity to generate Type 1 responses against *S. aureus*.

Very strongly resistant bacteria such as *Mycobacterium avium* to most non-toxic compounds are often isolated from AIDS patients as a co-infection factor. The investigation of Viveiros et al. [126], concerning the in vitro activity of chlorpromazine (CPZ), thioridazine, promazine (PZ), promethazine and desipramine against referential and clinical strains of *M. avium*, showed that thioridazine at concentrations ranging from 10 to more than 50 mg/L was the most active derivative of this group against AIDS. Their findings suggested that thioridazine could be applied in vivo as a complementary therapy of co-infected AIDS patients.

The discovery of calmodulin-like protein in *M. leprae* opened new access to diminish the growth of this microorganism. Indeed, *M. leprae* was treated in vitro by six phenothiazines. Among the six compounds under study, trifluoperazine appeared to be the most potent for inhibiting the microorganism, chloropromazine, triflupromazine and thioridizine being less active in inhibition, while acetopromazine and fluphenazine were completely ineffective [127].

The effectiveness and mechanism of phenothiazine compounds such as chlorpromazine (CPZ) and trifluoperazine against *Candida albicans* were investigated in vitro by Sharma et al. [128]. The authors explained that the anticandidal effect of these phenothiazine compounds might be due to their interactions with Ca^{2+}/calmodulin, which depended on the protein phosphorylation. The phenothiazines also produced a significant decline in the cellular lipids and phospholipids. It was proposed that CPZ and trifluoperazine might affect the cell cycle through DNA synthesis (S phase) and cell division phases, which are governed by calmodulin and Ca^{2+}/calmodulin-dependent protein phosphorylation. Moreover, these results demonstrated the important role of phenothiazine compounds in developing new anticandidal drugs.

For the peroxidase/H_2O_2/system, phenothiazines were found to irreversibly inhibit *Trypanosoma cruzi* dihydrolipoamide dehydrogenase (LADH). Gutierrez-Correa et al. [129, 130] discovered that the inactivation of the parasite enzyme depended on: (a) the phenothiazine structure, (b) the peroxidase nature, (c) the incubation time, and (d) the presence of a cation-radical scavenger. In the case of the myeloperoxidase/H_2O_2/system, promazine, trimeprazine, thioridazine, promethazine, prochlorperazine, chlorpromazine (CPZ) and perphenazine were the most effective derivatives among 12 phenothiazines under study. It was found that their electron-acceptor substituents, such as Cl, trifluoromethyl, propionyl and nitrile groups, which are located at the phenothiazine ring 2-position, decreased or suppressed the phenothiazine activity. The peroxidase systems activated by phenothiazines produced the corresponding cation radicals, which had an essential role in the phenothiazine chemotherapeutic action. This production of phenothiazine cation radicals was demonstrated by optical spectroscopy.

The trypanothione located in the redox defences of pathogenic trypanosomal and leishmanial parasites, in contrast to glutathione for their mammalian hosts, might play the important role of selective inhibitors of trypanothione reductase. Chan et al. [131] have designed a model including tricyclic neuroleptic molecular frameworks as structures leading to the development of selective inhibitors for trypanothione reductase over host glutathione reductase. From a homological-modelled structure for trypanothione reductase, replaced in the later stages of the study by the X-ray coordinates for the enzyme from *Crithidia fasciculate*, a series of inhibitors based on phenothiazines was designed. These phenothiazines were shown to constitute reversible inhibitors of trypanothione reductase from *T. cruzi*, linearly competitive with trypanothione as substrate, but non-competitive with reduced nicotinamide adenine dinucleotide phosphate (NADPH). Also, the same authors discovered an additional hydrophobic enzymatic region available to suitable N-substituents of the promazine nucleus. The selected compounds were tested for in vitro activity against three strains of *T. brucei*, *T. cruzi*, and

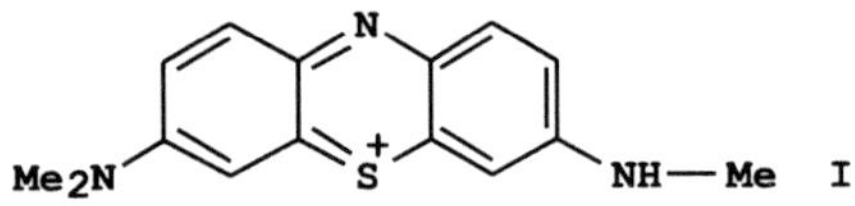

Fig. 10 Phenothiazinium derivatives with antiparasitic and biological activities

Leishmania donovani, and showed selective activities. As a consequence, the promazine analogues were recommended as potentially active drugs against trypanosomiasis and leishmaniasis [131].

The phenothiazine derivatives and related tricyclic neuroleptic compounds also exhibited inhibitory and lytic effects on *Entamoeba histolytica* HK9 and HMI trophozoites. Ondarza et al. [132] found that clomipramine (KD002), the most potent in vitro inhibitor of trypanothione reductase among 30 tested tricyclic compounds, caused a substantial decrease in the number of *E. histolytica* HK9 trophozoites. A substantial inhibitory effect on the cell proliferation was also demonstrated with metronidazol, which was used clinically against amoebiasis. Under similar experimental conditions, other tricyclic and phenothiazine derivatives which were designed originally to inhibit the trypanothione reductase of trypanosomatides also presented an inhibitory effect of 16 to 95% against *E. histolytica* HK9. The cell proliferation under anaerobiosis was higher in strain *E. histolytica* HK9 than in strain *E. histolytica* HM1 which is highly virulent, although metronidazol and clomipramine were less effective against HM1. The inhibitory activity on the cell proliferation and the lytic effects on the human parasite of the tricyclic derivative clomipramine, as well as the phenothiazines under study, suggested that these compounds were potentially anti-amoebic drugs.

Several newly synthesized dye-type phenothiazinium derivatives (see Fig. 10) were proposed as drugs having a significant antiparasitic activity against *Plasmodium*, *Babesia*, *Toxoplasma*, *Trypanosoma*, *Onchocerca*, *Filaria*, *Leishmania*, nematodes, plathelminthes and nemathelminthes [133]. However, up to now no other data are available on these compounds which were patented in 2001.

3.2 Neurological Effects of Phenothiazines

For a long time, some phenothiazine derivatives have been used as neuroleptic drugs, although their precise action on the central nervous system (CNS) at the molecular level still remains an important matter of investigation. Phenothiazines belong to a group of atypical antipsychotics which exhibit a neuromuscular transmission in the case of vertebrates. Nguyen and Miledi [134] studied the dissection of the cutaneous pescatoris muscles and nerves from *Rana temporaria* and *Rana pipenes* frogs. The effects of clozapine, sulpiridine, olanzapine and risperidone on the nicotinic synaptic transmission at the frog neuromuscular junction were examined. The results indicated that the contribution of different cytochrome P-450 isoenzymes to the N-demethylation and 5-sulphoxidation reactions might be involved. Therefore, these atypical antipsychotics could derive some of their therapeutic effects not only from their inhibitory action on dopaminergic receptors, but also from their pre- and post-synaptic modulation of the nicotinic neurotransmission.

Moreover, a clinical experience suggested the possibility of treating peripheral neuropathy by administering to symptomatic patients, who were suffering from pain and/or burning symptoms in the legs or feet and soles of the feet, a combination of two drugs, including a substituted phenothiazine and a tricyclic antidepressant. The substituted phenothiazines either exhibited a specific activity, or acted synergistically with the tricyclic antidepressants by providing effects which otherwise were not obtainable with one medication alone at normal dosage levels. The substituted phenothiazines were chosen from a group consisting of chlorpromazine hydrochloride, mesoridazine besylate, thioridazine hydrochloride, acetophenazine maleate, fluphenazine, fluphenazine hydrochloride, fluphenazine enanthate, fluphenazine decanoate, perphenazine, trifluoperazine hydrochloride, and their pharmaceutically acceptable free forms, and acid additional salts and esters thereof. The selected substituted phenothiazines might also be taken alone, but the most preferred combination of antidepressant and substituted phenothiazine was desipramine hydrochloride plus fluphenazine hydrochloride. The applied dosage level was found to depend on the severity of the neuropathy [135].

Daniel et al. [136] investigated the effects of phenothiazine neuroleptics on the rate of caffeine demethylation and hydroxylation in the rat liver. The influence of several phenothiazine neuroleptics, such as chlorpromazine, levomepromazine, thioridazine and perazine, on the cytochrome P-450 activity was measured by means of the caffeine oxidation in rat liver microsomes. The obtained results demonstrated that all of the neuroleptics under study competitively inhibited their caffeine oxidation in the rat liver, although their potency to inhibit the particular metabolic pathways was not equal. Levomepromazine exerted the most potent inhibitory effect on the caffeine oxidation pathways, and had the most pronounced influence on 8-hydroxylation.

The investigation of Taniguchi et al. [137] opened new hope in the search for medicaments in Alzheimer's disease and other taupathies such as Down's syndrome. They tested 42 compounds belonging to nine different chemical classes for their ability to inhibit the heparin-induced assembly of tau protein into filaments in vitro. Some of these compounds, including phenothiazines (methylene blue, azure A, azure B and quinacrine mustard), polyphenols (myricetin, epicatechin 5-gallate, gossypetin and 2,3,4,2′,4′-pentahydroxybenzophenone) and the porphyrin ferric deuteroporphyrin IX,

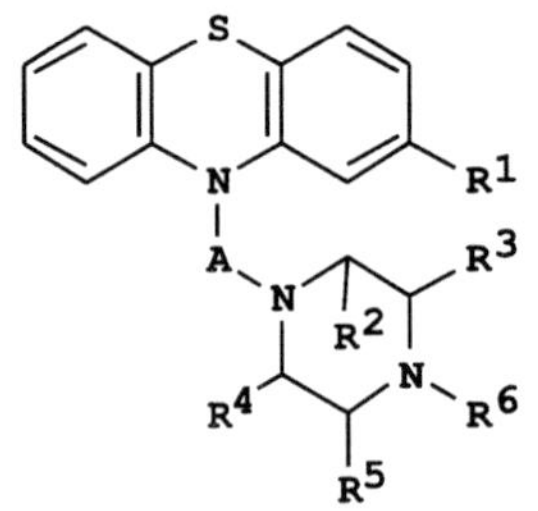

I

Fig. 11 General structure of piperazine phenothiazine derivatives with neuroprotective and/or neurotrophic effects on the central and/or peripheral nervous system

were found to inhibit the tau filament formation. The disassembly of tau filaments was observed in the presence of the porphyrin phthalocyanine. Compounds that inhibited tau filament assembly also inhibited the formation of Aβ fibrils. The compounds under study did not affect the ability of tau to interact with microtubules. The identification of small molecular inhibitors of heparin-induced assembly of tau could be a starting point for the development of mechanism-based therapies for Alzheimer's disease and other taupathies.

The manufacture of a phenothiazine-based medicament with neuroprotective and/or neurotrophic effects on the central and/or peripheral nervous system was recently patented [138]. The starting substances were piperazine phenothiazine derivatives, and their pharmaceutically acceptable salts or esters. One of the compounds included in the patent is flufenazine.

The general formula of the piperazine phenothiazine derivatives under study is presented in Fig. 11. In the structure of **I**, A = (un)branched $C_{2\text{--}6}$ alkylene separating N atoms linked thereto by at least two carbon atoms; R1 = H, halo, lower alkyl, lower alkoxy, etc.; R2, R3, R4 and R5 = H, Me or Et; R6 = H, lower alkyl, hydroxy lower alkyl, etc.

A combination of an antipsychotic compound belonging to the phenothiazine series and a device allowing the rapid delivery of this phenothiazine antipsychotic was proposed for the acute treatment of headache [139].

3.3
Antitumour Activities, Cytotoxicity and Cancer Multidrug Resistance Properties of Phenothiazines

It is very well known that the phenothiazines alone possess no antitumour and/or antineoplasma activity. But when they are applied in combination with other drugs, the phenothiazines can increase the anticancer activity and provoke a decrease of mutagenesis, diminishing the cytotoxic effects caused by radiation, carcinogenic chemicals and multidrug resistance (MDR). The positive effects of a simultaneous chemotherapy of cancers by combining some phenothiazines and other drugs could be especially remarkable in the brain, where the accumulation of phenothiazines is generally greater than in other organs [140].

3.3.1 Antitumour Activities

About 10 years ago, the first in vitro antitumour screening system based on half-mustard type phenothiazines was developed by Wuonola et al. [141]. The authors studied the antitumour effect of phenothiazines on 57 different tumour cell lines, including leukaemias, non-small lung, colon, CNS, ovarian, renal, breast, and prostate cancers, as well as melanoma cell cultures. Alkylurea derivatives of phenothiazine exhibited in vitro antitumour activity, but the phenothiazine phthalimido derivatives were not active on the majority of cancer cell cultures. In contrast, propylurea phenothiazine derivatives were active against some leukaemia cell types. Only two phenothiazine molecules possessing a butylene [$(CH_2)_4$] linker were active against non-small lung cancer cells, whereas phenothiazine compounds with a propylene linker were less effective. The same compounds were efficient on colon cancer lines, tumour cells from the CNS and melanoma cells, and it was supposed that their effectiveness was due to the presence of substituents at the phenothiazine 2-position. The majority of ovarian cancer cell lines (except one type, IGROVI) and five among eight renal cancer lines were not sensitive to these phenothiazine derivatives. The two butylene-linked phenothiazine ureas had a moderate antiproliferating action on two renal cancer cell lines. The prostate cancer and some breast cancer cell lines were not sensitive. Nevertheless, some breast cancer cell lines were apparently sensitive to CF_3-substituted phenothiazine alkylureas. In addition, the selectivity or organ cell specificity of the efficient phenothiazines had to be targeted for their improvement, in order to avoid the general cytotoxic effects of the half-mustard type phenothiazines.

Tanaka et al. [142] showed that the antitumour activity of phenothiazines was connected to their reaction with DNA, shielding DNA base pairs, which was verified by the UV absorption spectra. Indeed, phenothiazine, promethazine, promazine and chloropromazine formed coordination complexes with copper, palladium or gold. In the presence of metallic ions, these phenothiazine–metal coordination complexes increased the antimutagenic activity when compared to the parent compounds. The investigation was carried out on the *Salmonella typhimurium* TA98 strain which was treated with a strain-specific, direct-acting mutagen, 4-nitro-*O*-phenylenediamine, and the ligand–DNA interaction was analysed by UV absorption spectrometry [142].

Harris et al. [143] performed an investigation oriented towards the search for more efficient, non-classical cancer treatments, accompanied by photodynamic therapy (PDT), on EMT-6 tumour cells and erythrocytes. A series of phenothiazinium-based photosensitizers (PhBPs) were tested as PDT agents. PhBPs were incubated with EMT-6 tumour cells and erythrocytes, respectively, under dark and light conditions. Then, both the comet assay and haemolytic assay were used to assess the cellular photodamage induced by these PhBPs. In vitro assays were applied to determine the light absorption characteristics, singlet oxygen yields and lipophilicity of these PhBPs. The results suggested that DNA might be the primary target for the photodynamic antitumour activity of all the PhBPs under study, with the exception of 1,9-dimethylmethylene blue (DMMB) which might potentially also have the tumour cell membranes as the target.

Nichols et al. [144] patented some combinations of drugs including phenothiazines for the treatment of neoplasms. This invention provided formulations and structural modifications for phenothiazine compounds, resulting in altered biodistribution, thereby reducing the occurrence of adverse reactions associated with this class of drugs.

Phenothiazines exhibited antiproliferation effects on several tissue cultures. Gil-Ad et al. [145] studied in vitro the potential anticancer activity of perphenazine, fluphenazine, thioridazine, trifluoperazine and chlorpromazine in the wild-type and MDR B16 mouse melanoma cell lines. At the same time, the authors determined in vivo the antitumour effect of an in vitro selected, highly potent phenothiazine (thioridazine) in a murine melanoma model, using female C57/Bl mice. They concluded that all these agents induced a dose-dependent decrease in the cell viability in wild-type and MDR B16 melanoma cells, but that thioridazine displayed the highest antiproliferation activity. Apoptosis was confirmed by the co-staining of thioridazine-treated B16 cells with propidium iodide and Hoechst 33342 reagents and with Caspase-3 expression, a typical mediator of apoptosis. It was found that apoptosis was markedly increased following a 4-h exposure of B16 cells to thioridazine. The laboratory animals were inoculated with wild-type B16 cells by intravenous injection (i.p.) into the tail vein. Mice were treated with thioridazine, while the control animals received saline solution. The mice were monitored for 21–30 days, and their body weight was recorded. After autopsy, the lung weight and number of pulmonary melanoma colonies were determined. Thioridazine administration (i.p. or p.o.) resulted in the reduction of the lung tumour burden and an increase in mice survival. Therefore, it was concluded that several phenothiazines, particularly thioridazine, induced apoptosis of B16 melanoma cells and demonstrated a significant in vivo antitumour activity.

3.3.2
Cytotoxicity

Some phenothiazines, the so-called half-mustard type, stimulated T-cell blast formation, presented a natural killer cell activity, via possibly the activation of monocytes and macrophages, and an antibody-dependent cellular cytotoxicity of human peripheral blood mononuclear cells, and also showed cytotoxicity against several human cancer cell lines. These phenothiazines also might induce an in vivo antimicrobial activity by possibly their host-mediated immuno-potentiation [146]. Phenothiazines did not demonstrate any apparent mutagenic activity, but they were rather antimutagenic. These data suggest the possible applicability of half-mustard type phenothiazines and benzo[*a*]phenothiazines to cancer chemotherapy.

It is well known that phenothiazines can inhibit the typical shape changes displayed by activated lymphocytes and thereby their migration through polycarbonate filters. De Braekeleer et al. [147] studied the effect of phenothiazine on lymphocyte migration in an environment with living solid tissue cells. The effects of trifluoperazine (TPZ) and chlorpromazine (CPZ), which are strong inhibitors of the lymphocyte motility, and of pimozide (PIM), which is a much weaker inhibitor of the lymphocyte motility but a strong inhibitor for calmodulin, were investigated. This study was performed by examining the effects of these phenothiazines on the invasion of human Molt-4 T cells across pre-cultured fibroblast monolayers. The invasion of T cells across the cellular monolayers could be inhibited by PIM as well as by TPZ and CPZ via two distinct mechanisms. Indeed, TPZ and CPZ inhibited the lymphocyte motility via a calmodulin-independent pathway, whereas PIM impaired the monolayer tolerance to the invasion, most likely via a calmodulin- and a CamKII-dependent pathway.

Hawtrey's research group demonstrated the stimulation of a luciferase pRSVL DNA uptake and expression in HeLa cells at low concentrations (10^{-5} M) of CPZ and related antipsychotic phenothiazine compounds [148]. Higher concentrations of phenothiazines were toxic to the cells, and could not be tested. However, at concentrations higher than 10^{-4} M, the phenothiazines strongly stimulated the luciferase expression and activity by chloroquine. The result of this assay (free base) suggests that the major contribution to the binding might occur through an intercalation process. Furthermore, the interaction of phenothiazines and chloroquine was studied by using the ethidium bromide (EB)/calf thymus (CT) DNA intercalation method. The intercalated CT–DNA complexes with EB were examined for their possible dissociation into free DNA and EB upon addition of either chloroquine-SO_4 or chlorpromazine-HCl. A partial dissociation was observed with both compounds. Moreover, an alternative method of formation of pBR 322 DNA–polylysine complexes was used to give a better explanation of this reaction. Further studies of chloroquine and chlorpromazine on their interactions with plasmid DNA as well as with DNA–polylysine complexes were reported.

The cultured human glioma cell line U-251 was treated with trifluoperazine (TFZ) and phenothiazine, and their antitumour activities and radiosensitizing effects were observed in vitro [149]. The clone-forming test and flow cytometry were used as methods to follow the radiosensitivity and cell cycle of U-251 cells, respectively. The inhibition of growth in U-251 cells increased with higher dosages of TFZ and phenothiazine and also increased with the time of administration. It is worthwhile noting that TFZ exhibited a stronger effect than phenothiazine. In addition, TFZ increased the radiosensitivity of U-251 cells. There was no change in the cell cycle of U-251 cells treated with TFZ only.

The antipsychotic phenothiazines are also known to undergo a peroxidase-catalysed oxidation to form cation radicals that are stable at an acid pH, but not detected at a neutral pH. Eghbal et al. [150] found that catalytic amounts of phenothiazines were oxidized by the peroxidase system, and also caused the ascorbate, glutathione (GSH) and NADH co-oxidation. NADH and GSH co-oxidation was accompanied by oxygen uptake and was increased by the addition of catalytic amounts of superoxide dismutase (SOD), indicating that superoxide radicals were formed. In the presence of phenothazines, the NADH, ascorbate or GSH co-oxidation was faster at pH 6.0 than at pH 7.4, thereby partly reflecting the cation-radical stability. The order of catalytic efficiency was: promazine > CPZ > TFZ. Peroxidase also markedly increased the phenothiazine cytotoxicity in the case of isolated rat hepatocytes at non-toxic phenothiazine concentrations. At both pH values, the same order of phenothiazine catalytic efficiency was observed. The authors proposed the hepatocyte/peroxidase system as a useful model for studying the drugs which induced the idiosyncratic hepatic injury enhanced by inflammation.

Cytotoxicity could also be caused by the aggregation of the degradation proteins resulting from the same disease. A method and compounds for reducing the aggregation of either neurodegenerative proteins associated with neurotoxicity or other proteins were the object of a recent patent [151]. The compounds included a first domain or targeting element for binding to the target proteins, which might be linked to a second domain or recruiting element that was bound to an aggregation-inhibiting protein such as a prolyl isomerase. The patented compounds have the formula R1-L-R2, wherein R1 is chosen from the group consisting of 1-Br or I-2,5-bis(3-hydroxycarbonyl-4-hydroxy)stilbene, thioflavin, thioflavin T, chrysamine G, X-34, congo red, IMPY, imipramine, carbamazepine, phenazine, phenothiazine, promazine, chloropromazine, haloperidol, clozapine, 2-chlorophenothiazine, promethazine, chloroprothixen, acepromazine, deoxydoxorubicin, rifamycin, acridone and acridone derivatives, flavonoids and alkaloids, benzofurans, 9-substituted acridines, pamaquine, chloroquine and amacrine, and methylene blue; L is a bond or linking group of no more than about 36 atoms other than hydrogen; and R2 is selected from the group consisting of SLF, FK506 and rapamycin. By associating either the aggregation-forming proteins or neuronal cells under conditions where aggregating proteins are produced, with the patented compounds and an aggregation-inhibiting protein, the aggregation could be reduced. The patented agents might be used in assays for investigating the aetiology of the neuronal diseases as well as for prophylaxis and therapy.

3.3.3 Modulators of Cancer Multidrug Resistance

Multidrug resistance (MDR) remains as one of the biggest problems for the successful application of cancer chemotherapy. The mechanism of MDR is still not fully understood. In fact, until now most investigations have provided only parcellar information on the MDR mechanism, which might be considered as a small part of a very complicated puzzle. A number of experimental studies have concerned the detection of substances which are modulators of cancer MDR – also called substances with a reversal MDR effect or chemosensitizers. Phenothiazines belong as well to a group of drugs known to modify the MDR phenomenon occurring for different types of cells, from the cancer cells up to various kinds of microorganisms.

Motohashi et al. [152] investigated the drug resistance reversal, antimutagenic and antiretroviral effects of substituted phenothiazines, such as phthalimido- and chloroethyl-phenothiazines, for three different systems: bacteria, cancer cells and reverse transcriptase enzyme of Moloney leukaemia virus. These phenothiazine derivatives were also synergistic with tetracycline for bacteria and presented an antimutagenic effect. No mutagenic effect was observed in the case of the TA98 strain of *S. typhimurium*. The chloroethyl-substituted phenothiazines exhibited an antimutagenic behaviour equivalent to that of the parent compounds. P-glycoprotein, responsible for MDR, was also inhibited in the tumour cells. Some of the substituted phenothiazines were effective P-glycoprotein blockers, whereas some compounds had moderate activity but others were without effect when compared to 5 μM verapamil. Correlations were established between the biological activity and the dipole moment, as well as the entropy of the molecules under study [152]. The results suggest that the inhibition of Hly+ plasmid replication and P-glycoprotein function might partly depend on similar electronic properties of these phenothiazine derivatives.

Nacsa et al. [153] confirmed that the MDR gene could be induced by different environmental stresses ("SOS gene"). The authors tested the inhibitory action of coordinating complexes of Cu(II) with TFZ and CPZ on megacin encoding bacterial gene induced by mitomycin C (MMC) as an example of "SOS induction", and on the efflux pump of mouse lymphoma cells. The interaction of compounds with DNA was evaluated by measuring the thermal stability of DNA. It was found that the addition of these complexes before the MMC administration had an inhibitory action on megacin induction at a lower concentration than TFZ alone. The inhibitory effect of some metal coordinating complexes, such as TFZ-Cu(II) and TFZ-V(IV), was found to exceed the action of TFZ alone on efflux pump. It was proposed that these compounds might form a complex with the regulatory protein or DNA, resulting in the inhibition of the "SOS response" and of the MDR function by inactivating the P-glycoprotein as well [153].

Most MDR modulators such as antipsychotic phenothiazine derivatives are mainly cationic amphiphiles. The molecular mechanism of their effect involved interactions with protein transporters as well as with the lipid membrane. The interactions between anionic phospholipids and MDR modulators might be crucial for their action. Wesolowska et al. [154] investigated the interactions of 2-trifluoromethyl-10-[4-(methane-sulphonylamide)butyl]-phenothiazine (FPhMS) with neutral and anionic lipids. Using microcalorimetry as well as steady-state and time-resolved fluorescence spectroscopy, the authors showed that FPhMS interacts with the group of lipids under study and that the drug location in the membrane depends on the lipid class. Presumably, the electrostatic attraction between the drug and lipid head groups keeps the phenothiazine derivative molecules closer to the surface of the negatively charged membranes with respect to the neutral membranes. The effects of FPhMS on bilayer properties were found to be not proportional to the phosphatidylserine content in lipid mixtures.

Hendrich et al. [155] investigated the mechanism of incorporation of phenothiazines in the membrane bilayer lipids. They studied the influence of a particular phenothiazine derivative, TFZ, on the thermal properties of dimyristoyl phosphatidylcholine and dimyristoyl phosphatidylethanolamine by microcalorimetry. The main phase transition of both lipids was affected by this drug, depending on its concentration. The results suggest that TFZ was probably incorporated into both dimyristoyl phosphatidylcholine and dimyristoyl phosphatidylethanolamine bilayers. The phase separation was presumably induced by the different modes of the drug–bilayer interactions of protonated and unprotonated forms of TFZ. Only phosphatidylcholine, which possesses polar heads less densely packed in bilayers than phosphatidylethanolamine ones, was able to distinguish between the different protonated forms of TFZ.

Recently, Michalak et al. [156] presented a review on the interactions of phenothiazines with lipid bilayers and their role in MDR reversal. The drug partition between water and the lipid phase was discussed. The molecular size and polarizability of phenothiazines were found to correlate satisfactorily with the reduction of MDR. The most important structural features required for a high anti-MDR activity of phenothiazines seem to be their hydrophobic ring system, the type of substituent located at the 2-position, the presence of a carbonyl group and of an alkyl bridge connecting the ring system with side groups, and the protonable nitrogen [156].

Some bacteria, such as *Mycobacterium tuberculosis*, also exhibited a resistance to drugs of the MDR type. A comparative study of the activity of five phenothiazines against the MDR strains of *M. tuberculosis* (MDRTB) was carried out by Bettencourt et al. [157], using the Bactec 460 system. The order of their antimycobacterial activity was as follows: CPZ = thioridazine > promethazine > promazine = desipramine. These phenothiazines might be expected to be used as adjuvants for treating freshly diagnosed tuberculosis in patients from populations with a high prevalence of MDRTB. A further investigation on the *M. tuberculosis* resistance to rifampicin and isoniazid indicated that CPZ could enhance the antibiotic activity against *M. tuberculosis*, except in the case of ethambutol [158]. This might result from a reduction of the dose of some or all of the antibiotics under use, without sacrificing the integrity of the treatment. Upon combining CPZ, thioridazine and promethazine at concentrations that were minimally effective when applied separately, it was found that this combination of phenothiazines could enhance the activity of rifampicin and streptomycin against clinical strains of *M. tuberculosis* resistant to two or more antibiotics (MDRTB). In contrast, these phenothiazines had no effect on the activity of isoniazid against resistant MDRTB [158].

3.4
Mechanisms of Action of Phenothiazines and of MDR Overcoming

The investigation of the mechanisms of action of drugs such as phenothiazines and of MDR overcoming represent a very important research field. The recently gathered information could be considered as part of the mosaic pieces of the real picture of the reactions occurring in organs, cells and subcellular particles during the resorption, distribution, connection, chemical modification, degradation and elimination of drugs. The culture of human cells or of selected, well-differentiated types of cell lines was used in the investigations on the mechanisms of phenothiazines actions and MDR overcoming.

A comparative molecular analysis study based on three-dimensional quantum structure–activity relationships was performed by Pajeva and Wiese [159] on 40 phenothiazines and structurally related drugs to predict their MDR modification. More than 350 theoretical models were derived and evaluated using steric, electrostatic and hydrophobic fields alone and in combination. All examined fields were found to contribute to MDR reversing activity, and their hydrophobic fields improved the correlative and predictive power of the models in all cases. The results point out the role of hydrophobicity as a space-directed molecular property to explain the differences in anti-MDR activity of the drugs under study.

The interactions of several phenothiazine drugs with DNA were studied by electronic absorption and fluorescence spectra, using ethidium bromide (EB) as a probe. According to the results, both CPZ-HCl and TFZ-HCl could intercalate into DNA, but promethazine-HCl, which possesses a similar alkyldimethylamino group to CPZ, could weakly bind to DNA by mixed modes. These interactions could be inhibited at high ionic concentrations [160].

Second-derivative electronic absorption spectrophotometry, which is a method applied without any separation procedure, was used to determine the partition coefficients of three N-monodemethylated phenothiazine drugs, such as CPZ, triflupromazine and promazine, between the phosphatidylcholine bilayers of small unilamellar vesicles (SUVs) and water, in order to evaluate their affinity to biomembranes. The second-derivative spectra exhibited distinct isobestic points, confirming the entire elimination of the residual background signal effects of the SUVs which were observed in the conventional absorption spectra [161].

Hu et al. [162] investigated the stimulatory response to the oxidation of benzidine and other xenobiotics mediated by lipoxygenase in the presence of phenothiazines. The metabolic interactions were studied in vitro between CPZ and other phenothiazines, and benzidine or six other xenobiotics which were mediated by soybean lipoxygenase (SLO) or a purified human term placental lipoxygenase (HTPLO) in the presence of hydrogen peroxide (H_2O_2). The CPZ cation radical generated by lipoxygenase triggered a rapid oxidation of benzidine to benzidine diimine cation. Indeed, benzidine oxidation in the presence of phenothiazines was about 42 times greater than when benzidine was alone. The magnitude of stimulation of benzidine oxidation depended on the pH, amount of enzyme, and CPZ, benzidine and hydrogen peroxide concentrations. Both SLO and HTPLO could mediate the stimulatory response to the oxidation of xenobiotics in the presence of phenothiazines.

The hypothesis that HTPLO and SLO were able to mediate the N-demethylation of selected phenothiazines and insecticides in the presence of polyunsaturated linoleic acid (LA) was proposed by Hover and Kulkarni [163]. This N-demethylation reaction might be limited by the incubation time, pH of the medium and concentration of the enzyme and substrate. The reaction was followed by measuring the formaldehyde production. The results confirmed that the polyunsaturated free fatty acids could support the N-demethylation of xenobiotics via the lipoxygenase pathway.

A combination of chemotherapeutic agents and some phenothiazines changed their physicochemical properties. The neuroleptic protonated phenothiazine derivatives, including promethazine, promazine, triflupromazine, methotrimeprazine, propiomazine, trifluoperazine and fluphenazine, were complexed with water-insoluble chemotherapeutic agents such as 5-fluorouracil (5-FU), methotrexate (MTX) and sulindac, as well as with components of biomembrane and synthetic phospholipids considered as possible models of cancer and microbial cells. In all cases, water-soluble micellar-type inclusion adducts were formed, which exhibited an electron charge-transfer complex (CT complex) behaviour with the appearance of thiazine free radicals. The thiazines sequestered the drugs and phospholipids in well-defined molar ratios parabolically dependent on the dipole moments (μ) of the protonated phenothiazine derivatives. The pH comparison studies showed that the inclusion adducts followed a model in which the compounds were enveloped in the lipophilic cavity of the thiazine aggregates, while the side chains of the latter were in contact with the aquatic environment. The physicochemical measurements indicated that new thiazine drugs could be designed, to be used either alone or as a carrier in anticancer therapy [164].

The interaction of two phenothiazine derivatives, CPZ and TPZ, with phospholipid monolayers was investigated by Hidalgo et al. [165], as a model of a drug-interaction membrane system. The changes in monolayer behaviour caused by the incorporation of a few molar ratios of drug molecules were practically within the experimental dispersion for the zwitterionic dipalmitoyl phosphatidyl glycerol (DPPG). Therefore, more refined analyses were required to unequivocally probe these interactions. For both drugs, the changes were more prominent at the lift-off of the surface pressure, i.e. at the gas-condensed phase transition, and the surface pressure and surface potential isotherms became more expanded with the drug incorporation. The changes caused in the surface pressure and dipole moment by small concentrations of TPZ or CPZ could only be explained by some cooperative effects through which the contribution of DPPG molecules would change considerably, i.e. even DPPG molecules that are not neighbours to a CPZ or TPZ molecule are also affected. These types of changes might occur either through a significant reorientation of the DPPG molecules or a modification in their hydration state. They are probably implicated in biological activities.

The cytochrome P-450 isoenzymes could play an important role in contributing to the metabolism of phenothiazine neuroleptics. The optimum conditions for studying the promazine and perazine metabolism in rat liver microsomes, and the influence of specific cytochrome P-450 inhibitors on 5-sulphoxidation and N-demethylation of these neuroleptics, were determined by Daniel et al. [166]. These authors studied the metabolism of neuroleptics in liver microsomes, and showed the existence of a linear dependence of the product formation with time, and the protein and substrate concentrations. A Dixon analysis of the metabolism of neuroleptics indicated that quinine (a CYP2D1 inhibitor), metyrapone (a CYP2B1/B2 inhibitor) and α-naphthoflavone (a CYP1A1/2 inhibitor) affected the phenothiazine neuroleptic biotransformation, whereas erythromycin (a CYP3A inhibitor) and sulphaphenazole (a CYP2C inhibitor) did not change it. These results revealed the effect of drug structure and species differences on the contribution of cytochrome P-450 isoenzymes to the metabolism of phenothiazines.

Several phenothiazines and other types of compounds, including TFZ, fluphenazine (FP) and CPZ as well as calmidazolium and *N*-(6-aminohexyl)-5-chloro-l-naphthalenesulphonamide (W-7), are calmodulin-binding peptide antagonists on the Ca^{2+}-ATPase activity, and their effect on sarcoplasmic and endoplasmic reticulum was investigated [167]. The results showed that calmidazolium and calmodulin-binding peptide were the most potent inhibitors of skeletal muscle sarcoplasmic reticulum Ca^{2+}-ATPase activity. Moreover, these results suggested that the effects of these calmodulin antagonists were independent of calmodulin and that they could inhibit the Ca^{2+}-ATPase in an isoform-specific manner.

Two types of cells, such as *Saccharomyces cerevisiae* and *Schizosaccharomyces pombe*, were used to investigate their proliferation resulting from the effects of phenothiazines as calmodulin inhibitors [168]. The stimulation of proliferation of *Schizosaccharomyces pombe* was more evident upon lowering the calcium ion concentration in the culture medium. Although the proliferation of *Schizosaccharomyces pombe* was not affected by low concentrations of TFZ, this proliferation was partly inhibited when the TFZ concentration was increased to 100 μmol/L, and was even completely inhibited for a TFZ concentration up to 200 μmol/L. When applying the same method to *Saccharomyces cerevisiae*, the same inhibition effect appeared for a lower TFZ concentration of about 20 μmol/L. These different concentration effects could be attributed to differences of penetration of TFZ and of stimulation by Ca^{2+} ions in both kinds of cells.

MDR appeared when the drug accumulation in cells was decreased as a result of its ATP-dependent efflux across the membrane by P-glycoprotein (P-gp) transport [169]. Different mechanisms were proposed for the action of MDR modulators. The widely supported mechanism is based on the concept that modulating agents bind to P-gp, thus inhibiting either competitively or non-competitively the drug binding or the ATPase activity of protein [170]. At the same time, the membrane–drug interactions would provoke some modifications of the membrane physicochemical properties and a perturbation of the P-gp lipid environment [171].

It seems that the most important MDR mechanism in cancer cells is the inhibition of the MDR1 carrier, responsible for the efflux of different xenobiotics out of the cells. Therefore, the inhibition of the MDR1 carrier should diminish the importance of MDR. Konya et al. [172] proposed the use of some novel phenothiazine derivatives, possessing a strong in vitro MDR1 inhibitory activity. Some of these derivatives were more active than verapamil, and one of them was at least as active as cyclosporin A for inhibiting MDR. These interesting results suggest that the new structural elements incorporated into these novel phenothiazines might increase their MDR1 inhibitory activity, and they also may serve as a basis for developing an effective MDR1 inhibitor and reversal drug.

3.5
Other Biomedical Applications of Phenothiazines

Wainwright [173] described the use of well-established phenothiazinium methylene blue (MB) derivatives as photodynamic antimicrobial agents, which represents a simple method for the inactivation or destruction of pathogens contained in donated blood and blood products. The procedure could be carried out in blood bags, by using basic low-power light sources. It could be easily applied because bacterial, yeast and protozoal classes were susceptible to these phenothiaziniums. The photoantimicrobial mode of action generally took place via oxidative damage to their cellular components, either by the action of reactive oxygen species generated in situ by the photodynamic effect, or by the redox reactions occurring between the photoantimicrobial agent and a biomolecule target.

The anti-inflammatory effects of FP and triflupromazine, either alone or in combination with triamcinolone, were studied in rats after one single dose and 8 days, measuring histamine, serotonin, prostaglandin (PGE2), volume of exudate and histaminase enzyme [174]. The results indicated that the administration of FP and triflupromazine in these conditions induced a significant anti-inflammatory activity which could be attributed to the inhibition of prostaglandin E2.

In order to understand how the quaternary, tertiary and secondary structural organizations of certain proteins could regulate their binding behaviour with phenothiazines, a model was proposed by Bhattacharyva et al. [175]. The authors investigated the reaction of haemoglobin (Hb) and myoglobin with antipsychotic phenothiazine drugs such as CPZ and TFZ. In the case of Hb, a NaCl-induced alteration in the quaternary structure of this protein was shown to influence its mode of binding with phenothiazines. Minor alterations of the tertiary structure of thermally denatured myoglobin in the temperature range 30–70 °C did not affect its affinity, nor its binding behaviour with the phenothiazine drugs, whereas alterations in the secondary structure modified the binding when Hb was denatured at higher temperatures between 70 and 80 °C.

Chapman et al. [176] patented methods and an apparatus for the inactivation of contaminants in biological fluid, by using phenothiazine dyes and light. A photochemical agent such as MB was added to the biological fluid, which was submitted to illumination with a high-pressure sodium lamp. In these conditions, the high-intensity light was able to inactivate some of the contaminants.

Another interesting patent deposited by Boland and McDonough [177] described the efficiency of the promethazine (+) enantiomer in inhibiting bone resorbing cells (osteoclasts), thus providing new methods for reducing bone losses and treating periodontitis and osteoporosis. This invention gave a range of the uses, methods, medical and associated compounds, and kits. The preparation of the enantiomer was also described in the patent.

Takikawa et al. [178] compared the effects of colchicine and phenothiazine, applied as substrates for P-glycoprotein, on the vesicular transport and biliary excretion of some organic anions in rats. The colchicine treatment slightly inhibited the biliary excretion of indocyanine green, dinitrophenyl glutathione and pravastatin, but this treatment had no effect on the biliary excretion of sulphobromophthalein and dibromosulphophthalein. In contrast, the phenothiazine treatment did not affect the biliary excretion of indocyanine green and pravastatin, but it significantly increased the sulphobromophthalein-glutathione biliary excretion. These findings suggest that P-glycoprotein might play a supplementary role in the biliary excretion of some organic anions in addition to the canalicular, multispecific organic anion transporter.

A method to prevent ischaemia and re-perfusion injury by using agents that inhibit cytochrome P-450 enzymes was patented [179]. This method was based on administering to a mammal at risk of injury, associated with ischaemia and/or re-perfusion, an effective amount of one or more agents that inhibit cytochrome P-450 enzymes. The agents were selected from a group consisting of an H2-receptor antagonist, an H^+/K^+-ATPase inhibitor, an antimicrobial, an antifungal, a tricyclic antidepressant, a serotonin re-uptake inhibitor, a phenothiazine, a benzodiazepine, a calcium channel blocker and a non-steroidal anti-inflammatory drug. The method could be applied to myocardial ischaemia, cerebral ischaemia, intestinal ischaemia, renal ischaemia, or general tissue ischaemia. It was also applicable to the re-perfusion injury associated with myocardial infarction or vascular intervention procedures, angioplasty or coronary artery bypass graft surgery. Moreover, this method could be used to inhibit ischaemia and re-perfusion injury in a donor organ by treating the donor organ with the agents described in the patent.

Creutzfeldt–Jakob disease (CJD) and its new variant (nvCJD) are incurable, and, although the diagnosis has become clinically more precise, it is only certain at autopsy. Some phenothiazine derivatives were found to inhibit the production of prions, the disease-causing agent, in cultured neuroblastoma cells, and an advanced case of nvCJD was recently brought to remission by using these derivatives in combination with an antimalarial drug [180]. Korth et al. [181] showed that it was possible to treat cultured cells chronically infected with prions by using tricyclic acridine and phenothiazine derivatives. The authors found that these derivatives produced a half-maximal inhibition of prion formation at effective concentration (EC50) values ranging between 0.3 and 3 μM. The EC50 for CPZ was 3 μM, whereas quinacrine was ten times more potent than CPZ. Since quinacrine and CPZ have been used for many years in humans as antimalarial and antipsychotic drugs, respectively, and are known to pass the blood–brain barrier (BBB), the authors suggested that these substances might constitute good candidates for the treatment of CJD and other prion diseases.

In the past two decades, researchers have sought to identify new, alternative oestrogen actions. These novel actions, widely known as non-genomic effects, are not directly related to nuclear transcription events, but they might depend on the interactions of oestrogens with sites present at the plasma membrane or cytosolic locations. The produced effects could range from the modulation of plasma membrane ion channel activity to the regulation of different intracellular signal cascades. Valverde et al. [182] investigated the modulation of a large conductance chloride channel (called Maxi Cl^- channel) by oestrogens, non-steroidal triphenylethylene antioestrogens and phenothiazines in NIH3T3 fibroblasts, as well as the effect of guanosine 5′-triphosphate (GTP) on the Maxi Cl^- channel activation. Their data indicated that the non-steroidal antioestrogens toremifene and tamoxifen, and phenothiazines, such as CPZ and triflupromazine, were activators of the Maxi Cl^- channel. In contrast, 17β-oestradiol and cAMP, added prior to the exposure to antioestrogens, were able to prevent the channel activation.

Some phenothiazines exhibited a direct antibacterial activity in vitro, but until the last 5 years no phenothiazine immunopotentiation activity was reported. Andrieu et al. [183] patented an invention for rebuilding the immune defences. This patent concerned more particularly the use of a mixture of compounds capable of acting on the monocytes/macrophages and/or dendritic cells, in order to provide them with a signal inducing the proliferation of lymphocytes including T lymphocytes. This might permit a possible regulation of the immune defences to be established in the treated organism. The active substances were selected among phenazine, phenoxazine and phenothiazine derivatives, such as aminoperazine 2-amino-10-[3′-(1-methyl-4-piperazinyl)-propyl]phenothiazine. The patented drug was recommended in the treatment of tumours, infections and other immune deficiencies.

In a clinical study, several patients with neuroleptic malignant syndrome associated with myoglobulinemic acute renal failure were treated with phenothiazine, butyrophenone (haloperidol), benzamide, iminomide, benzisoxazole, antidepressants and hypnotics (benzodiazepine and barbiturate) for the treatment of schizophrenia [184]. The clinical manifestations of neuroleptic malignant syndrome were characterized by altered consciousness, muscle rigidity and weakness, fever and excessive perspiration. All patients were successfully cured of acute renal failure by haemodialysis or haemodiafiltration.

3.6 Unwanted Effects of Phenothiazines

The extensive and long-term use of phenothiazine derivatives might result in unwanted or side effects. Moreover, their application as antihistaminic and/or neuroleptic drugs in the therapy of schizophrenia, organic psychoses or other psychic illnesses might induce various changes at different cellular levels through interactions with proteins, lipids or DNA. The manifestation of these interactions very often produced some unwanted effects, such as skin photosensitization and disorders in the lipid status.

For example, changes in the lipid status were observed in the serum of 138 patients with chronic schizophrenia who were compared to 100 normal subjects. Significantly higher triglyceride (TG) and lower HDL lipoproteinaemia levels were found in the sera of patients treated with phenothiazine drugs [185]. This situation could lead to arteriosclerosis, especially in those patients having schizophrenia with brain morphological changes. In addition, the negative correlation of high TG and low HDL-C might represent a risk factor of ischaemic disease for the heart and brain in the case of the schizophrenia patients.

Shin and Kim [186] investigated the effects of three phenothiazines, thioridazine, CPZ and TFZ, on the cardiac action potential duration in rabbits. Purkinje fibres were excised from the rabbit left ventricle and stored in a chamber super-fused with a normal Tyrode solution. The phenothiazine drugs displayed different effects on the duration of cardiac action potential. The obtained results indicated that it would be necessary to propose an assay for the cardiac action potential duration, in order to elucidate the electrophysiological effects of these phenothiazines on the heart.

Another study relative to the photo-oxidative damage provoked by phenothiazines to rat liver isolated mitochondria was carried out by Rodrigues et al. [187]. The UV irradiation of rat liver isolated mitochondria, performed in the presence of TR, TFZ and FP, demonstrated the absence of a dose-dependent respiratory control ratio. These effects were not accompanied by a significant swelling and oxidation of the protein thiol groups, but by a lipid peroxidation. The addition of lycopene and sorbate was ineffective at protecting the mitochondria lipids against damage promoted by the excited-state phenothiazines, which suggests that the photochemically produced cation radicals were probably the photo-oxidant species. These results constituted a significant contribution to the understanding of the photochemical properties of phenothiazines in the biological systems.

Mio et al. [188] examined the augmentation of the UVB light-induced histamine release from peritoneal rat mast cells by certain phenothiazines, such as promethazine, trimeprazine, mequitazine, chlorpromazine, TFZ, ethopropazine and TR. Simultaneously, lactate dehydrogenase (LDH) was released. This suggested that the cell membrane permeability to low molecular weight substances was increased by UV exposure. Moreover, the light-induced histamine release was inhibited by ascorbic acid, indicating the involvement of a radical-type step in the reaction mechanism. The authors concluded that the phototoxicity of these phenothiazines might be partly due to an enhancement of UVB light-induced histamine release from the mast cells.

The excited-state properties and in vitro phototoxicity of three phenothiazines, namely perphenazine, FP and TR, were investigated by Elisei et al. [189]. The excited-state properties were studied by both stationary and time-resolved fluorimetry, and by laser flash photolysis. After a preliminary study of the photo-behaviour in organic solvents and in water, the phototoxicity of the three drugs was examined on various biological substrates through a series of in vitro assays under UVA irradiation. Photohaemolysis of mouse erythrocytes and phototoxicity on cultured murine fibroblasts were observed for all three phenothiazines. Then, the authors investigated the lipid photoperoxidation by using a linoleic acid as the unsaturated lipid model, and isolated red blood cell membranes. The drug-induced photodamage on the proteins was also evaluated by measuring the photosensitized crosslinking in the erythrocyte ghosts. This combined approach was valuable for elucidating the mechanism by which these phenothiazine derivatives induced the skin photosensitization. The photophysical properties of the phenothiazines under study and the results of their phototoxicity study appeared to be in good agreement with a radical-cation mechanism.

James et al. [190] carried out an epidemiological study of phenothiazine, butyrophenone and other psychotropic medication poisonings in children and adolescents. Decreasing levels of consciousness and dystonia were the most commonly presented signs, appearing in 90.7 and 51.2% of the patients, respectively. Miosis occurred in only 13.9% of the patients. Fluid boluses were administered to 28.7% of the patients, but about 25% of these had co-ingested potentially cardiotoxic drugs. The patients with dystonias were more likely to be submitted to more extensive diagnostic testing for neurological disease than those presenting no sign of dystonia. The effects of phenothiazine and butyrophenone ingestion in children and adolescents might be non-specific and provoked by co-ingestants. This suggested that physicians might not recognize the dystonias as a clinical finding characteristic of phenothiazine or butyrophenone exposure.

The elimination of an excessive dose of phenothiazines from plasma was analysed by using the novel polymeric solid-phase extraction (SPE) sorbent, strata-X-CW [191]. Strata X-CW was found to bind phenothiazines strongly, facilitating an intense organic wash and consequently allowing delivery of very clean extracts of these phenothiazines from biological matrixes.

One of the side effects of some phototoxic phenothiazine drugs, such as promazine, promethazine and chlorpromazine, was to depolymerize an aqueous solution of sodium hyaluronate (HA) upon exposure to light [192]. The depolymerization rate constants were found to increase at low drug concentrations below 0.05 mmol L^{-1}. In contrast, the molecular weight of HA was practically unchanged without UV irradiation in the presence of these phenothiazines, as well as with UV irradiation in the absence of these drugs. This indicated that the phenothiazines required photo-energy to yield any type of damaging chemical species active in the HA depolymerization reaction. An involvement of active oxygen radicals in the effects of promazine and promethazine was demonstrated by the inhibition of the depolymerization reaction under anaerobic conditions. Furthermore, the addition of mannitol was found to control the reaction in the presence of oxygen, pointing to hydroxyl radicals as the damaging agent. Chlorpromazine (CPZ) might preferably depolymerize the HA under anaerobic conditions, suggesting the participation of hydrated electrons.

Several phenothiazine derivatives have been shown to cause reproductive toxicity. The biochemical mechanisms responsible for these effects are not fully understood at present. In the in vitro study of Rajadhyaksha et al. [193], hydrogen peroxide-dependent oxidation of six phenothiazines by purified lipoxygenase from soybean and human term placenta was investigated. CPZ was used as the prototype phenothiazine drug because, during incubation at pH 7.0 and 6.5 with soybean and human term placenta, respectively, CPZ was easily demethylated, releasing formaldehyde in the presence of hydrogen peroxide. The results suggested the enzymatic nature of the reaction. Besides CPZ, soybean was also able to oxidize PZ, triflupromazine, TFZ, trimeprazine and PP, although at different rates, in the presence of hydrogen peroxide. The evidence gathered in this in vitro study showed that several phenothiazines might undergo a peroxidative N-demethylation reaction via a lipoxygenase pathway. However, the precise role of this biochemical mechanism in the in vivo developmental toxicity of phenothiazines still remained to be investigated. The reproductive toxicity of phenothiazines was also investigated by Yang and Kulkarni [194]. The authors studied the hydrogen peroxide-dependent oxidation of seven phenothiazines by purified human term placental peroxidase. Promazine was used as the prototype phenothiazine drug. In acetate buffer of pH 3.5, promazine was easily oxidized to a radical cation when incubated with the human term placental peroxidase in the presence of hydrogen peroxide. The results indicated that the reaction might have an enzymatic nature. Indeed, reduced glutathione, ascorbate and NADH inhibited the promazine radical formation probably by reducing it back to promazine. The radical-cation formation and NADH oxidation rates significantly declined in the presence of potassium cyanide and sodium azide, which are well-known peroxidase inhibitors. This enzyme could also oxidize other phenothiazines, such as CPZ, triflupromazine, TFZ, trimeprazine, FP and PP, in the presence of hydrogen peroxide. This study strongly suggests that the peroxidative bioactivation of phenothiazines to their corresponding radical-cation species by human placental peroxidase probably represents one of the biochemical mechanisms responsible for the reported developmental toxicity of these phenothiazines.

The most severe unwanted effects of phenothiazine derivatives are arrhythmias and sudden death. According to Kim and Kim [195], these relatively rare phenomena could be due to the blockage of cardiac potassium channels, such as the human ether-a-go-go-related gene (hERG) channel, which plays a central role in the arrhythmogenesis. But the effects of many of the phenothiazine drugs on the hERG channel expressed in mammalian cell lines still remained unknown. Therefore, these authors investigated the effects of four phenothiazines, TR, CPZ, TFZ and PP, on the hERG channel. This channel was expressed in Chinese hamster ovary (CHO) cells, and the ion currents were measured by means of the patch-clamp technique. The inhibition of the hERG channel by TR was characterized by important changes in voltage dependence, whereas no significant changes were noted in the case of the other phenothiazines.

3.7 Biomedical Applications of Benzo[*a*]phenothiazines

The biological and biomedical properties of benzo[*a*]phenothiazine derivatives were widely investigated in the last two decades. However, the precise interpretation of the biomedical properties and applications of these large molecules remains the matter of future research.

3.7.1 Antibacterial, Antifungal, Antiviral and Antiparasitic Activities

The synthesis of some novel fluoro-substituted benzo[*a*]phenothiazines and their nucleosides as possible antimicrobial agents was reported [196]. 6-(2-Methoxy-5-methyl/2-methoxy-4-chloro-5-methyl-anilino)-9-fluoro-5*H*-benzo[*a*]phenothiazine-5-ones/thiones, 12*H*-benzo[*a*]phenothiazine-5-ols, 5-acetoxy-12*H*-benzo[*a*]phenothiazines, 5-methoxy-12*H*-benzo[*a*]phenothiazines and nucleosides, viz *N*-(2,3,5-tri-*O*-benzoyl-β-D-ribofuranosyl)-6-(2-methoxy-5-methyl/2-methoxy-4-chloro-5-methyl-anilino)-9-fluoro-5-acetoxy benzo[*a*]phenothiazines and 9-fluoro-5-methoxy benzo[*a*]phenothiazines, were synthesized and tested for their antimicrobial activity. The structures of novel compounds were proposed on the basis of elemental analyses, IR and ^{1}H and ^{19}F NMR studies.

Mucsi et al. [197] studied the combined antiviral effects of some benzo[*a*]phenothiazines and 9-[2-hydroxy(ethoxy)methyl]guanine (acyclovir, ACV, and acycloguanisine) on the multiplication of herpes simplex virus type 2 (HSV-2). Vero cells were infected with HSV-2 and treated with a combination of ACV and selected benzo[*a*]phenohiazines, including 5-oxo-5*H*-benzo[*a*]phenothiazine and 6-methyl-5-oxo-5*H*-benzo[*a*]phenothiazine. The authors found that such a treatment decreased the infective virus population, probably by reduction of the mutagenic rate in the virus populations.

3.7.2 Neurological Activities

The well-known antipsychotic properties of phenothiazines opened the door to the investigation of the use of benzo[*a*]phenothiazines for the same purpose. A recent patent of Cuine et al. [198] described the combination of a short-acting hypnotic agent which was selected from among a modulator of receptors GABA-A, a benzodiazepine, a phenothiazine, a melatonin derivative and a melatonin receptor agonist, and a long-acting hypnotic agent which was chosen from a modulator of receptors GABA-A, a benzodiazepine, an antagonist of receptors 5HT2A and a calcium ion modulator. This invention was designed for the treatment of sleep disorders.

3.7.3 Antitumour Activities, Cytotoxicity and Effects on Multidrug Resistance

As already seen in the previous section, the efficiency of chemotherapy is often decreased by the development of resistance of cancer cells to the cytostatic drugs (MDR). Since it has been shown that, generally, the phenothiazines have the effect of diminishing the MDR and increasing the cytostatic efficiency, it seemed worthwhile to examine whether the benzo[*a*]phenothiazines presented the same type of properties.

Several antiplasmid and anticancer compounds were tested for their ability to reserve the MDR of lymphoma cells. Some well-known anticancer drugs, such as platidiam, novantrone, fluorouracil, bleomycin and methotrexate, were ineffective, while vinca alkaloids exerted a strong reversal effect on the MDR of lymphoma cells. Molnar et al. [199] investigated the ability of some selected antitumour phenothiazines, such as TFZ and benzo[*a*]phenothiazines, to inhibit the P-glycoprotein (Pgp) function. This fact is independent of the antiproliferation or differentiation inducing effects. Because the polylactosamine specific tomato lectin could prevent the action of the chemosensitizers under study, it was supposed that the site of action of phenothiazines and benzo[*a*]phenothiazines might be located at the first loop in the glycoprotein transmembrane. The efflux pump activity of Pgp in the brain capillary endothelium, which is responsible for the blood–brain barrier, might also be inhibited by some phenothiazines and benzo[*a*]phenothiazines. However, the tomato lectin sensitivity of Pgp was different in mouse lymphoma and human brain capillary endothelial cells. The MDR gene expression of the mouse lymphoma cells (which were transfected with the human MDR-1 gene) could be reduced by phenothiazines, such as promethazine and TFZ, when the cells were cultured in the presence of 0.5 μg/mL of these phenothiazines. Further synergism was found between two resistance modifiers, verapamil and TFZ, on the inhibition of MDR glycoprotein.

Benzo[*a*]phenothiazines induced the monocytic differentiation and apoptotic cell death (characterized by inter-nucleosomal DNA fragmentation) only in human myelogenous leukaemic cell lines, but not in other cancer cell lines [146]. As already noted in the case of phenothiazines (Sect. 3.3.2), these compounds also yielded an antimicrobial activity in vivo, possibly by host-mediated immunopotential. In addition, benzo[*a*]phenothiazines exhibited an antimutagenic activity. These data open the possible application of half-mustard type phenothiazines as well as of benzo[*a*]phenothiazines to cancer chemotherapy. In another study, Cincotta et al. [200] showed, both separately and in combination, the antitumour activity of a BPHT derivative such as 5-ethylamino-9-diethylamino-benzo[*a*]phenothiazinium chloride (EtNBPHT) and a benzoporphyrin derivative monoacid ring A (BPD-MA), acting as photosensitizers in photochemotherapy. By means of this approach, the authors showed significantly the increase of the photodynamic therapy (PDT) efficiency of both compounds, due to their various mechanisms of action. It is worthwhile noting that the application of this PDT permitted the decrease of large murine sarcoma tumours which were implanted subcutaneously in BALB/c mice. Moreover, in a fluorescence microscopic study of EMT-6 cells and tumour tissues, the in vitro and in vivo administration of the photosensitizers revealed a greater intracellular localization for EtNBPHT than for BPD-MA.

3.7.4
Mechanisms of Biological Action

The mechanisms of biological action and chemical reactivity of benzo[*a*]phenothiazines might be explained at least partially by their physicochemical properties. For example, Satoh et al. [201, 202] suggested that the induction of human leukaemic cell differentiation by benzo[*a*]phenothiazines might be initiated by their radical-mediated reactions. It was shown by ESR spectroscopy that 12*H*-benzo[*a*]phenothiazine, 9-methyl-12*H*-benzo[*a*]phenothiazine, 10-methyl-12*H*-benzo[*a*]phenothiazine, 11-methyl-12*H*-benzo[*a*]phenothiazine and 5-oxo-5*H*-benzo[*a*]phenothiazine might induce the differentiation of human myelogenous leukaemic cell lines into the maturing macrophages against the produced radicals under alkaline conditions. On the other hand, 6-hydroxy-5-oxo-5*H*-benzo[*a*]phenothiazine, 6-methyl-5-oxo-5*H*-benzo[*a*]phenothiazine and 5*H*-benzo[*a*-1,4]benzothiazino-[3,2-*c*]-phenothiazine as well as 13 phenothiazines had little or no differentiation-inducing activity with no detectable amounts of radicals. Seven of the eight benzo[*a*]phenothiazines under study significantly enhanced their radical intensity of sodium L-ascorbate and sodium 5,6-benzylidene-L-ascorbate (SBA), whereas only three out of 13 phenothiazines showed similar effects.

As already pointed out in Sect. 2.2, Motohashi et al. established relationships between the biological activity and dipole moments of several benzo[*a*]phenothiazines [91]. 9-Benzo[*a*]phenothiazine, 12*H*-benzo[*a*]phenothiazine, 9-methyl-12*H*-benzo[*a*]phenothiazine, 10-methyl-12*H*-benzo[*a*]phenothiazine and 11-methyl-12*H*-benzo[*a*]phenothiazine induced the monocytic differentiation of human myelogenous leukaemic cell lines and displayed antitumour activity. These benzo[*a*]phenothiazines exhibited a significantly smaller value of the calculated ground-state dipole moment (μ_g) and a larger value of the first excited-state dipole moment (μ_e). The authors proposed that the μ_g and μ_e values might be useful parameters for prediction of the biological activity of benzo[*a*]phenothiazines.

The L5178Y mouse lymphoma cell (parent) and its MDR sub-line were used to investigate the effect of some resistance modifiers on apoptosis induction by benzo[*a*]phenothiazines [203]. 12*H*-Benzo[*a*]phenothiazine induced apoptosis both in the parent cells and in the MDR cells. The apoptosis induction by 12*H*-benzo[*a*]phenothiazine was not affected significantly by a post- or pre-treatment with resistance modifiers, whereas in the cells treated by verapamil before and after apoptosis induction with 12*H*-benzo[*a*]phenothiazine the apoptosis was somewhat higher. The resistance modifier compounds alone also induced apoptosis and it was slightly higher in the parent cells than its MDR1/A gene-transformed sub-line.

4 Concluding Remarks

In this chapter, we have reviewed recent progress in spectroscopic, photophysical, photochemical and analytical studies as well as the biological and biomedical properties of a number of bioactive phenothiazines and benzo[*a*]phenothiazines.

Particular emphasis was placed on recent developments in the studies concerning the luminescence properties of 2-substituted, 10-alkylated phenothiazines and 4′-substituted *N*-phenylphenothiazine derivatives, including their fluorescence and phosphorescence spectral characteristics, triplet-state lifetimes and quantum yields, and the marked dependence of these properties on the nature and position of substituents and solvent effects. Overall, the fluorescence quantum yield values of most substituted phenothiazines were found to be very weak, whereas their chemical or photochemical oxidation resulted in the formation of strongly fluorescent reaction products. The physicochemical aspects of the complexation, aggregation and various interactions of phenothiazine derivatives occurring in organized media, such as CDs and micelles, were also the object of detailed investigation in connection with their possible pharmacological and biological applications. Various analytical methods, including HPLC, direct or derivatized (either chemically or photochemically) spectrofluorimetry and FIA, were discussed for the determination of pharmaceutically important phenothiazines. Most of these methods have been shown to be successful from the standpoint of selectivity, sensitivity and detection of phenothiazines, at rather low concentrations in the ng/mL range. Moreover, a number of analytical applications relative to the pharmaceutical formulations and biological media (urine, human serum ...) containing phenothiazines were proposed with satisfactory recovery values, generally between 90 and 110%.

BPHTs were found to present markedly different luminescence properties when compared to simple substituted phenothiazines. In particular, most BPHTs presented a strong native fluorescence with significantly red-shifted emission spectra when compared to the simple phenothiazine derivatives. Moreover, their fluorescence spectra and quantum yields markedly depended on the kind of investigated media (organic solvents of various polarities, water and aqueous solutions of CDs). The dipole moment values of BPHTs in both their ground state and singlet excited state were also discussed, and the establishment of linear solvatochromic relationships suggested the existence of important solute–solvent interactions in the excited state of these BPHTs. Moreover, some structure–biological property correlations have been obtained between the dipole moments and various types of biological activities (tumour antigen expression, antimutagenicity, antitumour activity). The formation of inclusion complexes of several BPHTs with CDs were marked by an important fluorescence enhancement. Photophysical and thermodynamic studies revealed that weak 1 : 1 inclusion complexes were formed, and the use of semi-empirical AM1 calculations led to the proposal of a molecular structure for these inclusion complexes. Also, several applications of BPHTs have been described in a variety of fields, such as the relationships between their antitumour activity and the π-electron density calculated on the basis of molecular orbital methods (Hückel method, MNDO and Aihara aromaticity theory), the use of BPHTs as photosensitizers to develop a photoresist technology, as well as their determination at the ng/mL level in biological media (human urine) by a CD-enhanced fluorescence analytical technique.

Recently, the biological and biomedical properties of phenothiazines and BPHTs have revealed a great importance and a number of applications. In fact, phenothiazine derivatives have been used as antibacterial, antiviral and antifungal substances for a long time, but generally, the quantity and frequency of their utilization were evaluated by empirical means. During the last decade, the investigation of the antibacterial, antiviral, antifungal and other therapeutic properties of phenothiazines was based on much more rigorous scientific parameters. The study of their properties was examined by using well-selected biological systems, such as laboratory animals, cell lines with well-defined characteristics, bacteria, viruses, parasites, sub-cellular systems and monolayer or bilayer membranes.

In the case of the antibacterial properties, some experiments have shown that, for example, irradiation by UV–visible light of phenothiazines used as photosensitizers increased their antibacterial effects via the oxidative damage of bacteria cellular components. Also, the antimicrobial qualities of novel benzo[*a*]phenothiazines possessing fluoro substituents were investigated. Several phenothiazines, such as thioridazine, were found to induce an intracellular antibacterial activity in addition to that of the human peripheral blood monocyte macrophages against *S. aureus*. Thioridazine was also proposed as adjuvant therapy against highly resistant bacteria such as *M. avium*, isolated from AIDS patients. Another phenothiazine, namely trifluoperazine, was applied to the inhibition of *M. leprae*. The interactions of some phenothiazines with the purine bases in DNA had an influence on the metabolism of the antiviral drug famciclovir, increasing its activity against the herpes and hepatitis B viruses. The combined treatment by both acyclovir and some benzo[*a*]phenothiazines of Vero cells infected with HSV-2 gave good results in decreasing the viral infection. In the case of the antifungal activity, it has been suggested that some phenothiazines might be applied to develop new anticandidal drugs, because these phenothiazines were found to diminish the cellular lipids and phospholipids, and therefore affect the cell division phases. Several newly synthesized phenothiazine dyes showed a significant antiparasitic activity against parasites such as *Plasmodium*, *Babesia*, *Toxoplassa*, *T. brucei*, *T. cruzi*, *L. donovani*, *Onchocerca*, *Filaria*, nematodes, plathelminthes and nemathelminthes, and were recently patented.

In the case of vertebrates, phenothiazines have also been used as antipsychotic drugs for many years, before the beginning of the investigation of their action on a molecular level. Indeed, recently a number of works have appeared on the influence of phenothiazines on the nicotinic synaptic transmission, indicating that the contribution of cytochrome P-450 isoenzymes to N-demethylation and 5-sulphoxidation reactions was involved. The action of phenothiazine antipsychotics on the pre- and post-synaptic neurotransmission was experimentally examined on the tissues and enzymes extracted from the same neuronal tissues. Some phenothiazines were proposed to treat peripheral neuropathy. A new hope might also arise in the search for novel antipsychotic drugs based on phenothiazines to treat Alzheimer's disease and other taupathies, because these new compounds could inhibit tau filament formation.

The effect of phenothiazines on the antitumour activities, decrease of cytotoxicity and overlap of cancer MDR has been widely investigated. Generally, phenothiazines alone have no antitumour activity with MDR reversal, but the expressive effects of anticancer drugs were enhanced in the presence of phenothiazines, which diminished the mutagenesis and decreased the cytotoxic effects caused by radiations and carcinogenic chemicals as well as MDR, especially for brain tumours. Many experiments on the tissues and cell lines of different types of cancers showed that the combination of several drugs, including some phenothiazines, could reduce the development of the cancer cells. Although MDR still remains a serious problem in the success of cancer chemotherapies, the use of phenothiazine derivatives gives hope to preparing a combination with conventional anticancer agents, such as MMC and actinomycin D, in order to reach a higher anticancer potency. Also, some benzo[*a*]phenothiazines were found to overlap the MDR and diminish apoptosis.

Various interesting biomedical applications of phenothiazines were investigated, such as the anti-inflammatory effects of fluphenazine and triflupromazine, the use of chloropromazine and trifluoperazine to determine the secondary and tertiary structure of haemoglobin and myoglobin, and the efficiency of the promethazine (+) enantiomer to inhibit the bone resorbing cells, which led to the development of new methods for the treatment of periodontitis and osteoporosis. Recently, it was shown that some phenothiazine derivatives inhibited the production of prions in cultured neuroblastoma cells, suggesting that these phenothiazines might be applied to treat CJD.

Finally, it must be pointed out that the extensive and long-term therapeutic use of phenothiazines might often cause several unwanted or side effects, such as skin photosensitization, disorders in the lipid status, reproductive toxicity and cardiac arrhythmia. Therefore, it is very important to understand better the biochemical mechanisms responsible for the different side effects of phenothiazines and their toxicity.

In the future, many novel phenothiazines with specific biomedical characteristics should be synthesized and lead to novel drugs designed for the treatment of various diseases. Their spectroscopic and photochemical properties as well as their biochemical mechanisms of action should be carefully investigated in order to develop important and safe applications of bioactive phenothiazines in the fields of the pharmaceutical industry and public health.

Acknowledgements The authors wish to warmly thank the devoted, hardworking and clever efforts of their co-workers whose names appear in the cited references.

References

1. Gupta RR (1988) In: Phenothiazines and 1,4-benzothiazines: chemical and biological aspects. Elsevier, Amsterdam
2. Dahl SG (1981) In: Usdin E, Dahl SG, Gran LF (eds) Clinical and pharmacology psychiatry neuroleptic antidepressants research (Int Mtg), 2nd edn. Macmillan, London, p 125
3. Curry SM (1985) Drug Psychiatry 3:79
4. Garcia C, Oyola R, Pinero LE, Arce R, Silva S, Sanchez V (2005) J Phys Chem A 109:3360
5. Horn AS, Snyder SH (1971) Proc Natl Acad Sci USA 68:2325
6. Kaufman JJ, Kerman E (1974) Int J Quantum Chem Quantum Biophys Symp 1:36615
7. Fowler GJS, Reej RC, Devonshire R (1990) Photochem Photobiol 52:489
8. Párkányi C, Boniface C, Aaron J-J, Meuguelati F, Murray JS, Politzer P, RaghuVeer KS (1992) In: Keyzer H, Eckert GM, Forrest IS, Gupta RR, Gutmann F, Molnár J (eds) Thiazines and structurally related compounds. Krieger, Malabar, p 103

9. Borowicz P, Herbich J, Kapturkiewicz A, Anulwicz-Ostrowska R, Nowacki J, Grampp G (2000) Phys Chem Phys 2:4275
10. Laassis B, Aaron J-J (1997) Analusis 25:183
11. Seetharamappa J, Motohashi N, Kovala-Demertzi D (2006) Curr Drug Targets 7:1107
12. Barra M, Redmond MW, Allen MT, Calabrese GS, Sinta R, Scaiano JC (1991) Macromolecules 24:4972
13. Kurihara T, Motohashi N, Pang GL, Higano M, Kiguchi K, Molnár J (1996) Anticancer Res 16(5A):2757
14. Motohashi N, Kurihara T, Satoh K, Sakagami H, Mucsi I, Pusztai R, Szabó M, Molnár J (1999) Anticancer Res 19(3A):1837
15. Motohashi N, Sasaki Y, Wakabayashi Y, Sakagami H, Molnar J, Turihara T (1992) Anticancer Res 12(5):1423
16. Aaron J-J, Maafi M, Kersebet C, Párkányi C, Antonious MS, Motohashi N (1995) Spectrosc Lett 28(7):1111
17. Aaron J-J, Maafi M, Kersebet C, Párkányi C, Antonious MS, Motohashi N (1996) J Photochem Photobiol A 101:127
18. Gaye Seye MD, Prot C, Adenier A, Aaron J-J, Motohashi N (2001) New J Chem 25:1290
19. Maafi M, Mahedero MC, Aaron J-J (1997) Talanta 44:2193
20. Gaye Seye MD, Prot C, Adenier A, Aaron J-J, Motohashi N (2000) Luminescence 15:62
21. Maafi M, Aaron J-J, Mahedero M C, Salinas F (1998) Appl Spectrosc 52(1):91
22. Maafi M, Aaron J-J, Mahedero MC, Salinas F (1997) J Fluoresc 7(1):11S
23. Gaye Seye MD, Parain C, Aaron J-J, Motohashi N (2002) Luminescence 17:250
24. Parkanyi C, Boniface C, Aaron J-J, Maafi M (1993) Spectrochim Acta A 49:1715
25. Gaye Seye MD, Aaron J-J, Parkanyi C, Motohashi N (2006) Curr Drug Targets 7:1083
26. Kemp TJ, Moore P, Quick GR (1980) J Chem Soc Perkin Trans 2:291
27. Borowicz P, Herbich J, Kapturkiewicz A, Opallo M, Nowacki J (1999) Chem Phys 249:49
28. Garcia C, Smith GA, Grant M, Kochevar IE, Redmond RW (1995) J Am Chem Soc 17:10871
29. Elisei F, Latterini L, Aloisi GG, Mazzucato U, Viola G, Miolo G, Vedaldi D, Dall'Acqua F (2002) Photochem Photobiol 75:11
30. Barra M, Calabrese GS, Allen MT, Redmond RW, Sinta R, Lamola AA, Small RD, Sciano JC (1991) J Chem Mater 3:610
31. Turro NJ, Khudyakov IV, Willingen HV (1995) J Am Chem Soc 117:12273
32. Lhoste JM, Merceille JB (1968) J Chim Phys Chim Biol 65:1889
33. Lui YH, McGlynn SP (1978) Spectrosc Lett 11:47
34. Porter G, Suppan P (1965) Trans Faraday Soc 61:1664
35. Degani Y, Willner I, Mackay RA (1984) Chem Phys Lett 104:496
36. Dan P, Willner I, Dixit NS, Mackay RA (1984) J Chem Soc Perkin Trans 2:455
37. Hirai H, Toshima N, Uenoyama S (1985) Bull Soc Chem Jpn 58:1156
38. Lee C, Moon MS, Park JW (1996) J Electroanal Chem 407:161
39. Lee C, Moon MS, Park JW (1996) J Inclusion Phenom 26:219
40. Lee C, Sung YW, Park JW (1997) J Electroanal Chem 431:133
41. Lee C, Sung YW, Park JW (1999) J Phys Chem B103:893
42. Maafi M, Laassis B, Aaron J-J, Mahedero MC, Munoz de la Pena AM, Salinas F (1995) J Inclusion Phenom 22:235
43. Raj CR, Ramaraj R (1996) J Electroanal Chem 405:141
44. Caetano W, Tabak MJ (2000) Colloid Interf Sci 225:69 and references therein

45. Rabinowitch E, Epstein LJ (1941) J Am Chem Soc 63:69
46. Lewis GN, Bigelisen J (1943) J Am Chem Soc 65:1144
47. Schubert M, Levine A (1955) J Am Chem Soc 77:4197
48. Bergmann K, O'Konski C (1963) J Phys Chem 67:2169
49. Spencer W, Sutter JR (1979) J Phys Chem 83:1573
50. Frömming KH, Szejtli J (1994) In: Davies JED (ed) Topics in inclusion science: cyclodextrins in pharmacy. Kluwer, Dordrecht
51. Fridan O, Davidson AG (1988) J Chromatogr 442:363
52. Green GL, O'Haver TC (1974) Anal Chem 46:91
53. Cimbura G (1972) J Chromatogr 10:387
54. Kapadia AJ, Barber MA, Martin E (1970) J Pharm Sci 59:1476
55. Li S, Purdy WS (1991) J Pharm Biomed Anal 9:409
56. Prabhakar RG, Kollmorgen D, Kraut B (1998) LC GC 16:1006
57. Sato K, Mizuno Y, Lee X-P, Taguchi T, Katsumata Y (1997) In: Takatori T, Takasu A (eds) Current topics in forensic science: proceedings of the 14th meeting of the International Association of Forensic Sciences, Tokyo, August 26–30, 1996, 2:348. Shunderson, Ottawa
58. Tesarova E, Bosakova Z (2003) J Sep Sci 26:661
59. Guttierez MC, Gomez-Henz A, Perez-Bendito D (1994) Anal Chim Acta 290:1847
60. Ragland JB, Kinross-Wright VJ (1964) Anal Chem 36:1356
61. Mellinger TJ, Keeler CE (1963) Anal Chem 35:554
62. Zakari NA, Rizk M, Walsh MI, Ibrahim F (1985) Anal Lett 18:1405
63. Laassis B, Aaron J-J, Mahedero C (1994) Anal Chim Acta 290:27
64. Laassis B, Maafi M, Aaron J-J, Mahedero MC (1997) Anal Lett 30:1541
65. Martinez Calatayud J, Garcia Mateo V (1992) Anal Chim Acta 264:283
66. Martinez Calatayud J, Garcia Sanchez T (1992) J Pharm Biomed Anal 10:37
67. Perez-Ruiz T, Martinez-Lozano C, Tomas V, Sidarach de Cardona C (1993) Talanta 40:1341
68. Chen D, Rios A, Luque de Castro MD, Valcarcel M (1991) Analyst 116:171
69. Chen D, Rios A, Luque de Castro MD, Valcarcel M (1991) Talanta 38:1227
70. Martinez Calatayud J, Gomez Benito C (1992) Anal Chim Acta 256:105
71. Mellado Romero A, Gomez Benito C, Martinez Calatayud J (1992) Anal Lett 25:1889
72. Laassis B, Aaron J-J, Mahedero MC (1994) Talanta 41:1985
73. Aly FA, Alarfaj NA, Slwarthan AA (1998) Anal Chim Acta 358:255
74. Soriano T, Jurado C, Menendez M, Repetto M (2001) J Anal Toxicol 25:137
75. Petit C, Murakami K, Erdem A, Kilinc E, Ortiz Borondo G, Liegeois J-F, Kauffmann J-M (1998) Electroanalysis 10:1241
76. Madej K, Wozniakiewicz M, Kala M (2005) Chromatographia 61:259
77. Gessner T, Kowalsky W, Schildknecht C (2006) PCT Int Appl, p 91
78. Gessner T, Schmidt H-W, Thelakkat M, Baete M (2006) PCT Int Appl, p 111
79. Yonemura H, Tokudome H, Noda M, Hayashi K, Moribe S, Yamada S (2002) RIKEN Rev 44:66
80. Moribe S, Yonemura H, Yamada S (2006) C R Chim 9:247
81. Lukhtanov EA, Vorobiev AV, Reed MW, Vermeulen NMJ (2003) PCT Int Appl, p 103
82. Hinze WL, Srinivasen N, Smith TK, Igarashi S, Hoshino H (1991) In: Warner IM, McGown LB (eds) Advances in multidimensional luminescence, vol 1. JIA, Greenwich
83. Maafi M, Laassis B, Aaron J-J, Mahedero MC, Munoz de la Pena A, Salinas F (1995) J Inclusion Phenom Mol Recognit Chem 22:237
84. Baeyens W, Ling BL, Corbisier V, Raemdonk A (1990) Anal Chim Acta 234:187

85. Munoz de la Pena A, Salinas F, Gomez MJ, Acedo MI, Sanchez Pena M (1993) J Inclusion Phenom Mol Recognit Chem 15:131
86. Nigam S, Durocher G (1996) J Phys Chem 100:7135
87. Kamlet MJ, Abboud JL, Taft RW (1977) J Am Chem Soc 99:6027
88. Kamlet MJ, Abboud JL, Abraham MH, Taft RW (1983) J Org Chem 48:2877
89. Párkányi C, Aaron J-J (1998) In: Párkányi C (ed) Theoretical organic chemistry. Elsevier, Amsterdam, p 233
90. Párkányi C, Antonious MS, Aaron J-J, Maafi M, Gil O, Kersebet C, Motohashi N (1995) In: Barbe J, Keyzer H, Soyfer JC (eds) Biological and chemical aspects of thiazines and analogs. Enlight, San Gabriel, p 178
91. Motohashi N, Kurihara T, Yamanaka W, Satoh K, Sakagami H, Molnár J (1997) Anticancer Res 17:3431
92. Kurihara T, Motohashi N, Párkányi C, Aaron J-J, Wakabayashi H, Shindo K, Molnár J (1998) In: Chakrabarty AN, Molnár J, Dastidar AN, Motohashi N (eds) Non Antibiotics. NISCOM, New Delhi, p 19
93. Pusztai R, Motohashi N, Párkányi C, Aaron J-J, Rao BK, Molnár J (1996) Anticancer Res 16:2961
94. Tanaka M, Wayda K, Molnár J, Párkányi C, Aaron J-J, Motohashi N (1997) Anticancer Res 17:839
95. Motohashi N, Kurihara T, Ghosh S, Dastidar SG, Chakrabarti A, Chakrabarty AN, Molnár J (1998) In: Chakrabarty AN, Molnár J, Dastidar AN, Motohashi N (eds) Non Antibiotics. NISCOM, New Delhi, p 246
96. Pohlers G, Scaiano JC, Sinta R, Brainard R, Pai D (1997) Chem Mater 9:1353
97. Flamigni L (1993) J Phys Chem 97:9566
98. Will AY, Schuette-Parsons JM, Agbaria RA, Warner IM (1995) Appl Spectrosc 49:1762
99. Froehlich PM, Yeats M (1976) Anal Chim Acta 87:185
100. Hoshino M, Imamura M, Ikehara K, Hama Y (1981) J Phys Chem 85:1820
101. Smith VK, Ndou TT, Munoz de la Pena A, Warner I (1991) J Inclusion Phenom 10:471
102. Connors KA (1987) In: Binding constants: the measurement of molecular complex stability. Wiley, New York
103. Martin RB (1997) J Chem Educ 74:1238
104. Munoz de la Pena A, Ndou T, Zung JB, Warner I (1991) J Phys Chem 95:3330
105. Szejtli J (1982) In: Cyclodextrins and their inclusion compounds. Akadémiai Kiadó, Budapest
106. Li S, Purdy WC (1992) Chem Rev 92:1457
107. Kurihara T, Motohashi N, Pang GL, Higano M, Kiguchi K, Molnár J (1996) Anticancer Res 16:2757
108. Motohashi N, Kurihara T, Satoh K, Sakagami H, Mucsi I, Pusztai R, Szabó M, Molnár J (1999) Anticancer Res 19:1837
109. Motohashi N, Sasaki Y, Wakabayashi Y, Sakagami H, Turihara T (1992) Anticancer Res 12:1423
110. Berry AK, Feely WE, Thompson SD, Calabrese GS, Sinta R, Lamola AA, Thackeray JW, Orsula GW (1990) SPIE Adv Resist Technol Proc 1262:575
111. Mitchell SC (2006) Curr Drug Targets 7:1181
112. Harris F, Sayed Z, Hussain S, Phoenix DA (2004) Photodiagn Photodyn Ther 1:231
113. Komatsu N, Motohashi N, Fujimaki M, Molnar J (1997) In Vivo 11:13
114. Hussain S, Harris F, Phoenix DA (2006) FEMS Immunol Med Microbiol 46:124
115. Birgit KG (2005) PCT Int Appl, p 71
116. Kristiansen MM, Leandro C, Ordway D, Martins M, Viveiros M, Pacheco T, Kristiansen JE, Amaral L (2003) Int J Antimicrob Agents 22:250

117. Tegos GP, Hamblin MR (2006) Antimicrob Agents Chemother 50:196
118. Sayed Z, Harris F, Phoenix DA (2005) FEMS Immunol Med Microbiol 43:367
119. Tassaneeyakul W, Birkett DJ, Miners JO (1998) Xenobiotica 28:293
120. Kalkanidis M, Klonis N, Tilley L, Deady LW (2002) Biochem Pharmacol 63:833
121. Pignol B, Puel J-L, Auvin S, Chabrier de Lassaunière P-E, Wang J (2005) PCT Int Appl, p 33
122. Pignol B, Puel J-L, Auvin S, Chabrier de Lassaunière P-E (2005) Société de Conseils de Recherches et d'Applications Scientifiques - Consulting Society in Research and Scientific Applications (SCRAS) Fr Demande, p 21
123. Rashidi MR, Malek A, Hanaee J, Hamidi A (2004) Ulum-i Daroei 1:87
124. Wainwright M, Phoenix DA, Smillie TE, Wareing DRA (2001) J Chemother 13:503
125. Ordway D, Viveiros M, Leandro C, Arroz MJ, Amaral L (2002) Int J Antimicrob Agents 20:34
126. Viveiros M, Martins M, Couto I, Kristiansen JE, Molnar J, Amaral L (2005) In Vivo 19:733
127. Dhople AM (1999) Microbios 98(390):113
128. Sharma S, Kaur H, Khuller GK (2001) FEMS Microbiol Lett 199:185
129. Correa JG, Fairlamb AH, Stoppani AOM (2001) Rev Argent Microbiol 33:36
130. Gutierrez-Correa J, Stoppani AOM (2002) Rev Argent Microbiol 34:83
131. Chan C, Yin H, Garforth J, McKie JH, Jaouhari R, Speers P, Douglas KT, Rock PJ, Yardley V, Croft SL, Fairlamb AH (1998) J Med Chem 41:148
132. Ondarza RN, Hernandez E, Iturbe A, Hurtado G (2000) Biotechnol Appl Biochem 32:61
133. Galey L (2001) Fr Demande, p 13
134. Nguyen Q-T, Yang J, Miledi R (2002) Neuropharmacology 42:670
135. Conforti J (2005) US Pat Appl Publ, p 6
136. Daniel WA, Syrek M, Rylko Z, Kot M (2001) Pol J Pharmacol 53:615
137. Taniguchi S, Suzuki N, Masuda M, Hisanaga S, Iwatsubo T, Goedert M, Hasegawa M (2005) J Biol Chem 280:7614
138. Appert-Colin A, Callizot N (2004) (Neuro 3d, Fr) Eur Pat Appl, p 26
139. Hale RL, Lloyd PM, Lu AT, Munzar P, Rabinowitz JD, Skowronski R (2006) US Pat Appl Publ, p 37
140. Motohashi N, Kawase M, Satoh K, Sakagami H (2006) Curr Drug Targets 7(9):1055
141. Wuonola MA, Palfreyman MG, Motohashi N, Kawase M, Gabay S, Nacsa J, Molnar J (1997) Anticancer Res 17:3409
142. Tanaka M, Molnar J, Kidd S (1997) Anticancer Res 17:381
143. Harris F, Sayed Z, Hussain S, Phoenix DA (2004) Photodiagn Photodyn Ther 1:231
144. Nichols JM, Foley MA, Keith C, Padval M, Elliott P (2005) PCT Int Appl, p 92
145. Gil-Ad I, Shtaif B, Levkovitz Y, Nordenberg J, Taler M, Korov I, Weizman A (2006) Oncol Rep 15:107
146. Motohashi N, Kawase M, Saito S, Sakagami H (2000) Curr Drug Targets 1:237
147. Grabski R, Dewit J, De Braekeleer J, Malicka-Blaskiewicz M, De Baetselier P, Verschuerena H (2001) Biochem Pharmacol 61:1313
148. Hawtrey A, Joubert D, van Jaarsveld P, Pieterse A, van Zyl J, Ariatti M (2002) Drug Deliv 9:47
149. Yan Y, Sun W, Zhou P (2005) Zhonghua Fangshe Yixue Yu Fanghu Zazhi 25:305
150. Eghbal MA, Tafazoli S, Pennefather P, O'Brien PJ (2004) Chem Biol Interact 151:43
151. Graef IA, Crabtree GR, Gestwicki JE (2005) PCT Int Appl, p 36
152. Motohashi N, Kurihara T, Kawase M, Hever A, Tanaka M, Szabo D, Nacsa J, Yamanaka W, Kerim A, Molnar J (1997) Anticancer Res 17:3537

153. Nacsa J, Nagy L, Sharples D, Hever A, Szabo D, Ocsovszki I, Varga A, Konig S, Molnar J (1998) Anticancer Res 18:3093
154. Wesolowska O, Hendrich AB, Motohashi N, Kawase M, Dobryszycki P, Ozyhar A, Michalak K (2004) Biophys Chem 109:399
155. Hendrich AB, Wesolowska O, Michalak K (2001) Curr Top Biophys 25:67
156. Michalak K, Wesolowska O, Motohashi N, Molnar J, Hendrich AB (2006) Curr Drug Targets 7:1095
157. Bettencourt MV, Bosne-David S, Amaral L (2000) Int J Antimicrob Agents 16:69
158. Viveiros M, Amaral L (2001) Int J Antimicrob Agents 17:225
159. Pajeva IK, Wiese M (1998) J Med Chem 41:1815
160. Cao Y, Li Y, Gao Z, He X (1999) Yaoxue Xuebao 34:276
161. Takegami S, Kitamura K, Takahashi K, Kitade T (2002) J Pharm Sci 91:1568
162. Hu J, Wu Y, Kulkarni AP (2004) Zhonghua Lao Dong Wei Sheng Zhi Ye Bing Za Zhi 22:184
163. Hover CG, Kulkarni AP (2000) Placenta 21:646
164. Flores VC, Keyzer H, Kim HK, Molnar J (2002) Anticancer Res 22:959
165. Hidalgo AA, Caetano W, Tabak M, Oliveira ON (2004) Biophys Chem 109:85
166. Daniel WA, Syrek M, Haduch A (2002) Eur Neuropsychopharmacol 12:371
167. Khan SZ, Longland CL, Michelangeli F (2000) Biochem Pharmacol 60:1797
168. Lu L, Jiang AQ, Yuan S, Yin LH, Huang WY, Fan QS (2000) Shiyan Shengwu Xuebao 33:141
169. Tsakovska I, Pajeva I (2006) Curr Drug Targets 7:1123
170. Ambudcar S, Dey S, Hrycyna C, Ramachandra M, Pastan I, Gottesman M (1999) Annu Rev Pharmacol Toxicol 39:361
171. Seydal JK, Wies M (2002) In: Mannhold RK, Folkers G (eds) Drug membrane interactions. Wiley, Weinheim, p 245
172. Konya A, Andor A, Satorhelyi P, Nemeth K, Kurucz I (2006) Biochem Biophys Res Commun 346:45
173. Wainwright M (2000) Int J Antimicrob Agents 16:381
174. Saad S, El-Sayed M, Abdallah D (1998) Bull Fac Pharm (Cairo University) 36:151
175. Bhattacharyya J, Bhattacharyya M, Chakraborti AS, Chaudhuri U, Poddar RK (1998) Int J Biol Macromol 23:11
176. Chapman JR, Stark PRH, Swallow MV, Larson DN (2001) Baxter International Inc, USA, p 30
177. Boland EJ, McDonough J (2004) PCT Int Appl, p 48
178. Takikawa H, Sano N, Akimoto K, Ogasawara T, Yamanaka M (1998) J Gastroenterol Hepatol 13:427
179. Gottlieb R, Wentworth P, Johnson EF, Yeager MJ, Granville D (2004) PCT Int Appl, p 80
180. Amaral L, Kristiansen JE (2001) Int J Antimicrob Agents 18:411
181. Korth C, May BCH, Cohen FE, Prusiner SB (2001) Proc Natl Acad Sci USA 98:9836
182. Valverde MA, Hardy SP, Diaz M (2002) Steroids 67:439
183. Andrieu J-M, Lu W, Achour A (2002) PCT Int Appl, p 39
184. Sanai T, Matsui R, Hirano T, Torichigai S, Yotsueda H, Higashi H, Hirakata H, Iida M (2006) Renal Failure 28:51
185. Wen F, Cheng Z (2000) Zhongguo Yiyuan Yao Xue Za Zhi 20:389
186. Shin W-H, Kim E-J (2005) Basic Clin Pharmacol Toxicol 96:143
187. Rodrigues T, Faria PAF, Pessoto FS, Santos CG, Nantes IL (2005) Rev Cien Farm Basica Aplicada 26:19
188. Mio M, Yabuta M, Kamei C (1999) Immunopharmacology 41:55

189. Elisei F, Latterini L, Aloisi GG, Mazzucato U, Viola G, Miolo G, Vedaldi D, Dall'Acqua F (2002) Photochem Photobiol 75:11
190. James LP, Abel K, Wilkinson J, Simpson PM, Nichols MH (2000) J Toxicol Clin Toxicol 38:615
191. Huq S, Dixon A, Garriques M, Kallury K (2005) Phenomenex Inc, Torrance, CA, USA, LCGC North America (Suppl) 64
192. Miyazaki T, Yomota C, Okada S (2000) Colloid Polym Sci 278:84
193. Rajadhyaksha AV, Reddy V, Hover CG, Kulkarni AP (1999) Teratog Carcinog Mutagen 19:211
194. Yang X, Kulkarni AP (1997) Teratog Carcinog Mutagen 17:139
195. Kim KS, Kim EJ (2005) Drug Chem Toxicol 28:303
196. Singh G, Swati, Mishra AK, Prakash L (1999) J Fluorine Chem 98:37
197. Mucsi I, Molnar J, Motohashi N (2001) Int J Antimicrob Agents 18:67
198. Cuine A, Decobert M, Francon D, Saunal H (2007) PCT Int Appl, p 35
199. Molnar J, Szabo D, Mandi Y, Mucsi I, Fischer J, Varga A, Koenig S, Motohashi N (1998) Anticancer Res 18:3033
200. Cincotta L, Szeto D, Lampros E, Hasan T, Cincotta AH (1996) Photochem Photobiol 63:229
201. Satoh K, Sakagami H, Motohashi N (1997) Anticancer Res 17:2539
202. Satoh K, Sakagami H, Kurihara T, Motohashi N (1997) Anticancer Res 17:2465
203. Mucsi I, Varga A, Kawase M, Motohashi N, Molnar J (2002) Anticancer Res 22:2833

Subject Index